T5-CRP-170

M02 40004 50569

QC
311
B 4

4642

Bent, Henry A.
The Second Law

DATE DUE

FEB 24 '71			
JAN 17 1973			
DEC 29 '73			

Waubonsee Community College

keep for FL

The Second Law

Drawing by Steinberg;
© 1963 The New Yorker Magazine, Inc.

The Second Law

An Introduction to Classical and Statistical Thermodynamics

HENRY A. BENT
ASSOCIATE PROFESSOR OF CHEMISTRY
UNIVERSITY OF MINNESOTA

QC
311
B4

4642

NEW YORK OXFORD UNIVERSITY PRESS 1965

Copyright © 1965 by Oxford University Press, Inc.
Library of Congress Catalogue Card Number: 65-15608
Printed in the United States of America

Acknowledgments

I am grateful to Saul Steinberg and *The New Yorker* Magazine for permission to reproduce the drawing that appears as the frontispiece to this book; to Professor E. U. Condon and the American Institute of Physics for permission to include his observations on the early investigations of radioactivity; to Professor Peter Debye and the American Association for the Advancement of Science for permission to include his observations on the origins of quantum mechanics; to Professor W. F. Giauque and *The Journal of Chemical Physics* for permission to quote from his discussion of the word "pernt"; to the Harvard University Press for permission to include excerpts from Joseph Black's work on the equilibrium of heat and the melting of ice; to the University of California Press for permission to quote from a translation of Ludwig Boltzmann's *Lectures on Gas Theory;* to the Philosophical Library for permission to quote from Max Planck's *Scientific Autobiography;* to the staff of Oxford University Press for their expert assistance in seeing this book into print; to former students, for whom earlier versions of this book were written, for many perceptive questions and comments; and, most particularly, to Professors Luke E. Steiner and Henry E. Bent, godfathers to this book, for many years of constructive criticism and encouragement.

Minneapolis
March 1965

H. A. B.

Contents

The Second Law

Maximum disorder was our equilibrium

T. E. LAWRENCE
Seven Pillars of Wisdom

Introduction and Synopsis

> I was going home to dinner, past a shallow pool, which was green with springing grass, . . . when it occurred to me that I heard the dream of the toad. It rang through and filled all the air, though I had not heard it once. And I turned my companion's attention to it, but he did not appear to perceive it as a new sound in the air. Loud and prevailing as it is, most men do not notice it at all. It is to them, perchance, a sort of simmering or seething of all nature. That afternoon the dream of the toads rang through the elms by Little River and affected the thoughts of men, though they were not conscious that they heard it. How watchful we must be to keep the crystal well that we are made, clear!
>
> THOREAU

Some things we scarcely notice. Like the fresh air we breathe and the pure water we drink, they have no unusual quality, no special odor or flavor to commend them to our senses. Yet, "It frequently happens," Count Rumford has said, "that *in the ordinary affairs and occupations of life, opportunities present themselves of contemplating some of the most curious operations of nature*."

A passing car, a fire in a fireplace, an ice cube melting in a glass, a bouncing ball coming to rest on the floor, how often after childhood do we detect anything unusual or noteworthy in these events? On reflection we may realize, however, a fact often overlooked. Though common enough in outward appearances, these things, once experienced, can never be experienced again. Events have truly only one life—one occurrence.

A birch log, for example, cannot be burned twice. One cannot take the hot flue gases of a wood fire and the warmth of the fire and from these reconstitute an unburnt log, fresh air, and room chilliness. Passage from the burnt state—the hot flue gases and a warm room—to the preburnt state—a birch log, fresh air, and a cold room—is impossible, or at most, highly improbable.

Similarly, the onrush of a passing car cannot be reversed in all aspects. Many cars, of course, can be driven backward. But no car, in an exact reversal of its motion forward, can suck through its tailpipe exhaust fumes and from these produce in its cylinders an ignition spark, liquid gasoline, and fresh air.

Burning logs and onrushing cars are typical examples of irreversible events. Viewed in their entirety, such events always produce unalterable changes in the universe. The full scope of these changes—formation of wood ashes and room warmth, or car noise and smog—may not always be apprehended by the unaided senses, however.

Seldom noticed, for example, is the thermal energy produced as a bouncing ball comes to rest. Yet if energy is conserved during this event, and we believe energy is always conserved, the loss in potential energy of the system (the ball plus the earth) should appear somewhere in some form. Permanent distortion of the ball or the floor might account for part of the energy; even so, it would be difficult to escape production (through friction and sound waves) of some thermal energy. In time, this thermal energy would become distributed between the ball and its thermal surroundings (the floor, the air, the walls, the furniture of the room), to each according to its heat capacity.

Thus, merely bringing the ball back to its initial position above floor-level would not restore the universe (the ball plus its thermal surroundings) to its initial condition. That is impossible. How remarkable would be a complete restoration of the universe can be appreciated by looking at a movie of a bouncing ball coming to rest, run backward. Objects obviously are not in the habit of springing spontaneously into the air at the expense of the thermal energy of their surroundings. The idea that they might is absurd. They never do. To paraphrase Lord Kelvin, events whose net effects are equivalent to the raising of a weight and the cooling of a thermal reservoir do not occur.

This is because a falling weight on impact produces in the universe a significant increase in microscopic disorder—the kind of disorder, invisible to the eye, that stems from the random thermal motions of the universe's constituent atoms. The randomness of this thermal motion increases as the thermal energy in the system increases. The random character of the thermal motion in the system is in this case greater in the final (warmer) state than in the initial (cooler) state. The final state (the state of greater microscopic disorder) is intrinsically more probable than the initial state, and passage from the initial, less microscopically disordered state to the final, more microscopically disordered state can occur spontaneously. Weights do have the habit of falling downward. The reverse process—the raising or floating upward of a weight solely at the expense of the thermal energy of its surroundings—does not occur because the microscopic disorder, or as it is called, the entropy of the universe, never (well, hardly ever) decreases.

Ice melts in warm water for much the same reason that weights fall downward. The process increases the entropy of the universe. For although melting ice absorbs thermal energy from its surroundings (an effect that must somewhat diminish the entropy of the surroundings), at high temperatures ($T > 273°K$) the relatively small entropy decrease suffered by the surroundings is more than made up by the entropy increase of the melting water as it changes from the relatively ordered crystalline solid—ice—to a more disordered state—the liquid. Melting at high temperatures increases the entropy of the universe.

Ice does not melt, however, at low temperatures, for much the same reason that a resting ball does not spontaneously begin bouncing. The process would decrease the entropy of the universe. True, the positive entropy change in the melting water itself would still exist, for this change—the difference between the entropies of liquid and solid water—is not greatly affected by changes in temperature. On the other hand, the change in entropy of the surroundings is greatly affected by changes in temperature. This entropy change, in fact, is just the change in the thermal energy of the surroundings (here -80 calories/gram of ice melted) divided by the absolute temperature of the surroundings. At low temperatures (small T) the quotient (-80 cal/gm)/T is a large, negative number, a number more negative than the entropy difference between liquid and solid water is positive. Thus, at low temperatures the sum of the two entropy changes is a negative number. Cold ice, therefore, does not melt.

Cold water, however, can freeze. For although the entropy of the frozen product (ice) is less than the entropy of the liquid reactant, a factor by itself unfavorable to the freezing process, the thermal energy liberated to the surroundings by the freezing water produces in the surroundings the favorable entropy change of $+80/T$ calories per degree Kelvin for each gram of water frozen. At low temperatures ($T < 273°K$) this term is a relatively large positive number, more positive, in fact, than the difference between the entropy of solid and liquid water is negative. The sum, therefore, is a positive number. Freezing can occur. At sufficiently low temperatures the freezing of water, or indeed of any other liquid, increases the total entropy of the universe.*

The freezing point of water can be depressed by adding to it sugar, salt, alcohol, and other "antifreezes"—substances soluble in liquid water but insoluble in ice. The presence of these second components in the liquid phase increases the contribution that the other component, the solvent water, makes to the disorder of the liquid phase and stabilizes the presence

* Liquid helium is an exception to this rule.

of the water in that phase. The escaping tendency of the water from the liquid phase is thereby diminished and the temperature range over which the liquid phase is stable with respect to its transformation into other phases, through solidification or evaporation, is extended at both ends. Its freezing point is lowered and its boiling point (if the solutes are nonvolatile) is raised.

These examples suggest that the stability of a given arrangement of atoms and molecules with respect to a different arrangement of the same atoms and molecules depends on the energies of the two arrangements and, also, on their entropies—and the temperature.

Ice is the most stable arrangement of water molecules at low temperatures. At higher temperatures, however, ice melts spontaneously to form liquid water; at still higher temperatures—and not too high pressures—liquid water spontaneously changes to steam; at very high temperatures steam dissociates into hydrogen and oxygen; at still higher temperatures hydrogen and oxygen ionize to positive ions and electrons.

These facts can be conveniently expressed by a function called free energy. The free energy of a substance is equal to its internal energy less its entropy multiplied by the absolute temperature. Included also is a term, generally small, the substance's volume multiplied by the pressure. In symbols—G for free energy, E for internal energy, S for entropy, V for volume, T for absolute temperature, and P for pressure—

$$G = E - TS + PV.$$

A small free energy implies stability, a large free energy instability.

At low temperatures (small T) the free energy of a substance depends mainly upon its internal energy. At high temperatures (large T) the temperature-entropy term becomes the dominant term. Forms of matter stable at low temperatures tend to be low energy forms (ice compared to liquid water, or liquid water compared to steam). Forms of matter stable at high temperatures tend to be large entropy forms (liquid water compared to ice, or steam compared to liquid water).

A chemical reaction can occur only if the free energy of the reactants exceeds the free energy of the products. Only then, we shall see, does the chemical change produce a net increase in the entropy of the universe. Thus, water can evaporate only if $G^{\text{liq}}_{\text{H}_2\text{O}} > G^{\text{gas}}_{\text{H}_2\text{O}}$, and aqueous acetic acid can ionize according to the equation

$$\text{HAc} + \text{H}_2\text{O} = \text{H}_3\text{O}^+ + \text{Ac}^-$$

only if $(G_{\text{HAc}} + G_{\text{H}_2\text{O}}) > (G_{\text{H}_3\text{O}^+} + G_{\text{Ac}^-})$.

Ionization is never complete, however. For as acetic acid disappears, its concentration decreases, its entropy per mole increases, and its free energy decreases. Simultaneously, the concentrations of the ions increase, their entropies per mole decrease, and their free energies per mole increase. Eventually (rather quickly, in fact) the concentrations reach such values that $(G_{HAc} + G_{H_2O}) = (G_{H_3O^+} + G_{Ac^-})$. At this point the system is at equilibrium with respect to the transfer of protons between acetate ions and water molecules. For dilute aqueous solutions at atmospheric pressure and 25°C, $(G_{HAc} + G_{H_2O}) = (G_{H_3O^+} + G_{Ac^-})$ whenever the concentrations of the species HAc, H_3O^+, and Ac^- satisfy the condition

$$\frac{(H_3O^+)(Ac^-)}{(HAc)} = 1.8 \times 10^{-5} \text{ (moles/liter)}.$$

When this condition is satisfied, the chemical reaction $HAc + H_2O = H_3O^+ + Ac^-$ produces no net change in the total entropy of the universe. To be sure, ionization of acetic acid does alter the entropy of the reaction mixture; at equilibrium, however, this entropy change, like the entropy change in melting ice at 0°C, is exactly balanced by the entropy change, equal in magnitude but opposite in sign, that occurs in the thermal surroundings, owing to the exchange of thermal energy between the chemical system and its surroundings. Indeed, from a knowledge of two things, the thermal energy exchanged between the chemical system and its surroundings and the values at some reference concentration of the entropies of HAc, H_2O, H_3O^+, and Ac^-, the number 1.8×10^{-5} can be calculated from a simple formula (Chapter 26). This formula, we shall find, is a direct consequence of the First and Second Laws of Thermodynamics and the thermodynamic definition of the absolute temperature.

I

Introduction to Classical Thermodynamics

I

The First Law (I)

$$E_{\text{total}} = \text{constant}$$

There was once a young inventor who thought he could produce energy from nothing. "It is well known," said he, "that an electric motor converts electrical energy into mechanical energy and that an electric generator converts mechanical energy into electrical energy. The two machines taken together are like the bacterium *Rhizobium leguminosarum* and the sweet pea. What one needs the other makes. The pea needs fixed nitrogen (nitrates), and makes carbohydrates; the bacteria need fixed carbon (carbohydrates), and make nitrates. In the end, this association yields a net profit to man. Why not, then, use the motor to run the generator and the generator to run the motor (Fig. 1.1, dashed lines) and create, thereby, an endless supply of energy? This energy could be used to do useful mechanical work; it could be used, for example, to lift a weight (Fig. 1.1). The whole unit could be

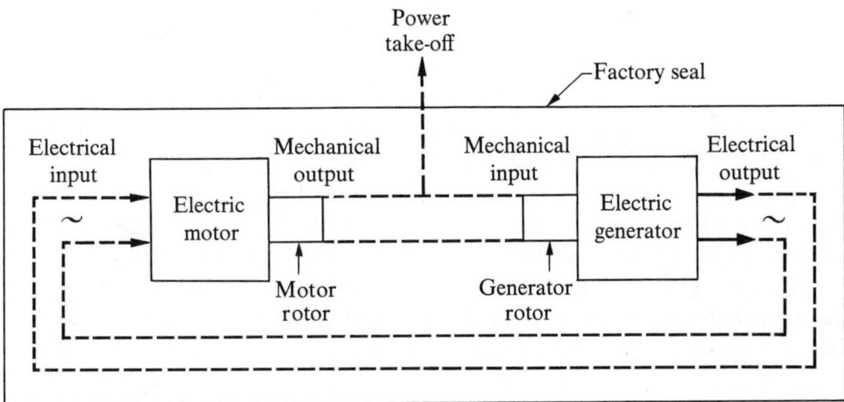

Fig. 1.1 Electric motor and generator. Dashed lines show proposed symbiotic coupling of these two units (see text).

factory-sealed and sold guaranteed to do work perpetually without the addition from outside of any mechanical, electrical, or chemical energy."

This proposal can be dismissed as impracticable. *No device for producing energy from nothing has ever worked.* And were the motor-generator to function in the aforedescribed manner (Fig. 1.1), energy would in fact be produced from nothing. For the net effect of the change from State 1 to State 2 is merely the raising of a weight; by hypothesis the condition of the motor-generator is unchanged. The energy changes in going from State 1 to State 2 would therefore be: for the weight, an increase in potential energy; for the motor-generator, no change; for the two together, an increase. In the notation of thermodynamics,

$$(E_{\text{Wt.}})_{\substack{\text{final}\\\text{state}}} > (E_{\text{Wt.}})_{\substack{\text{initial}\\\text{state}}}$$

$$(E_{\text{M.G.}})_{\substack{\text{final}\\\text{state}}} = (E_{\text{M.G.}})_{\substack{\text{initial}\\\text{state}}} .$$

Adding,

$$[E_{\text{Wt.}} + E_{\text{M.G.}}]_{\substack{\text{final}\\\text{state}}} > [E_{\text{Wt.}} + E_{\text{M.G.}}]_{\substack{\text{initial}\\\text{state}}} ,$$

or

$$(E_{\text{total}})_{\substack{\text{final}\\\text{state}}} > (E_{\text{total}})_{\substack{\text{initial}\\\text{state}}} .$$

This last statement is a direct contradiction of the First Law of Thermodynamics.*

Similarly, no one has ever devised a process whose net effect is the destruction of energy. Drop a weight to the floor, for example, or jump out a window, and energy equivalent to the decrease in potential energy of the system (weight plus earth) appears ultimately as heat.

The only processes known to man appear to be those for which

$$(E_{\text{total}})_{\substack{\text{final}\\\text{state}}} = (E_{\text{total}})_{\substack{\text{initial}\\\text{state}}} ;$$

i.e. processes for which

$$E_{\text{total}} = \text{constant}.$$

Processes whose *net* effects are equivalent to the raising (or lowering) of a weight have never been observed (Fig. 1.2).

Of course, weights can be raised or lowered. But always the gain or loss in energy of a weight whose altitude is altered is balanced exactly by a complementary change in energy in some other part of the universe.

The term E_{total} signifies the energy of everything that might conceivably be altered during an event. Those objects that might be altered during an event together make up a collection of objects that will be called a universe of the event.

* The production of heat by the motor-generator unit has been ignored in this discussion. Its inclusion, on the reasonable assumption (see later) that the motor-generator warms up during operation, would only make worse the violation of the First Law.

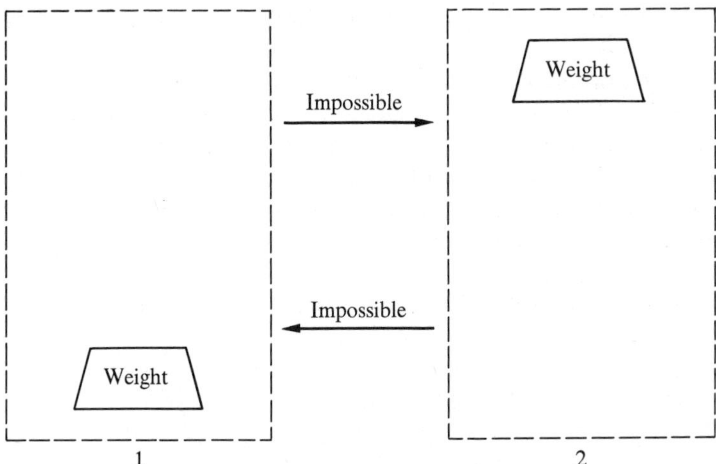

Fig. 1.2 Two impossible processes. Energy can be neither created (spontaneous passage from State 1 to State 2) nor destroyed (spontaneous passage from State 2 to State 1). The dashed line encompasses everything—here merely a weight—that might change in some fashion during the process under consideration.

The collection of objects composed of all stars and their satellites and the satellites of these satellites and the space between them is a universe of all events. Generally this is a larger universe than necessary. For most events a more modest collection of objects suffices. For a particular event the smallest collection of objects in terms of which every thermodynamically important aspect of the event can be described will be called *the* universe of the event. To be able to recognize *the* universe of a particular event is a skill which underlies most applications of thermodynamics to practical problems.

The universe of an event is, in effect, an isolated system. As far as the event is concerned, nothing outside its universe changes. Inside an isolated system, or universe, the sum of the energies of the parts remains constant irrespective of any changes that occur there. This fact is frequently expressed by saying that energy is conserved.

(Most people seem to believe this firmly; mathematicians because they believe it is a fact of observation; observers because they believe it is a theorem of mathematics; philosophers because they believe it is aesthetically satisfying, or because they believe no inference based upon it has ever been proven false, or because they believe new forms of energy can always be invented to make it true. A few neither believe nor disbelieve it; these people maintain that the First Law is a procedure for bookkeeping energy changes, and about bookkeeping procedures it should be asked, not are they true or false, but are they useful.)

To summarize, "energy" has two important properties. It is conserved, and it is additive.

Additive means merely that, like mass and volume, the energy of a system is the sum of the energies of its parts. Temperature, for example, is not additive, for the temperature of a system is not equal to the sum of the temperatures of its individual parts. The temperature of a house, for example, is not the sum of the temperatures of its individual rooms.*

Properties that are additive are called extensive properties. Energy, mass, and volume are extensive properties.

The additive property and the conservative property of energy form the theoretical basis for the assignment of molar energies to chemical substances from experimentally determined heats of reaction. The procedure for doing this is described in the following chapter.

Problems for Chapter 1

> One must learn by doing the thing; for though you think you know it, you have no certainty until you try.
>
> SOPHOCLES

Answers to the following problems, and to problems at the ends of subsequent chapters, are given at the back of the book. The purpose of these problems is to review and to summarize, in some cases to amplify and to extend, and in other cases to provide illustrative examples and alternative points of view to the discussion in the text.

1. Give several equivalent statements of the First Law.
2. Can the statement (E_{total}) = constant be applied to the earth?
3. Show that if energy can be neither created nor destroyed, $(E_{\text{total}})_{\substack{\text{final}\\\text{state}}}$ must be equal to $(E_{\text{total}})_{\substack{\text{initial}\\\text{state}}}$.
4. To bookkeep changes in energy accurately, what objects ought to be considered part of the universe of the following events?
   (a) Ammonium nitrate dissolves in water.
   (b) Hydrogen and oxygen explode in a closed bomb.
   (c) A solution of hydrochloric acid is titrated with a solution of sodium hydroxide.
   (d) Zinc pellets dissolve in aqueous hydrochloric acid.
   (e) A rubber band is rapidly extended by a hanging weight.
   (f) A rubber band is slowly extended by a hanging weight.

* Conservation and additivity are closely related. Indeed, it is difficult to conceive of a property being conserved if it is not additive. As we shall see, however, not all additive properties are conserved.

(g) The gas in a chamber is rapidly compressed by a weighted piston.
(h) The gas in a chamber is slowly compressed by a weighted piston.
(i) A glass shatters on the floor.
(j) Your mother-in-law steps on the bathroom scales.
(k) A tray of water freezes in the freezing compartment of an electric refrigerator.

5. Let the internal energies at a specified temperature and pressure of a mole of ice and a mole of liquid water be designated, respectively. $E_{H_2O}^{solid}$ and $E_{H_2O}^{liq}$.*

   (a) What is the internal energy in terms of $E_{H_2O}^{solid}$ and $E_{H_2O}^{liq}$ at the specified conditions of 54 grams of solid water? Of 54 grams of liquid water?
   (b) A beaker contains $n_{H_2O}^{solid}$ moles of ice and $n_{H_2O}^{liq}$ moles of liquid water at the previously specified temperature and pressure. In terms of the symbols $n_{H_2O}^{solid}$, $n_{H_2O}^{liq}$, $E_{H_2O}^{solid}$, and $E_{H_2O}^{liq}$, what is the energy of the beaker's contents?
   (c) To the beaker described in (b) is added sufficient energy from an outside source to melt one mole of the ice; the temperature and pressure of the system remain at their initially specified values. How many moles of ice remain? How much liquid water does the beaker contain? What is the energy of the beaker's contents?

6. Energy is liberated when hydrogen and oxygen explode. Does this constitute an exception to the First Law, which states that energy-producing processes are impossible?
7. Energy is absorbed when ice melts. Does this constitute an exception to the First Law, which states that energy-destroying processes are impossible?
8. A length of steel spring wire that has been wound into a tight coil is dissolved in hydrochloric acid. What happens to the mechanical energy spent coiling the wire?
9. Can hot kitchens be cooled by opening refrigerator doors?
10. The discovery of radioactivity produced a crisis in physics regarding the First Law. The situation has been aptly described by E. U. Condon (*Physics Today*, October 1962, p. 44).

They [the Curies] had not measured the half life, and the energy given off did not show any signs of weakening, and you know that physicists are great on extrapolation. They said that radium gives off energy *perpetually*—that was the word, perpetually.

So the question was, how could anything radiate perpetually at this tremendous

* In the notation of thermodynamics the subscript position after a letter such as E, V, S, or G is usually reserved for the symbol of the chemical in question, the superscript position for a description of the phase in which the chemical is present.

rate?—a rate unheard of when expressed in terms of energies of usual chemical reactions.

Kelvin had an idea. . . . Perhaps, he suggested, there was some kind of energy that one could not detect . . . floating around in space, and perhaps radium had the property of absorbing it, like a fountain, and shooting it out, and that was what was observed. Even in those days people were perfectly willing to balance the books on conservation of energy in such a manner.

Later, accurate measurements of the energies involved in beta decay produced yet another crisis in physics. The sum of the energies of the products of decay seemed to be less than the energy of the starting nucleus. What conjectures do you suppose this experimental result produced?

James P. Joule

If our present views regarding the First Law may be said to stem from one person, that person is probably James Joule, a Manchester brewer, who, in 1838, at nineteen, became interested in improving electric motors and discovered that mechanical work could be converted into heat, approximately 800 foot-pounds of mechanical work being required to raise the temperature of one pound of water one Fahrenheit degree. This result was communicated to the British Association in 1843. It was received with "entire incredulity" and "general silence."

In 1844 a paper by Joule on this same subject was rejected by the Royal Society.

In 1845 he discussed before the British Association further results in which it was suggested that the water at the bottom of a waterfall should be warmer than at the top, for Niagara falls, 160 feet high, about one-fifth of a Fahrenheit degree; also, from the thermal expansion of gases Joule deduced that there should be a "zero of temperature" 480°F below the freezing point of water. This was the first suggestion of absolute zero. These results again failed to provoke discussion.

Two years passed and once more, in 1847, Joule presented his work in more perfect form at an Oxford meeting of the British Association. As he later remarked, ". . . the communication would have passed without comment if a young man had not risen in the section, and by his intelligent observations created a lively interest in the new theory. The young man was William Thomson [later Lord Kelvin] who had two years previously passed the University of Cambridge with the highest honour, and is now [1885] probably the foremost scientific authority of the age."

In later years Thomson recalled that he was "tremendously struck" by Joule's paper, and added, "This is one of the most valuable recollections of my life, and is indeed as valuable a recollection as I can conceive in the possession of any man interested in science." Still, he found himself at first unable to accept Joule's views; he wrote to his brother James, "I enclose Joule's papers, which will astonish you. I have only had time to glance through them as yet. I think at present that some great flaws must be found."

Thomson has given this account of the historic 1847 Oxford meeting.

I made Joule's acquaintance at the Oxford meeting, and it quickly ripened into a life-long friendship. I heard his paper read at the section, and felt strongly impelled to rise and say that it must be wrong. . . . But as I listened on and on I saw that Joule had certainly a great truth and a great discovery, and a most important measurement to bring forward. . . .

Faraday was there and was much struck with it, but did not enter fully into the new views. It was many years after that before any of the scientific chiefs began to give their adhesion.

Miller and Graham, or both, were for many years quite incredulous as to Joule's results, because they all depended on fractions of a degree of temperature, sometimes very small fractions. His boldness in making such large conclusions from such very small observational effects is almost as note-worthy and admirable as his skill in extorting accuracy from them. I remember distinctly at the Royal Society, I think it was either Graham or Miller saying simply he did not believe in Joule because he had nothing but hundredths of a degree to prove his case by.

In the same account Thomson adds that about a fortnight after the Oxford meeting he was walking down from Chamonix to commence a tour of Mont Blanc, "and whom should I meet walking up but Joule, with a long thermometer in his hand, and a carriage with a lady in it not far off. He told me that he had been married since we parted at Oxford! and he was going to try for elevation of temperature in waterfalls."

Joule's paper "On the Mechanical Equivalent of Heat" was communicated by Michael Faraday to the Royal Society in 1849 and appeared in the *Philosophical Transactions* in 1850. The last paragraph of this historic paper ends with the statement,

I will therefore conclude by considering it as demonstrated by the experiments contained in this paper—

1st. That the quantity of heat produced by the friction of bodies, whether solid or liquid, is always proportional to the quantity of force extended. And,

2nd. That the quantity of heat capable of increasing the temperature of a pound of water (weighed in vacuo, and taken at between 55° and 60°) by 1° Fahr. requires for its evolution the expenditure of a mechanical force represented by the fall of 772 lb. through the space of one foot.

A third proposition, suppressed by a committee to whom the paper had been referred, stated that friction consists in the conversion of mechanical work into heat.

2

The First Law (II)

Calorimetry

What does it mean to say the heat of melting of ice is 80 calories/gram, or 1440 calories/mole?*

Two things.

It means the melting of one mole of ice will cool 1440 grams of liquid water 1° centigrade. (At room temperature removal of one calorie of energy from one gram of water lowers its temperature almost exactly one centigrade degree.)

But, in addition, the energy lost by the water surrounding the melting ice (1440 calories per mole of ice melted) must be accounted for; for according to the First Law, the energy of the "universe" (the melting ice plus the surrounding water) is constant. It might be supposed that the energy content of liquid water is greater than that of the solid by 1440 calories per mole. Indeed, there seems to be no other choice; together, experiment and the First Law appear to imply that

$$E_{H_2O}^{liq} = E_{H_2O}^{solid} + 1440 \text{ cal/mole}.$$

In this thermochemical equation, the symbols $E_{H_2O}^{liq}$ and $E_{H_2O}^{solid}$ stand for the energies of one mole of liquid and solid water, respectively. Note that it is only the difference between them that is determined ($E_{H_2O}^{liq} - E_{H_2O}^{solid} = 1440$ cal/mole). In ordinary applications of the First and Second Laws the absolute values of $E_{H_2O}^{liq}$ and $E_{H_2O}^{solid}$ cannot be determined and are not needed.

In a similar fashion, to say that the heat of reaction of hydrogen and oxygen at constant volume is 67,400 calories per mole of liquid water formed is to say that explosion of one mole of hydrogen gas and one-half a mole of oxygen gas in a closed vessel, a "bomb," liberates energy sufficient to warm 67,400 grams of liquid water almost exactly one centigrade degree after the

* $\left(80 \dfrac{\text{cal}}{\text{gm}}\right)\left(18 \dfrac{\text{gm}}{\text{mole}}\right) = 1440 \dfrac{\text{cal}}{\text{mole}}.$

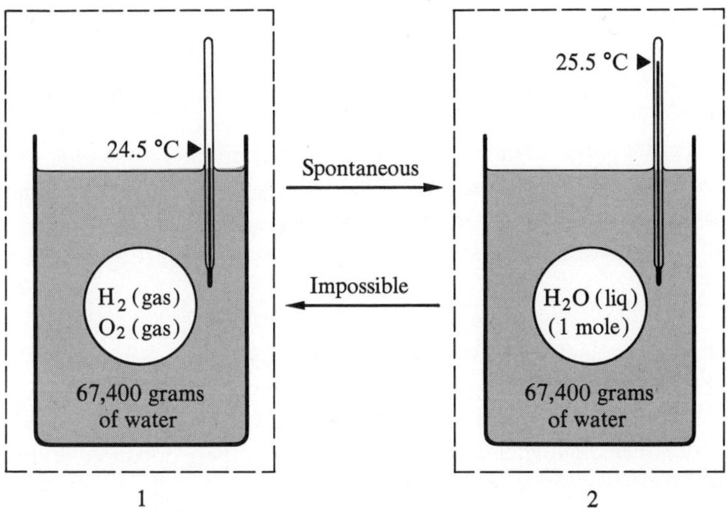

Fig. 2.1 Bomb calorimeter for measuring the thermal energy produced by the spontaneous reaction H_2(gas) + $\frac{1}{2}O_2$(gas) = H_2O(liq).

product of the reaction has cooled to the temperature of the surroundings (Fig. 2.1). The net effect of this change may be expressed as follows:

$$H_2(\text{gas}) + (1/2)O_2(\text{gas}) = H_2O(\text{liq}) + 67{,}400 \text{ cal (at } 25°\text{C});$$

or $$H_2(\text{gas}) + (1/2)O_2(\text{gas}) - 67{,}400 \text{ cal} = H_2O(\text{liq}).$$

But what is the source of the thermal energy gained by the bomb and the water surrounding the bomb? Other than a rise in temperature, the bomb itself, if a good one, suffers few, if any, significant changes. Evidently there is no alternative but to suppose that the energy gained by the bomb and its surroundings stems entirely from the fact that the energy of the final contents of the bomb, the products of the chemical reactions, H_2O(liq), is less than the energy of the initial contents of the bomb, the chemical reactants, H_2 and O_2. The thermochemical equation that summarizes this conclusion is similar to the last equation above read in reverse.

$$E^{\text{liq}}_{H_2O} = E^{\text{gas}}_{H_2} + (1/2)E^{\text{gas}}_{O_2} - 67{,}400 \text{ cal/mole (at } 25°\text{C}); \text{ or}$$

$$E^{\text{liq}}_{H_2O} - [E^{\text{gas}}_{H_2} + (1/2)E^{\text{gas}}_{O_2}] = -67{,}400 \text{ cal/mole}; \text{ or}$$

$$E_{\text{products}} - E_{\text{reactants}} = -67{,}400 \text{ cal/mole of } H_2O(\text{liq}) \text{ formed, where}$$

$$E_{\text{products}} = E^{\text{liq}}_{H_2O}$$

$$E_{\text{reactants}} = E^{\text{gas}}_{H_2} + (1/2)E^{\text{gas}}_{O_2}.$$

Note again that only an energy difference is given.

Often it proves convenient to take as a reference of energy the energies of the elements in their "standard states." The phrase "standard states" refers to the forms of the elements stable at 25°C and 1 atm.; for example, for hydrogen, hydrogen gas; for oxygen, oxygen gas; for sulfur, rhombic sulfur; for mercury, liquid mercury. This convention corresponds to setting $E_{H_2}^{0gas} = 0$ and $E_{O_2}^{0gas} = 0$ (also $E_S^{0rhombic} = 0$ and $E_{Hg}^{0liq} = 0$).* The superscript ^{0} stands for "at 25°C and 1 atm.". With this convention, and the measured heat of the reaction $H_2(gas) + (1/2)O_2(gas) = H_2O(liq)$ (Fig. 2.1), it follows that

$$E_{H_2O}^{0liq} = -67{,}400 \text{ cal/mole}.$$

This does not mean that the energy content of liquid water at 25°C and 1 atmosphere is actually negative. Rather, it signifies that at 25°C and 1 atm. the energy of one mole of liquid water is 67,400 calories less than the energy of its constituent elements in their standard states (here one mole of hydrogen gas and one-half a mole of oxygen gas).

One peculiarity about spontaneous reactions like the one illustrated in Fig. 2.1 may be noted. As surely as the change is spontaneous, it is irreversible. Once having passed from State 1 to State 2 spontaneously (Fig. 2.1), the isolated universe (the bomb, its contents, and its thermal surroundings) will never spontaneously revert back again to State 1. State 2 is evidently intrinsically more probable than State 1, at 25°C.

The melting of ice in warm surroundings is another example of a spontaneous, irreversible process. Ice melts in warm water; the reverse process, the freezing of warm water, does not occur. On the other hand, consider the effect of a change in temperature. Cold water does freeze. Also, it is a fact that a mixture of hydrogen and oxygen can be gotten too hot to burn; indeed, at very high temperatures the reverse reaction is spontaneous: water vapor at say 5000°C spontaneously absorbs energy from its surroundings and dissociates into hydrogen and oxygen.

Evidently chemical instability, or, what is the same thing, the spontaneity of a chemical reaction, depends on something more than the heat of reaction. The stable state of a chemical or a mixture of chemicals is not always the state of lowest energy. True, many exothermic reactions ($E_{\text{products}} < E_{\text{reactants}}$) are spontaneous; for example, the freezing of cold water and the explosion of cold hydrogen and oxygen; but then, again, there exist many endothermic reactions that are spontaneous: for example, the melting of hot ice and the dissociation of hot water vapor.

The additional factor that contributes to chemical stability is closely related to temperature. It is called entropy.

* Later we will find that it is usually more convenient to set equal to zero the standard enthalpies of the elements. The enthalpy is equal to $E + PV$. It is given the symbol H. At ordinary pressures, $H \approx E$.

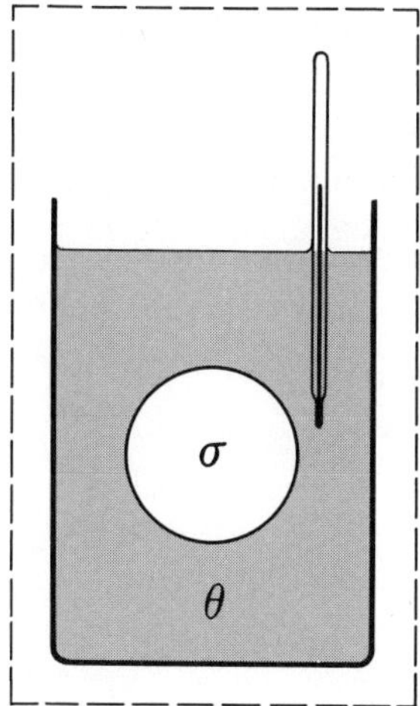

Fig. 2.2 Conceptually important parts of a bomb calorimeter.

A recapitulation in general terms. A system composed of a reaction vessel (e.g. a bomb) containing reactive chemicals (e.g. hydrogen and oxygen) and surrounded by a quantity of water (in which rests a thermometer) can be used to measure calorimetric changes; hence its name: calorimeter. From the thermochemical point of view a calorimeter is a self-contained universe (Fig. 2.2). Regardless of the specific character of the chemical change that occurs within the calorimeter, it will always be true for a bomb-type calorimeter that

$$(E_{\text{total}})_{\substack{\text{final}\\ \text{state}}} = (E_{\text{total}})_{\substack{\text{initial}\\ \text{state}}},$$

where for both initial and final states

$$E_{\text{total}} = E_{\substack{\text{bomb}\\ \text{contents}}} + E_{\substack{\text{bomb and the}\\ \text{surrounding water*}}}.$$

Chemists usually refer to the bomb contents as the "chemical system" and to the bomb and its surrounding water as the "thermal surroundings" of the chemical system. This partitioning of the universe in the mind's eye into two parts proves to be very useful in thermodynamic analyses of chemical events. It is therefore a considerable convenience, particularly in writing down formulas, to have for these two parts of the universe, the chemical system and its thermal surroundings, easily written and discernible symbols. The two symbols used in this book for these parts of the universe are σ and θ. σ comes from the *s* in "chemical *s*ystem," θ from the *t* in "*t*hermal surroundings"—or from the *t* in *t*emperature (in thermodynamics the only

* And the thermometer.

thing important about the thermal surroundings is, in fact, its temperature). Briefly,

$$\sigma = \text{chemical system}$$
$$\theta = \text{thermal surroundings of the chemical system.}$$

Introduction of this notation into the statement that one obtains by combining the two previous equations yields the following expression for a bomb-type calorimeter:

$$(E_\sigma + E_\theta)_{\substack{\text{final}\\\text{state}}} = (E_\sigma + E_\theta)_{\substack{\text{initial}\\\text{state}}}.$$

Written out in full, this reads

$$(E_\sigma)_{\substack{\text{final}\\\text{state}}} + (E_\theta)_{\substack{\text{final}\\\text{state}}} = (E_\sigma)_{\substack{\text{initial}\\\text{state}}} + (E_\theta)_{\substack{\text{initial}\\\text{state}}}.$$

To reflect correctly the fact that thermochemical measurements yield information regarding only the relation to each other of energy differences, this last expression may be written in some such form as this:

$$\left[(E_\sigma)_{\substack{\text{final}\\\text{state}}} - (E_\sigma)_{\substack{\text{initial}\\\text{state}}} + (E_\theta)_{\substack{\text{final}\\\text{state}}} - (E_\theta)_{\substack{\text{initial}\\\text{state}}}\right] = 0.$$

It is notationally convenient at this point to have a symbol that means "change in." The symbol commonly used in mathematics for this is Δ. If, then, one defines

$$\Delta(E_\sigma) \equiv \left[(E_\sigma)_{\substack{\text{final}\\\text{state}}} - (E_\sigma)_{\substack{\text{initial}\\\text{state}}}\right]$$

and

$$\Delta(E_\theta) \equiv \left[(E_\theta)_{\substack{\text{final}\\\text{state}}} - (E_\theta)_{\substack{\text{initial}\\\text{state}}}\right], \qquad (*)$$

the somewhat cumbersome expression above may be written

$$\Delta(E_\sigma) + \Delta(E_\theta) = 0 \qquad \text{or} \qquad \Delta(E_\sigma) = -\Delta(E_\theta).$$

The last equation may be regarded as still another statement of the conservation of energy: the energy one part of the universe gains another part loses.

Problems for Chapter 2

1. At 0°C what is the energy of H_2O(solid) with respect to the energy of H_2O(liq)?
2. At 25°C the energy of dissociation at constant volume of hydrogen gas is 103 kilocalories per mole of hydrogen gas consumed. This means that

* In general, $\Delta(X) \equiv X_{\substack{\text{final}\\\text{state}}} - X_{\substack{\text{initial}\\\text{state}}}$, where $X = E_\sigma, E_\theta, S_\sigma, S_\theta$, etc.

if a mole of hydrogen gas were to dissociate spontaneously at 25°C to produce in the original volume at the original temperature two moles of hydrogen atoms, it would be found that the thermal surroundings of the chemical system had lost 103 kilocalories of energy. What thermochemical inference follows from this fact? What is the energy of atomic hydrogen with respect to the energy of molecular hydrogen, at 25°C?

3. In the example $H_2(\text{gas}) + (1/2)O_2(\text{gas}) = H_2O(\text{liq})$, what are the values of $(E_\sigma)_{\substack{\text{initial}\\\text{state}}}$ and $(E_\sigma)_{\substack{\text{final}\\\text{state}}}$ in terms of $E^{\text{gas}}_{H_2}$, $E^{\text{gas}}_{O_2}$, and $E^{\text{liq}}_{H_2O}$?
4. For a bomb-type calorimeter,

$$\Delta E_\sigma = -\Delta E_\theta.$$

When ΔE_θ is positive $[(E_\theta)_{\substack{\text{final}\\\text{state}}} > (E_\theta)_{\substack{\text{initial}\\\text{state}}}]$, the chemical reaction occurring within the bomb is said to be ———(a). In such cases ΔE_σ must be ———(b).
5. When ΔE_θ for a bomb-type calorimeter is negative $[(E_\theta)_{\substack{\text{final}\\\text{state}}} < (E_\theta)_{\substack{\text{initial}\\\text{state}}}]$, the change occurring within the bomb is said to be ———(a). In such cases ΔE_σ must be ———(b).
6. What is the value of ΔV_σ for a reaction that occurs in a perfectly rigid bomb?
7. One might attempt to calculate the temperature of a hydrogen-oxygen torch that is operating under essentially adiabatic conditions (conditions under which the flame does not lose significant energy to its thermal surroundings) by supposing that the chemical reaction in the flame is quantitative conversion of hydrogen and oxygen to water vapor and that the entire difference between the energy of water vapor at ambient temperature and the energy of the hydrogen and oxygen at the same temperature goes into heating up the water formed in the flame reaction. One finds, however, that the temperature calculated in this way is significantly greater than the actual adiabatic flame temperature. How do you account for this fact?
8. Does the electrolytic dissociation of water into hydrogen and oxygen constitute an exception to the remark that the change from State 2 to State 1 in Fig. 2.1 is impossible?
9. Criticize this statement: At low temperatures the energy of a solid approaches zero, the energy of a gas—which at low temperatures is $(3/2)RT$—also approaches zero, and, therefore, the energy of sublimation approaches zero.
10. At low temperatures mixtures of hydrogen and oxygen explode and water freezes. At high temperatures, ice melts and water vapor dissociates into hydrogen and oxygen. What general rule do these facts suggest?

Joseph Black and the Melting of Ice

The magnitude of the energy changes that occur in the universe when ice melts was not appreciated until the experiments of an Englishman, Joseph Black (1728–99). The motivation behind these experiments, their nature, and their chief results, are described in Dr. Black's own words (with some modernizations) in the following paragraphs.*

Our experience of freezing of liquids when exposed to more or less powerful degrees of cold is almost universal. The exceptions are very few. The strongest spirit of wine and a few subtle and volatile oils are the only substances that have not yet been solidified by any degree of cold hitherto known. . . .

Quicksilver was, not long since, one of this small number of substances, which having never been seen in any other than a liquid state, was considered as naturally and essentially liquid, and incapable of being reduced to a solid form, until experiments were made with it, first in different parts of the Russian Empire, since the year 1760, and verified afterwards in other places. By these experiments, every person must be convinced that quicksilver is a metal that can become solid and malleable like the rest, but that it freezes at a lower temperature than has ever been observed over the greater part of the surface of this earth. In the same manner may we consider all other liquids as solids melted by heat.

Some philosophers, however, have offered many objections to this general proposition concerning the nature of liquids. They thought it necessary to suppose that water is an exception. They could not be persuaded that its liquidity is the effect of heat, but supposed this quality to be an essential one of the water, depending on the spherical form and polished surface of its particles, and that the freezing of it depended on the introduction of some extraneous, subtle matter. . . .

The propensity of many people to imagine water as naturally and essentially liquid is a prejudice contracted from the habit of seeing it much oftener in this state than in the solid state. . . .

In considering the effect of heat in producing liquifaction, we should first remark that innumerable experiments made with thermometers show that the change of a particular substance from solid to liquid occurs only when the temperature is increased to a certain value. Above this temperature the substance is a liquid. If the liquid is cooled back down to this temperature, it becomes solid, and it remains solid at all lower temperatures. This at least may be stated as the general fact. . . .

. . . [I]n general, each different kind of matter must be heated to a particular temperature to render it liquid, and below this temperature it is either solid or has some degree of solidity (beeswax, resin, tallow, glass, etc.). This temperature is therefore called the FREEZING or the MELTING POINT of the substance. It is called the

* Reprinted by permission of the publishers from *Harvard Case Histories in Experimental Science*, Vol. I, James B. Conant, ed., 1957. Copyright 1957 by the President and Fellows of Harvard College.

freezing point of such substances as exist commonly in the liquid state, and the melting point of those that are solid under ordinary circumstances. . . .

I must now add that the foregoing account of liquefaction as an effect of heat is not complete and satisfactory. . . .

Melting had been universally considered as produced by the addition of a *very small quantity* of heat to a solid body, once it had been warmed up to its melting point. . . . [italics mine]

This was the universal opinion on the subject, so far as I know, when I began to read my lectures in the University of Glasgow in the year 1757. But I soon found reason to object to it, as inconsistent with many remarkable facts, when attentively considered. . . .

The opinion I formed from attentive observation of the facts and phenomena is as follows. When ice or any other solid substance is melted, I am of the opinion that it receives a much larger quantity of heat than what is perceptible in it immediately afterwards by the thermometer. A large quantity of heat enters into it, on this occasion, without making it apparently warmer, when tried by that instrument. . . .

On the other hand, when we freeze a liquid, a very large quantity of heat comes out of it, while it is assuming the solid form, the loss of which heat is not to be perceived by the common manner of using the thermometer. . . .

If we attend to the manner in which ice and snow melt when exposed to the air of a warm room, or when a thaw succeeds to frost, we can easily perceive that, however cold they might be at first, they soon warm up to their melting point and begin to melt at their surfaces. And if the common opinion had been well founded—if the complete change of them into water required only the further addition of a very small quantity of heat—the mass, though of a considerable size, ought all to be melted within a very few minutes or seconds by the heat incessantly communicated from the surrounding air. Were this really the case, the consequences of it would be dreadful in many cases; for, even as things are at present, the melting of large amounts of snow and ice occasions violent torrents and great inundations in the cold countries or in the rivers that come from them. But, were the ice and snow to melt suddenly, as they would if the former opinion of the action of heat in melting them were well founded, the torrents and inundations would be incomparably more irresistible and dreadful. They would tear up and sweep away everything, and this so suddenly that mankind would have great difficulty in escaping their ravages. This sudden liquefaction does not actually happen. . . .

In order to understand better this absorption of heat by the melting ice, and concealment of it in the water, I made the following experiments. . . .

I chose two thin globular glasses, 4 inches in diameter, and very nearly the same weight. I poured 5 ounces of pure water into one of them, and then set it in a mixture of snow and salt until the water was frozen into a small mass of ice. It was then carried into a large empty hall, in which the air was not disturbed or varied in temperature during the progress of the experiment. . . .

I now set up the other globular glass precisely in the same way, and at the distance of 18 inches to one side, and into this I poured 5 ounces of water, previously cooled almost to the freezing point—actually to 33°F. Suspended in it was a very delicate thermometer, with its bulb in the center of the water, and its stem so placed that I could read it without touching the thermometer. I then began to observe the ascent of this thermometer, at suitable intervals, in order to learn with what celerity the water received heat; I stirred the water gently with the end of a feather about a minute before each observation. The temperature of the air, examined at a little distance from the glasses, was 47°F.

The thermometer assumed the temperature of the water in less than half a minute, after which, the rise of it was observed every 5 to 10 minutes, during half an hour. At the end of that time, the water was 7 degrees warmer than at first; that is, its temperature had risen to 40°F.

[Dr. Black next explains that the time to warm the ice-containing glass to 40°F was 10½ hours.]

It appears that the ice-glass had to receive heat from the air of the room during 21 half-hours in order to melt the ice and then warm the resulting water to 40°F. During all this time it was receiving heat with the same celerity (very nearly) as had the water-glass during the single half-hour in the first part of the experiment. . . . Therefore, the quantity of heat received by the ice-glass during the 21 half-hours was 21 times the quantity received by the water-glass during the single half-hour. It was, therefore, a quantity of heat, which, had it been added to liquid water, would have made it warmer by $(40 - 33) \times 21$, or 7×21, or 147 degrees. No part of this heat, however, appeared in the ice-water, except that which produced the temperature rise of 8 degrees; the remaining part, corresponding to 138 to 140 degrees, had been absorbed by the melting ice and was concealed in the water nto which it was changed. . . .

11. What is the heat of melting of ice as determined by Dr. Black's data?
12. In an article on machine guns in the eleventh edition of the *Encyclopaedia Britannica* there appears this statement.

    The great difficulty which has to be met in all single-barrel machine guns is the heating of the barrel. The 7½ pints of water in the water-jacket of the Maxim gun are raised to boiling point by 600 rounds of rapid fire—i.e. in about 1½ minutes—and if firing be continued, about 1½ pints of water are evaporated for every 1000 rounds.

    Estimate from these data the energy required to evaporate water in terms of the energy required to heat the same quantity of liquid water one degree centigrade.

3

The Second Law (I)

Further Limitations on Weightlifting

Performance of useful mechanical work—the lifting of a weight, for example—requires energy in some form or other. Electric motors use electrical energy; diesel engines, rocket motors, and professional weightlifters use chemical energy; and so forth.

It has been suggested that mechanical energy might be delivered more economically by drawing upon the thermal energy of the surroundings.* Why not operate an ocean-going vessel, for example, off the virtually inexhaustible supply of thermal energy in the oceans? Pump aboard warm water, extract energy from it to operate the ship's mechanical machinery, and when finished eject the cold water, or icebergs, overboard. There being no violation of the First Law, who, except the Coast Guard, could possibly object?

As it turns out, this is rather like asking a cold man to warm himself on the energy of an icicle. Nothing in the First Law prohibits the transaction *icicle colder – man warmer*; nonetheless, it is well known that energy transfers of this kind never occur.

Similarly, processes whose net effects are equivalent to the cooling of a body and the raising of a weight (or the performance of some other kind of useful mechanical work, as, for example, the operation of an ocean-going vessel) have never been observed (Fig. 3.1).

These two hypothetical processes, the flow of energy from an icicle to a warmer body and the complete conversion of thermal energy to mechanical energy, have several features in common. In both processes, energy is conserved. Nevertheless, neither process actually does occur. Finally, the reverse of each process is a process that does occur in nature. Energy does flow from hot to cold, as, for example, when a warm-blooded man holds

* The thermal energy of the surroundings is the energy it has in excess of what it would have at absolute zero.

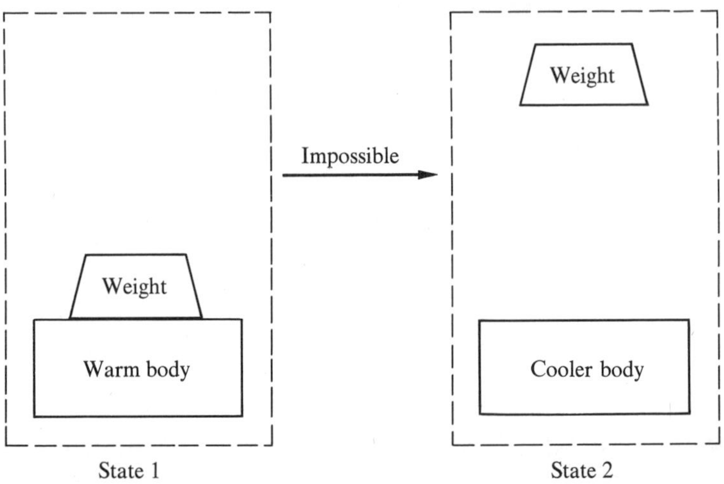

Fig. 3.1 A limitation on weightlifting. Though energy is conserved, the transition from State 1 to State 2 is impossible.

an icicle in his hand. And mechanical energy can be converted completely into thermal energy, by friction, for example, or by dropping a leaden weight on the floor.

Problems for Chapter 3

1. List several characteristics common to the following processes: the melting of very cold ice; the freezing of hot water; the thermal dissociation at room temperature of water into hydrogen and oxygen; the flow of heat from a cold body to a warmer one; the raising of a weight solely at the expense of the thermal energy of a body.
2. Consider a chamber containing a gaseous mixture of nitrogen, hydrogen, and ammonia, pressurized by a movable piston, and suppose that the mixture is at equilibrium with respect to the rearrangement of atoms

$$2N_2(\text{gas}) + 3H_2(\text{gas}) = 2NH_3(\text{gas}).$$

Suppose, further, that introduction of a catalyst shifts the equilibrium to the left by increasing the rate of the backward reaction (the formation of nitrogen and hydrogen from ammonia) more than it increases the rate of the forward reaction (the formation of ammonia from nitrogen and hydrogen) with a consequent increase in the number of moles of gas in the chamber and rise in the piston's equilibrium position, and that on removing the catalyst the system returns to its original state.

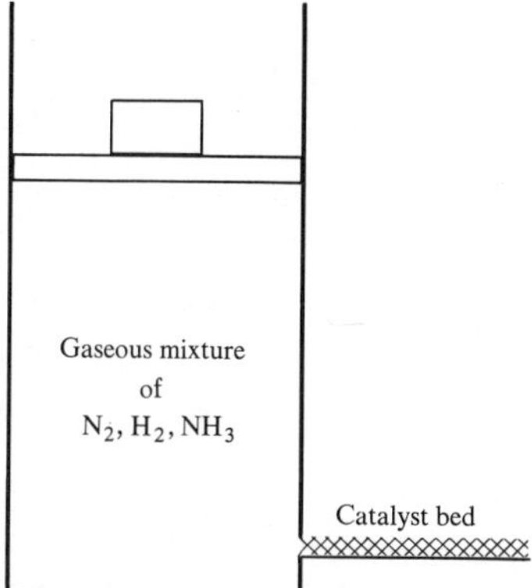

Apparatus for testing the effect of a solid catalyst on a chemical system that is at equilibrium with respect to the reaction $N_2 + 3H_2 = 2NH_3$.

The cycle could then be repeated: catalyst in, piston up; catalyst out, piston down. In effect one has at hand a little engine that should be capable of doing useful mechanical work. To keep the engine from cooling off—presumably this might interfere with its efficient operation, and according to the First Law the energy that appears as work must come from somewhere—it is allowed to remain in thermal contact with its surroundings. This costs nothing. Would not this be a good way to operate ocean-going vessels, particularly in the lower latitudes?

3. In the Introduction it was stated that weights do not spring spontaneously into the air at the expense of the thermal energy of their surroundings. In this chapter it was stated that processes whose net effects are equivalent to the raising of a weight and the cooling of a body have never been observed. Are these two statements equivalent?

4

The Second Law (II)

The Downfalling Habits of Weights

The small child who tosses a toy or a blanket or some other imperfectly elastic object from his crib soon learns what to expect. The tossed object will fall to the floor, and there it will come to rest, eventually. The process is entirely spontaneous *and* completely irreversible. Never does the object at some later time come back spontaneously into the crib.

In common with other spontaneous processes—the freezing of cold water and the melting of warm ice, the explosion of cold hydrogen-oxygen mixtures and the dissociation of hot water vapor, and the flow of thermal energy from hot objects to colder ones—the process *weight lower – surroundings warmer* results in a net increase in the disorder of the universe.

This disorder cannot be seen with the unaided eye, however, for it mainly concerns the chaotic character of the random thermal motions of the individual atoms and molecules of the weight and its thermal surroundings. As illustrated explicitly, and with exaggeration, in Fig. 4.1, the disorder repre-

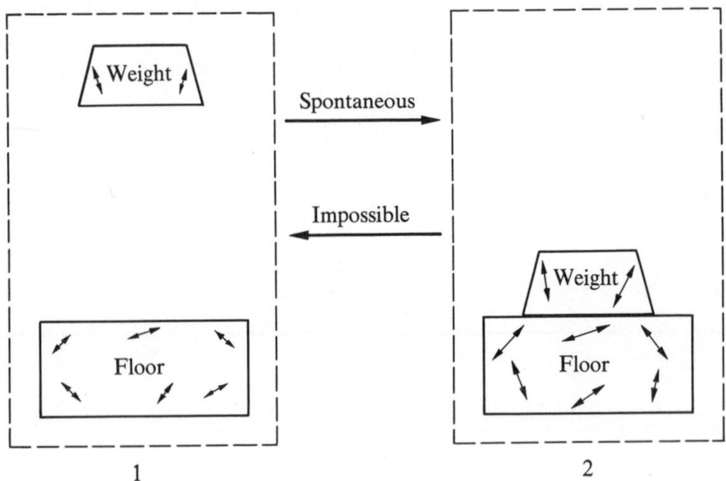

Fig. 4.1 A spontaneous process: the conversion of potential energy into random molecular motion. The reverse process is impossible: the energy of random molecular motion can never be converted completely into potential energy.

sented by the random motions of the atoms and molecules of the weight and its surroundings is greater in final State 2 than in initial State 1. State 2, the state of greater microscopic disorder, is intrinsically more probable than State 1 and passage from the less microscopically disordered State 1 to the more microscopically disordered State 2 can, given time, occur spontaneously.

The reverse process, *weight higher – surroundings cooler*, would correspond to passage from a more microscopically disordered state to a less microscopically disordered state and in common with other disorder-decreasing processes—the freezing of warm water and the melting of cold ice, the explosion of very hot hydrogen-oxygen mixtures and the dissociation of cold water vapor, the flow of thermal energy from cold objects to hotter ones—does not occur. An extreme example of an improbable process would be for the weight to rise spontaneously to such heights that, to conserve energy, the temperature of the weight and its thermal surroundings would have to drop to absolute zero. Events as improbable as this are generally termed impossible. At room temperature, for example, conversion of a single calorie of thermal energy completely into potential energy is a less likely event than the production of Shakespeare's complete works fifteen quadrillion times in succession without error by a tribe of wild monkeys punching randomly on a set of typewriters.

Spontaneous, irreversible processes (the freezing of cold water or the melting of warm ice) are processes in which the entropy of the universe increases. Impossible processes (the freezing of warm water or the melting of cold ice), were they to occur, are processes in which the entropy of the universe would decrease. Reversible processes (the freezing of water or the melting of ice at 0°C, and 1 atm.) are processes in which the entropy of the universe does not change.

These remarks may be summarized as follows:

$$(S_{\text{total}})_{\substack{\text{final}\\ \text{state}}} > (S_{\text{total}})_{\substack{\text{initial}\\ \text{state}}}$$

$\Delta S_{\text{total}} > 0\,.$ Process may occur. Will be irreversible.

$$(S_{\text{total}})_{\substack{\text{final}\\ \text{state}}} < (S_{\text{total}})_{\substack{\text{initial}\\ \text{state}}}$$

$\Delta S_{\text{total}} < 0\,.$ Process highly improbable. Probably will never occur.

$$(S_{\text{total}})_{\substack{\text{final}\\ \text{state}}} = (S_{\text{total}})_{\substack{\text{initial}\\ \text{state}}}$$

$\Delta S_{\text{total}} = 0\,.$ Process reversible. May occur in either direction.

Entropy, unlike energy, is not conserved. The statement that the entropy of a universe tends to increase is known as the Second Law of Thermodynamics.

But entropy, like energy, is additive. The entropy of a system is equal to the sum of the entropies of its individual parts. This property of entropy, we shall see, makes it possible to predict whether the total entropy of the universe will increase, decrease, or remain the same during a chemical reaction if three things are known: the heat of the reaction, the absolute temperature, and the molar entropies of the reactants and the products.

Problems for Chapter 4

1. Give several equivalent statements of the Second Law.
2. List several characteristics common to all spontaneous processes.
3. List several spontaneous, irreversible, entropy-producing processes.
4. List several reversible, entropy-conserving processes.
5. On what important point is the First Law entirely silent?
6. On what important point is the Second Law entirely silent?
7. What factors determine the shape adopted by a flexible rope hung between two horizontal supports?
8. Show that if thermal energy could be converted entirely to mechanical energy, heat could be made to flow from a cold body to a warmer one without there occurring in the universe any other change.
9. It has been pointed out that a fried egg can be unfried by grinding it up with mash and feeding it back to a hen. In time the hen will produce a fresh egg. Does this constitute an exception to the statement that spontaneous processes are never reversible?
10. In a steam engine thermal energy of the boiler appears ultimately as work. Is this a violation of the statement that processes whose net effects are equivalent to the raising of a weight solely at the expense of the thermal energy of a body are impossible?

Addendum to Chapter 4

To restore every part of a universe to its original condition after a weight, say a brick, has fallen on a floor, the brick's elevation must be increased and the floor's temperature must be decreased.

It has been suggested that the latter change, the cooling of the floor, might be accomplished with a piece of ice.

But in cooling a warm floor, ice initially at 0°C will begin to melt.

That could be rectified by placing the product, the liquid water, in a refrigerator—provided, of course, the refrigerator is plugged in, and the electricity is on. Which raises another question.

What is the source of the refrigerator's electricity?

It might come from a generator at the base of Niagara Falls—in which case, to complete the reverse process (no more and no less than the raising of a weight and the cooling of an object), the water at the bottom of Niagara Falls that supplied the electricity to the refrigerator that froze the water that cooled the floor that was heated by the brick must somehow be gotten back to its original elevation. Here nature might help.

Let the water evaporate into the atmosphere and condense as a rain cloud over Lake Erie.

In fact, this only makes matters worse. Evaporation occurs best when water is warm, condensation when water is cold. The net effect of the two processes is to restore the water to its original elevation and, simultaneously, to transfer thermal energy from a warm spot, the spot that provided the heat of evaporation, to a colder spot, the spot that received the heat of condensation. Energy transfers of this type are difficult, in fact impossible, to reverse. Heat simply does not flow spontaneously from cold objects to warmer ones, although it can be forced in that direction with appropriate devices, one example of which is a refrigerator.

A refrigerator transfers heat from its cooling coils (the freezing compartment) to its thermal surroundings (the kitchen). To provoke this uphill transfer of heat, however, the refrigerator needs help. It must be plugged in.

In summary, it seems impossible to precisely undo the effects produced by a falling brick. The brick can be raised again, and the object upon which it landed can be cooled. But inevitably this produces in the universe other changes (the melting of ice) which in turn cannot be undone without producing still further changes (water over a waterfall), and so forth ad infinitum.

As Lord Kelvin said, "Any process whose *net* effect is the raising of a weight and the cooling of an object is impossible."

Bricks cannot be sprung into the air solely at the expense of the thermal energy of their surroundings. To do so would be to bring about a most unusual event: the diminution of the disorder in the universe.

5

Survey of Molar Entropies (I)

Flatten 'em With Platinum

I shoot the Hippopotamus
  with bullets made of platinum,
Because if I use leaden ones
  his hide is sure to flatten 'em.

BELLOC, *The Bad Child's Book of Beasts*

Science, it has been said, begins in observation. The best way to become acquainted with birds, for example, is to look at some birds. A good way to become acquainted with entropy is to look at some entropies.

Substance	*Molar entropy at room temperature and atmospheric pressure.**
Diamond	0.6
Platinum	10.0
Lead	15.5
Laughing gas	52.6

These numbers suggest that entropy is related to hardness. Indeed, as a rule, hard, gem-like, abrasive, and refractory materials such as diamond, garnet, topaz, quartz, fused zirconia, silicon carbide, and boron nitride, in which the individual atoms are bound to each other in nearly infinite, three-dimensional lattices by genuine chemical bonds that severely limit random thermal motions of atoms, have small measured entropies. On the other hand, soft substances, especially gases, usually contain large amounts of thermal disorder at room temperature and have, correspondingly, large measured entropies. Lamellar crystals are an interesting exception. While hard substances always have small entropies, not all soft substances have large entropies. Graphite, which is hard in two dimensions, but soft in the third, has an entropy at room temperature of only 1.34 entropy units per

* The units of entropy will be considered in the following chapter. They are the same as the units of heat capacity: calories per degree per mole.

mole. Complex substances generally have larger entropies than simple substances of similar hardness. The entropies at room temperature of copper, sodium chloride, and zinc chloride, for example, are 8.0, 17.3, and 25.9 entropy units per gram formula weight. Another example is sodium sulfate and its decahydrate; their entropies at 25°C are, respectively, 35.7 and 141.7. It is interesting to note that one-tenth the difference between these two numbers is close to the molar entropy of ice extrapolated from its melting point to 25°C.

Melting and vaporization lead always to an increase in molar entropy (with a concomitant, but never overbalancing, decrease in entropy of the surroundings). For water, the molar entropies at 0°C and 1 atm. of the relatively ordered solid, the less ordered liquid, and the highly disordered vapor are, respectively, 9.8, 15.1, and 44.4.*

The high-temperature modification of a substance is always the form with the largest molar entropy. Typically the intermediate entropy of the liquid phase lies numerically closer to the small entropy of the solid than it does to the larger entropy of the vapor. For non-hydrogen-bonded substances—ethane, benzene, mercury, or chlorine, for example—the entropy of melting, $S(\text{liquid}) - S(\text{solid})$, is generally about 2 to 3 entropy units per mole and the entropy of vaporization at the normal boiling point is generally about 20 to 23 entropy units per mole. For hydrogen-bonded substances, like water, the corresponding values are somewhat larger.

From the figures cited above for water, one infers that liquids are more solid-like than gas-like. This inference agrees well with the familiar view that in solids and liquids the individual atoms and molecules are constantly in contact with their nearest neighbors. By contrast, gases are highly dispersed phases. In a dilute gas, the free volume—the volume not actually occupied by the molecules themselves—constitutes more than 99.9 per cent of the total volume.

The existence of enormous free volumes in gases explains many of their most characteristic properties: their low densities; their high compressibilities; their low absorption and reflection of light; their poor conduction of sound; their complete miscibility in each other; their nearly identical equations of state; their large entropies; and their property of having thermal energies and entropies, which (like all Gaul) can be divided into three parts—one for translational motion of the centers of mass of the molecules, one for unhindered rotation of the molecules as rigid bodies about their centers of mass, and one for thermally excited bond-bending and bond-stretching vibrations (Fig. 5.1). This division is illustrated below for the entropies of nitrous oxide and water vapor.

*This is an extrapolated value. Water vapor at 1 atm. is unstable with respect to condensation at all temperatures below 100°C.

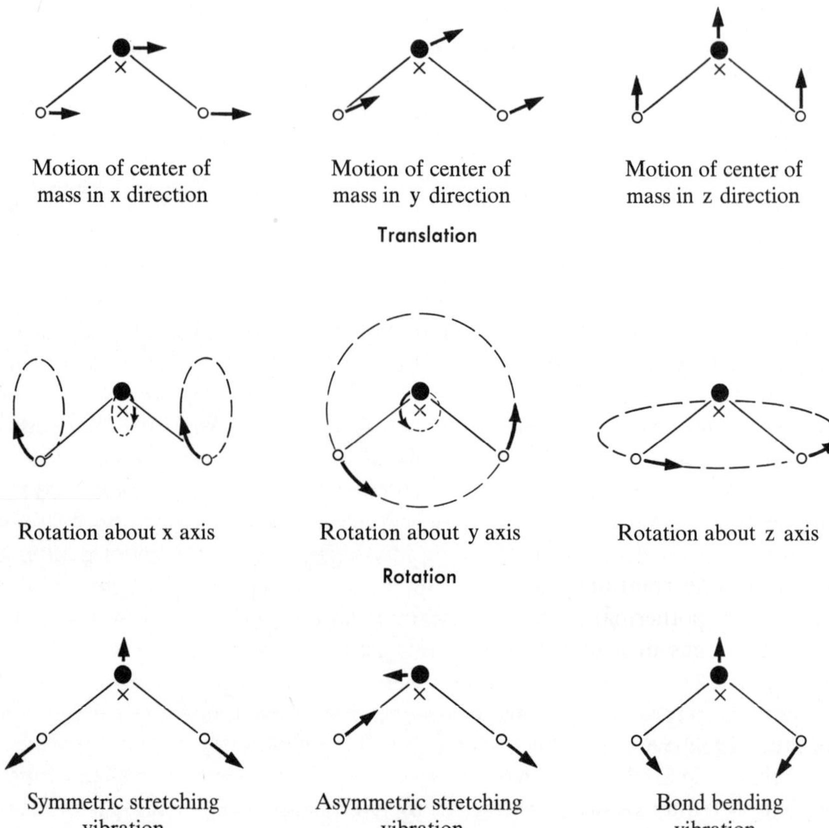

Fig. 5.1 Translational, rotational, and vibrational degrees of freedom of a water molecule. ● = oxygen nucleus; ○ = hydrogen nucleus; x = molecule's center of mass, which remains stationary in pure rotational and vibrational motion.

	Translational entropy (25°C, 1 *atm.*)	*Rotational entropy* (25°C)	*Vibrational entropy* (25°C)	*Total entropy* (25°C, 1 *atm.*)
N_2O(gas)	37.3	12.9	2.4	52.6
H_2O(gas)	34.6	10.5	0.0	45.1

The small contributions that vibrational degrees of freedom make to the entropies of water vapor and nitrous oxide at room temperature stem from the fact that at this temperature the vibrational degrees of freedom of these molecules are not easily excited. This fact is revealed, also, in the heat capacities of the two gases. The translational heat capacity of an ideal gas is (3/2)R. For water (a non-linear molecule), the rotational heat capacity at room temperature is (3/2)R; for nitrous oxide (a linear molecule), (2/2)R.* For H_2O(gas), the sum of the translational and rotational heat capacities is (6/2)R = 6.0 cal/deg-mole; for N_2O(gas), (5/2)R = 5.0 cal/deg-mole. The actual heat capacities at constant volume (for an ideal gas the heat capacity at constant volume is R less than the heat capacity at constant pressure) are (in cal/deg-mole): for H_2O, 6.038 (compared to 6.0); for N_2O, 7.265 (compared to 5.0). From this, one concludes (correctly) that vibrational degrees of freedom contribute very little to the heat capacity (and entropy) of H_2O at room temperature. For N_2O, there is a small contribution. The vibrational heat capacities of gases are discussed quantitatively in Chapter 22. The question as to why the electrons in the molecules do not contribute to the heat capacity and entropy at room temperature is discussed in Chapter 6.

The external pressure on a gas has a marked effect on the molar volume of the gas and, through this, on the molar entropy of the gas. A tenfold decrease in pressure at constant temperature increases the volume involved in the translational motion of the molecules of an ideal gas tenfold and results in a 4.6 entropy unit per mole increase in the translational entropy of the gas. The rotational and vibrational entropies of a dilute gas are not affected by isothermal changes in volume. Thus the net effect on the entropy of a dilute gas of a tenfold change in pressure and volume is a change in entropy of 4.6 entropy units per mole. A change in molar entropy of this magnitude may have a profound effect on the stability of the vapor. This is because the driving force behind evaporation (an endothermic process) lies entirely in the large entropy of the vapor. If the molar entropy of the vapor is diminished by allowing its pressure (or partial pressure, if other gases are present) to increase, as happens in a pressure cooker, evaporation may be prevented. On the other hand, by continually decreasing the partial pressure of a vapor a point can always be reached where the entropy of the vapor is sufficiently great to make evaporation of the corresponding solid or liquid, however cold, a spontaneous (although not necessarily rapid) process. Snow will evaporate, even on cold days, if the air is very dry.

The vapor pressure of a substance is the pressure (or partial pressure) of its vapor that makes the entropy of the vapor such that during evaporation or condensation the entropy of the universe does not change.

Problems for Chapter 5

1. Estimate the molar entropies at 25°C and 1 atm. of boron, barium, and ammonia.

* See Chapter 22, problem 7.

2. The entropy of $CaSO_4$ at 25°C is 25.5 cal/deg-mole. Estimate the entropy of $CaSO_4 \cdot 2H_2O$.
3. Which has the larger entropy, rhombic or monoclinic sulfur?
4. Calcium carbonate crystallizes in two forms called calcite and aragonite. At room temperature their entropies are, respectively, 22.2 and 21.2 cal/deg-mole. Which is the more stable form at high temperatures?
5. Compare the entropies of a liquid and its vapor at the normal boiling point. At the critical point.
6. Molecular motion that produces a change in the position of the center of mass of the molecule without altering bond angles and bond lengths and with which there is associated no angular momentum is called pure ——— motion. Molecular motion that produces no change in either the center of mass of the molecule or its bond angles and bond lengths but that does have associated with it angular momentum is called pure ——— motion. Molecular motion that produces no change in the position of the center of mass of the molecule and with which there is associated no angular momentum but that does produce changes in bond angles and/or bond lengths is called pure ——— motion.
7. Translate into technical terms the phrase "dry air."
8. The entropy of an ideal gas at a certain pressure and volume and temperature (P_1, V_1, T) is related to its entropy at another pressure and volume and the same temperature (P_2, V_2, T) in the following manner:

$$S(P_2, V_2, T) = S(P_1, V_1, T) + nR \ln (V_2/V_1).$$

If $V_2 > V_1$ (in which case $P_2 < P_1$ if $T_2 = T_1 = T$), the last term is positive and $S(P_2, V_2, T) > S(P_1, V_1, T)$. Show that for an ideal gas

$$S(P_2, V_2, T) = S(P_1, V_1, T) - nR \ln (P_2/P_1).$$

$R = 1.987$ cal/deg-mole, n = number of moles of gas, and $\ln (\ ) = \log_e (\ )$. What is the value of $[S(P_2, V_2, T) - S(P_1, V_1, T)]$ when $n = 1$ and $P_2/P_1 = 0.1$?
9. Why is there always a decrease in entropy of the surroundings when a substance melts or evaporates spontaneously? Why is this decrease never greater in magnitude than the increase in the entropy of the substance itself?

6

Survey of Molar Entropies (II)*

Thermal Excitation of Quantized Motion

Motion in nature may be classified as free or constrained. Constrained motion includes the motion of nucleons within atomic nuclei, the motion of electrons within the electron cloud of an atom, the motion of atomic nuclei during a molecular vibration or rotation, and the motion of the molecules of a gas that is confined to a container of finite dimensions.

An interesting feature of constrained motion, a feature not noticed at the everyday level of experience with macroscopic objects, is its quantization. This is a way of saying that the energy associated with constrained motion does not assume a continuous spectrum of values. The energy can have only discrete values, which are separated from one another by finite (although sometimes very small) intervals. Perhaps the most familiar example of this quantization is the set of electronic energy levels of the hydrogen atom. The lowest energy a hydrogen atom can have (the energy of the 1*s* state) is separated from the next allowed energy level of the system (the energy of the 2*s* state) by a large energy gap.

The spacing between the allowed energy levels of a system determines how easily the system will absorb thermal energy. This in turn determines what its entropy will be. If the spacing between allowed energy levels is small, thermal excitation of the system is relatively easy and its entropy at a given temperature will be relatively large. On the other hand, if the spacing between allowed energy levels is large (as in the hydrogen atom), thermal excitation is relatively difficult and the system's corresponding entropy will be small.

The spacing between the allowed energy levels in a system critically

* This chapter is in some respects like descriptive chemistry. In it are discussed facts that seem important but that are not at the moment explained. Surveyed and summarized here are the principal conclusions, as they bear on entropy, of statistical mechanics and quantum mechanics. The connection between these disciplines and entropy is discussed in more detail in Part III.

depends upon the distance the individual molecules, atoms, electrons, or nucleons within the system move when excited in the type of motion considered. If the characteristic amplitude of motion is small (as in atomic vibrations), the spacing between allowed energy levels is large and the associated entropy is small. On the other hand, if the characteristic amplitude is large (as in the translational motion of the individual molecules of a gas), the spacing between allowed energy levels is small and the associated entropy is large. In general, the larger the amplitude of motion, the larger the entropy.

It may seem paradoxical that small amplitude motion is the most difficult motion to excite. In fact, a moment's reflection reveals that this statement corresponds closely to experience. It is well known, for example, that the translational motion of gas molecules can be altered with an ordinary tire pump; whereas to explore after the fashion of a chemist the properties of individual atoms and molecules frequently requires the energy of a bunsen burner; while to produce changes in the tiny nucleus of an atom usually requires the use of immense equipment and enormous energies.

If the characteristic amplitude of motion is small, say only 10^{-4} to 10^{-5} angstroms, as in the motion of the individual nucleons within an atomic nucleus, thermal excitation from one allowed energy level to the next is exceedingly difficult and requires the million-degree temperatures of stellar interiors and nuclear explosions. On the other hand, when the characteristic amplitude of motion is large, say 10 cm, as in the translational motion of the individual molecules of a dilute gas, thermal excitation of the now very closely spaced energy levels occurs readily at temperatures only a small fraction of a degree above absolute zero. For this reason, translational degrees of freedom make a large contribution to the entropies of gases at room temperature. Molecular rotations, where the characteristic amplitude of motion is 1 to 3 angstroms, make a significant, but smaller, contribution. Molecular vibrations, whose amplitudes range typically from 0.1 to 0.5 angstroms, generally contribute little to the entropies of simple gases at room temperature. And at room temperature, contributions to the entropy from the vibrations or rotations of nucleons within atomic nuclei can usually be neglected.

Failure of electronic motion to contribute significantly to the room-temperature entropies of nitrous oxide and water vapor (Chapter 5) stems from the fact that electrons are very light. This is important because the ease of excitation of quantized motion depends, in fact, upon two things: the amplitude of the motion, as already discussed above, and the masses of the particles involved in the motion. The lighter the particles, the more widely spaced are the quantum-mechanically allowed energy levels for the system, and the more difficult its thermal excitation.* For this reason the values for the translational, rotational, and vibrational entropies of nitrous oxide (molecular weight 14 + 14 + 16) are slightly greater than the corresponding values for water vapor (molecular weight 1 + 1 + 16).

Addition of thermal energy to a substance generally increases its temperature and its entropy.

Similarly, removal of thermal energy from a substance decreases its entropy. This effect continues all the way down to the lowest temperatures that can be reached. At absolute zero the entropy of a perfect crystal, regardless of its chemical composition, may be taken as zero. This statement is known as the Third Law of Thermodynamics.

The increase in entropy of a substance when thermal energy is added to it is proportional to the amount of thermal energy added. The entropy of a cold substance, however, is altered more by the addition of one calorie of thermal energy than is the entropy of a hot substance. These facts may be read from the following equation, which is valid for all substances at all temperatures:

$$\Delta S = \left(\frac{1}{T}\right) \Delta E .$$

ΔS and ΔE are the changes in entropy and thermal energy, respectively (strictly speaking under the assumption that there is no change in the volume of the substance†); T is the absolute temperature.

Problems for Chapter 6

1. Estimate the molar entropy at 25°C and 1 atm. of neon.
2. The entropies at 25°C of NaCl and NaBr are 17.3 and 20.0 cal/deg-mole, respectively. Do these seem like reasonable values?
3. Estimate the entropy of KBr. The entropy of KCl is 19.7 cal/deg-mole.
4. Estimate the entropy of $PbBr_2$. The entropy of $PbCl_2$ is 32.6 cal/deg-mole.
5. Does the value 35.7 cal/deg-mole seem reasonable for the entropy of anhydrous sodium sulfate at 25°C?
6. Why is the molar entropy of oxygen (MW 32) larger than the molar entropy of xenon (At. Wt. 131)?

* The spacing between allowed energy levels is inversely proportional to the product of the characteristic mass involved in the motion and the square of the length of this mass's orbit (Chapter 24).

† And that during the act of absorption or desorption of thermal energy there occur within the substance no irreversible, entropy-producing chemical or physical changes.

7. Suggest an explanation for the fact that the entropy of NO_2 is slightly over 5 cal/deg-mole greater than the entropy of N_2O.
8. Why is the molar entropy of hydrogen fluoride gas (MW 20) less than the molar entropy of water vapor (MW 18)?
9. Why is the molar entropy of HCl less than the molar entropy of O_2?
10. Why are not the rotational and vibrational entropies of a dilute gas affected by isothermal changes in volume and pressure?
11. Comment on the qualitative changes that seem likely to occur in the translational, rotational, and vibrational components of the entropy of a polyatomic gas on liquefaction.
12. What are the changes in entropy of a 400°K body and a 300°K body that
    (a) gain 1200 calories of thermal energy?
    (b) lose 1200 calories of thermal energy?
    Assume that the bodies are so large the addition or removal of 1200 calories has no noticeable effect on their temperatures.
13. What is the change in the entropy of the universe when 1200 calories of energy flow from a 400°K body to a 300°K body?
14. Can the relation $\Delta S = \Delta E/T$ be used to calculate the entropy change that occurs in a substance whose temperature changes as energy is added to it?
15. List some of entropy's more characteristic properties.

Addendum to Chapter 6

DETERMINATION OF ABSOLUTE ENTROPIES

The entropy of a substance may be determined in a conceptually straightforward manner through use of the relation $\Delta S = \Delta E/T$, the Third Law, and calorimetric data. Simply cool the substance whose entropy is to be determined to a temperature T_1 close to 0°K; then add small measured increments of energy and measure, simultaneously, the substance's temperature as it warms up in small stages to a final temperature T_2. To compute the substance's entropy, divide each increment of added energy by the instantaneous value of the substance's absolute temperature, and sum. Add to this sum an estimate of the amount missed between 0°K and T_1. The total is the entropy of the substance at temperature T_2 if its entropy at 0°K is zero. Entropies determined in this manner are called Third Law entropies.

The computation of Third Law entropies from warm-up data is usually performed in a manner mathematically equivalent to, though not identical with (cf. references listed in the Bibliography), the graphical procedure described below.

The change in entropy of an object that absorbs ΔE calories of energy at temperature T may be represented graphically by the area of a rectangle

$1/T$ deg^{-1} high and ΔE calories wide (Fig. 6.1). Continued absorption of thermal energy by an object will in general cause the object's temperature to increase and the reciprocal of its temperature to decrease. The rectangles that represent the entropy increments of an object that absorbs successively ΔE_1, ΔE_2, ΔE_3, ... calories of energy at a series of increasing temperatures T_1, T_2, T_3, ... will, therefore, as shown in Fig. 6.2, become shorter and shorter. At the normal melting point, T_m, T and $1/T$ remain constant while energy equal to the energy of melting ΔE_m is absorbed. (For the compound H_2O, the rectangle representing the entropy of melting has an area of $(1/273°K)$ (80 cal/gram) = 0.293 cal/deg-gram = 5.28 cal/deg-mole.) Placed edge-to-edge, the separate rectangles produce a step-like figure (Fig. 6.3). As the separate energy increments ΔE_1, ΔE_2, ΔE_3, ... become very small the area of this figure approaches the area beneath the continuous curve formed by plotting the instantaneous value of the reciprocal of the object's

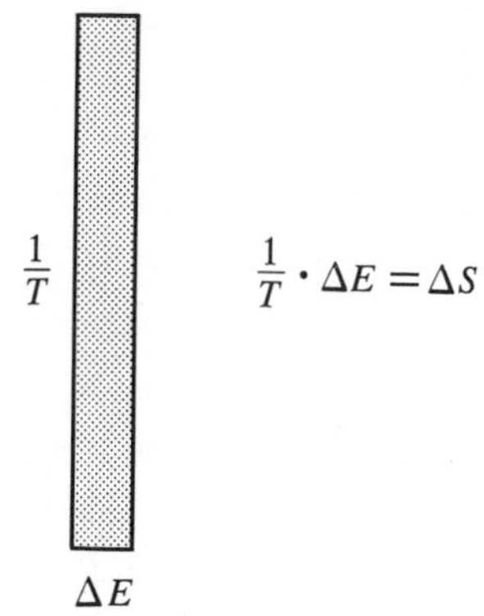

Fig. 6.1

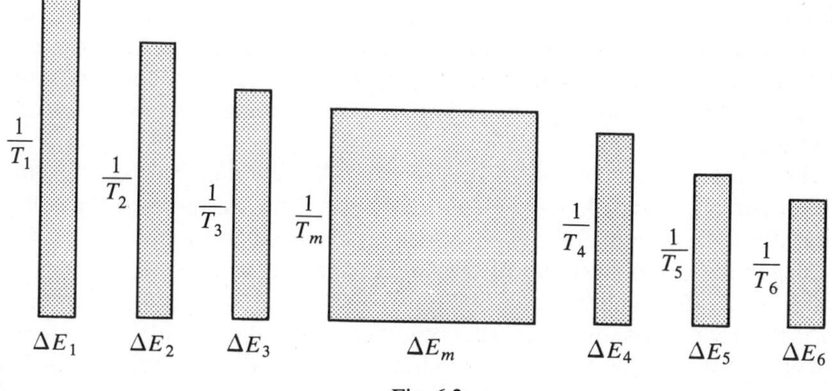

Fig. 6.2

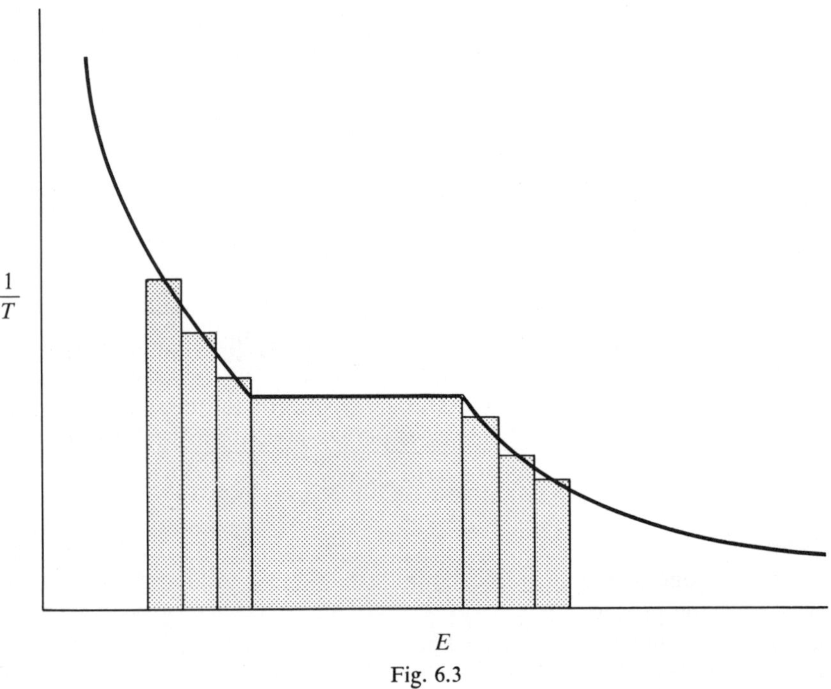

Fig. 6.3

absolute temperature against the thermal energy absorbed by the object during warm-up.

The area beneath this curve accurately gives the change in entropy of the warmed-up object.

(If during warm-up the parameter held constant is the pressure, not—as assumed here—the volume, the quantity plotted horizontally should be H, not E.)

The area beneath a $(1/T) - E$, or $(1/T) - H$, curve that has been carried experimentally to low temperatures and then extrapolated to absolute zero represents the entropy the object has in excess of the entropy it has at absolute zero.

The Third Law entropy of liquid water at 25°C (assuming that one begins with a perfect crystal) has these contributions: from warm-up of the solid from absolute zero to 0°C, 9.86 cal/deg-mole; from melting at 0°C, 5.28 cal/deg-mole; from warm-up of the liquid from 0°C to 25°C, 1.58 cal/deg-mole; total, 16.72 cal/deg-mole. Vaporization of the liquid at 25°C and its normal vapor pressure contributes another 35.3 cal/deg-mole to water's entropy. This number is equal to the heat of vaporization of water at 25°C, 10,514 cal/mole, divided by 298°K.

7

Temperature (I)

$$\frac{\Delta E}{\Delta S}$$

One of the simplest of all phenomena, and by the same token one of the most fundamental, is the flow of thermal energy from a hot object to a cold object. This process is one that requires no outside help. Whenever a hot object and a cold object are in thermal contact, migration of thermal energy from the hot one to the cold one occurs spontaneously. The spontaneous character of the process means—if the Second Law is applicable—that the entropy of the universe (hot object plus cold object) is greater after the migration of energy has occurred than it was before. And this is despite the fact that loss of energy by the hot object decreases its entropy, because of course the cold object in gaining thermal energy gains in entropy. Evidently, the loss in entropy of the hot object is more than made up by the gain in entropy of the cold object (Fig. 7.1).

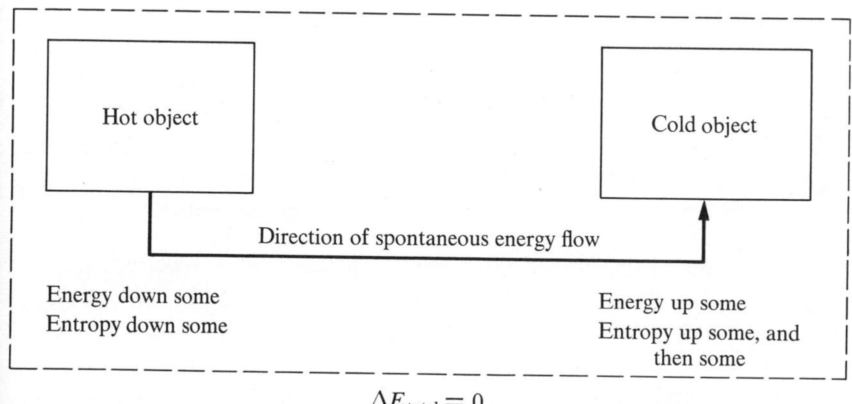

$$\Delta E_{\text{total}} = 0$$

$$\Delta S_{\text{total}} > 0$$

Fig. 7.1 Energy and entropy changes in the universe when thermal energy flows from a hot object to a cold object.

Further, it is seen that the direction of flow of thermal energy between two bodies depends not on the absolute value of the energy or the entropy of either body, or on their chemical composition. From the standpoint of the Second Law the direction of migration of thermal energy depends solely on the rate at which the entropy of each body changes as its thermal energy changes. If this rate is large (large change in entropy for given change in energy), the body's thermal behavior corresponds to that of cold objects. It will act as a good acceptor of thermal energy (large increase in entropy for given addition of energy) and as a poor donor of the same commodity (large decrease in entropy for given loss of energy). If, on the other hand, the rate of change of entropy of a body with respect to its thermal energy is small (small change in entropy for given change in energy), its thermal behavior will correspond to that of hot objects. Such objects are relatively poor acceptors of thermal energy (small increase in entropy for given addition of energy), but, by the same token, they are relatively good donors of this commodity (small decrease in entropy for given loss of energy).

To summarize, if reception of a given quantity of energy ΔE alters by ΔS the entropy of an object and ΔS is large, the object is a good acceptor and a poor donor of energy; in a word, cold. If, on the other hand, for the same change in energy, ΔS is small, the object is a poor acceptor and a good donor of energy; in a word, hot. Thus, the ratio

$$\frac{\Delta E}{\Delta S}$$

is an index of thermal behavior. For cold objects (large ΔS for given ΔE) its value is small. For hot objects (small ΔS for given ΔE), its value is large.

Problems for Chapter 7

1. Name two phenomena that probably played an important historical role in the formulation of the Second Law.
2. Describe the significance in thermodynamics of the terms "hot" and "cold."
3. Let the value of the ratio $\Delta E/\Delta S$ for an object be designated τ.

$$\tau = \frac{\Delta E}{\Delta S}$$

(a) If an object absorbs Q calories of thermal energy, what is its change in entropy in terms of Q and τ? Assume that Q is very small, or that the thermal capacity of the object is very large, so that during the absorption of energy there is no significant change in τ.

(b) Show that the flow of Q calories of energy from an object A to an object B (Q is taken to be a positive number) produces a net increase in the entropy of the universe if $\tau_A > \tau_B$.

(c) Show that if the flow of Q calories of energy from object A to object B produces a net increase in the entropy of the universe, τ_A must be greater than τ_B.

(d) Summarize parts (b) and (c) of this problem in a single statement.

4. Express in words the meaning of the expression $(\Delta S/\Delta E) \cdot Q$.

Dr. Black's Discovery Concerning the Distribution of Heat

The role temperature plays in the distribution of thermal energy (or "heat") was not understood until, some two hundred years ago, Dr. Joseph Black (see Chapter 2), a practicing physician, thought to examine the temperatures of objects that had been thermally equilibrated with each other. What he discovered is described below in his own words.*

An improvement in our knowledge of heat, which has been attained by the use of thermometers, is the more distinct notion we have now than formerly of the *distribution* of heat among different bodies. Even without the help of thermometers, we can perceive a tendency of heat to diffuse itself from any hotter body to the cooler ones around it, until the heat is distributed among them in such a manner that none of them is disposed to take any more from the rest. The heat is thus brought into a state of equilibrium.

This equilibrium is somewhat curious [italics mine]. We find that, when all mutual action is ended, a thermometer applied to any one of the bodies undergoes the same degree of expansion. Therefore the temperature of them all is the same. No previous acquaintance with the peculiar relation of each body to heat could have assured us of this, and we owe the discovery entirely to the thermometer. We must therefore adopt, as one of the most general laws of heat, the principle that *all bodies communicating freely with one another, and exposed to no inequality of external action, acquire the same temperature, as indicated by a thermometer.* All acquire the temperature of the surrounding medium.

By the use of thermometers, we have learned that, if we take a thousand, or more, different kinds of matter—such as metals, stones, salts, woods, cork, feathers, wool, water and a variety of other fluids—although they be all at first of different temperatures, and if we put them together in a room without a fire, and into which

* Reprinted by permission of the publishers from *Harvard Case Histories in Experimental Science*, Vol. I, James B. Conant, ed., 1957. Copyright 1957 by the President and Fellows of Harvard College.

the sun does not shine, the heat will be communicated from the hotter of these bodies to the colder, during some hours perhaps, or the course of a day, at the end of which time, if we apply a thermometer to them all in succession, it will give precisely the same reading. [Admittedly their temperatures would not *feel* the same. Owing to differences in thermal conductivity, a copper tray or metal lamp stand at 70°F feels colder than a rug or chair cushion at the same temperature.*] The heat, therefore, distributes itself upon this occasion until none of these bodies has a greater demand or attraction for heat than every other of them has; in consequence, when we apply a thermometer to them all in succession, after the first to which it is applied has reduced the instrument to its own temperature, none of the rest is disposed to increase or diminish the quantity of heat which that first one left in it. This is what has been commonly called an "equal heat," or "the equality of heat among different bodies"; I call it the *equilibrium of heat.*

The nature of this equilibrium was not well understood until I pointed out a method of investigating it. Dr. Boerhaave imagined that when it obtains, there is an equal quantity of heat in every volume of space, however filled up with different bodies. . . .

But this is taking a very hasty view of the subject. It is confounding the quantity of heat in different bodies with its intensity [temperature], though it is plain that these are two different things, and should always be distinguished, when we are thinking of the distribution of heat.

* Problem 5. Why is a person a poor thermometer?

8

Temperature (II)

The Absolute Temperature

The index of thermal behavior $\Delta E/\Delta S$ has all the properties of temperature. As described in the previous chapter, thermal energy can flow spontaneously from body A to B if and only if the value of this ratio for A is greater than its value for B.*

Furthermore, the value of the ratio $\Delta E/\Delta S$ for any object—like the temperature of an object—increases or decreases according as thermal energy is added to or taken from the object. Were this not so, thermal contact between a hot object and a colder one would make the hot one hotter and the cold one colder, which is absurd. The two could never possibly reach a state of thermal equilibrium in this way.

An additional similarity exists between temperature and $\Delta E/\Delta S$. Both are intensive properties. This fact about the temperature has already been commented upon (Chapter 2). That $\Delta E/\Delta S$ is an intensive property follows from the fact that energy and entropy are extensive properties, and the ratio of two extensive properties is always an intensive property (the ratio of the mass of an object to its volume, for example, is the object's density, an intensive property). The same remark holds true for the ratio of the changes in two extensive properties. If, for example, a small addition of energy ε alters the entropy by the small amount η, addition of 2ε will alter the entropy by 2η; however, the ratio of ΔE to ΔS, ε/η in the first case and $2\varepsilon/2\eta$ in the second, remains unchanged.†

* From the standpoint of the Second Law, thermal equilibrium between two bodies A and B exists if, and only if, $(\Delta E/\Delta S)_A = (\Delta E/\Delta S)_B$. In addition, if bodies A and C are found to be in thermal equilibrium with each other, it may be added that $(\Delta E/\Delta S)_A = (\Delta E/\Delta S)_C$. And since things equal to the same thing are equal to each other, it follows that $(\Delta E/\Delta S)_B = (\Delta E/\Delta S)_C$. This implies that bodies B and C if tested against each other would be found to be in thermal equilibrium, as is indeed the case. This fact is sometimes referred to as the Zeroth Law of Thermodynamics.

† In using the ratio $\Delta E/\Delta S$ as an index of temperature, the change in E that is used to test how S changes should be small so as not to alter sensibly the temperature of the body. Ideally, the value chosen for the temperature index should be the limiting value of the ratio of ΔE to ΔS as ΔE (or ΔS) approaches zero. This value is generally referred to as dE/dS, or $(\partial E/\partial S)_{V,M}$. The subscripts V and M are added to indicate explicitly that the volume and mass are to be held constant as the quantities E and S within the parentheses are varied.

Also, like the absolute temperature, the ratio $\Delta E/\Delta S$ is generally positive. And, like the absolute temperature, removal of thermal energy from a substance causes $\Delta E/\Delta S$ to approach the value zero.

In thermodynamics, the index of thermal behavior $\Delta E/\Delta S$ and the absolute temperature T are set equal to each other.

$$T = \frac{\Delta E}{\Delta S}$$

The absolute temperature *is* the value of the ratio $\Delta E/\Delta S$. And the value of this ratio *is* the absolute temperature.

The absolute temperature of a system may be viewed as the quantity of energy that must be added to the system to alter its entropy one unit. For example, if addition of 298 calories of thermal energy to a system increases the entropy of the system 1 entropy unit, the system's temperature is 298 calories per entropy unit, abbreviated 298°K. If this same energy addition increases the system's entropy 2 entropy units, half 298, or 149 calories, would increase the entropy 1 entropy unit; therefore, the system's temperature is 149°K. On the other hand, if addition of 298 calories to the system increases its entropy only 1/2 an entropy unit, twice 298, or 596 calories, would be required to increase the entropy 1 entropy unit; hence, $T = 596$°K. More generally, if a change in thermal energy ΔE (preferably small) alters the entropy of a system by the amount ΔS, the system's absolute temperature is ΔE divided by ΔS. In the examples cited above, the temperatures were, respectively, 298 cal/1 e.u. (e.u. for "entropy unit") = 298°K, 298 cal/2 e.u. = 149°K, and 298 cal/(1/2) e.u. = 596°K. Briefly stated, $T = \Delta E/\Delta S$.

In many of the mathematical expressions that occur in thermodynamics, T is commonly written in place of $\Delta E/\Delta S$. This substitution, widely favored by printers and persons who make frequent use of thermodynamic formulas, sacrifices clarity for notational simplicity. The expression $Q/(\Delta E/\Delta S)$, or $(\Delta S/\Delta E) \cdot Q$, for example, is generally written Q/T.

Usually the units of entropy are selected to make the thermodynamic temperature scale correspond to the temperature scale established by an ideal gas thermometer. When the freezing point of water is defined as 273.15°, the scale is called the Kelvin scale. If the absolute temperature is expressed in degrees Kelvin and the energy in calories, the units of entropy are of necessity calories per degree K.*

The figure 400°K signifies that to change by one unit the entropy of a substance whose temperature is this value requires the addition or removal of 400 calories of thermal energy.† This statement assumes that the substance

* On a molar basis, cal/deg K-mole.
† $\Delta E = T$ when $\Delta S = 1$.

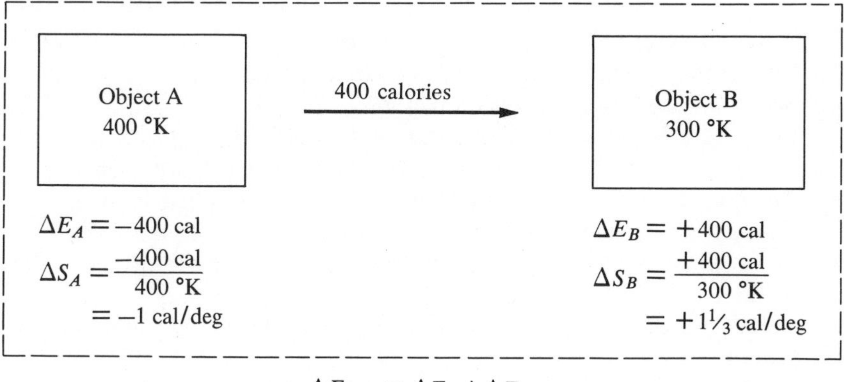

$$\Delta E_{\text{total}} = \Delta E_A + \Delta E_B = 0$$

$$\Delta S_{\text{total}} = \Delta S_A + \Delta S_B = +\tfrac{1}{3}\ \text{cal/deg}$$

Fig. 8.1 Energy and entropy changes in the universe when 400 calories of thermal energy flow from a 400°K object to a 300°K object.

is large enough to absorb or lose 400 calories of energy without any noticeable change in its temperature, and that in this process there is no net change in the substance's mass, volume, or chemical composition. A colder substance, one whose temperature is only 300°K, for example, would require for the same change in entropy (1 cal/deg) the addition or removal of only 300 calories of thermal energy.

Clearly the flow of 400 calories from a 400°K substance to a 300°K substance results in a net increase in the entropy of the universe (Fig. 8.1). This "downward" flow of energy is a spontaneous process. It is, as such, a process that in principle can be harnessed to do useful mechanical work. Of the 400 calories removed from the hot reservoir, only 300 need be delivered to the colder one; this is enough to make the increase in entropy of the cold reservoir just balance the decrease in entropy of the hot one. The remainder of the energy, one-quarter of the amount initially removed from the hot reservoir,* is available for the performance of useful mechanical work.

A machine that can produce useful mechanical work from the natural flow of thermal energy from high temperatures to lower temperatures is called a heat engine.

* $\dfrac{400 - 300}{400} = \left(\dfrac{T_{\text{hot}} - T_{\text{cold}}}{T_{\text{hot}}}\right) = \dfrac{1}{4}.$

Problems for Chapter 8

1. State the Zeroth Law of Thermodynamics.
2. List several ways in which T and $\Delta E/\Delta S$ are alike.
3. Might T be called a conversion factor?
4. List several ways in which the relation $T = \Delta E/\Delta S$ might be used.
5. Addition of 1 calorie of thermal energy to a substance increases its entropy 10 cal/deg. Is the substance hot or cold? What is its temperature on the Kelvin scale?
6. Estimate the heat of vaporization of a substance whose normal boiling point is 160°C. (In reversible evaporation the entropy of vaporization is equal to the energy of vaporization divided by the absolute temperature.)
7. What is the entropy change in the universe if 1200 calories of thermal energy flow from a 400°K body to a 400°K body? Assume both bodies are so large that addition or removal of 1200 calories of energy has no noticeable effect on their temperature. Is this process reversible?
8. What is the entropy change in the surroundings when a mole of ice melts at 0°C? When a mole of water freezes at 0°C? At −10°C?
9. What is the entropy change in the surroundings when a mole of hydrogen and one-half a mole of oxygen react in a bomb to form a mole of liquid water at 25°C?
10. Is a substance with a negative absolute temperature hot or cold?

Planck's Statement of Clausius's Hypothesis

Clausius, in 1865, stated the two laws of thermodynamics, the constancy of energy and the increasing property of entropy, in these words:

> "Die Energie der Welt ist constant.
> Die Entropie der Welt strebt einem Maximum zu."

These statements made a strong impression upon a young student, Max Planck. In the paragraphs quoted below, Planck describes how he happened to read Clausius's work, and his reaction to it.*

After my graduation from the Maximilian-Gymnasium, I attended the University, first in Munich for three years, then in Berlin for another year. . . . [I]t was in Berlin that my scientific horizon widened considerably under the guidance of Hermann von Helmholtz and Gustav Kirchhoff, whose pupils had every opportunity to follow their pioneering activities, known and watched all over the world.

* Max Planck, *Scientific Auiobiography and Other Papers*, trans. by Frank Gaynor, New York: Philosophical Library, 1949. Copyright 1949 by the Philosophical Library.

I must confess that the lectures of these men netted me no perceptible gain. It was obvious that Helmholtz never prepared his lectures properly. He spoke haltingly, and would interrupt his discourse to look for the necessary data in his small note book; moreover, he repeatedly made mistakes in his calculations at the blackboard, and we had the unmistakable impression that the class bored him at least as much as it did us. Eventually, his classes became more and more deserted, and finally they were attended by only three students. . . .

Kirchhoff was the very opposite. He would always deliver a carefully prepared lecture, with every phrase well balanced and in its proper place. Not a word too few, not one too many. But it would sound like a memorized text, dry and monotonous. We would admire him, but not what he was saying.

Under such circumstances, my only way to quench my thirst for advanced scientific knowledge was to do my own reading on subjects which interested me; of course, these were the subjects relating to the energy principle. One day, I happened to come across the treatises of Rudolf Clausius, whose lucid style and enlightening clarity of reasoning made an enormous impression on me, and I became deeply absorbed in his articles, with an ever increasing enthusiasm. I appreciated especially his exact formulation of the two Laws of Thermodynamics, and the sharp distinction which he was the first to establish between them. . . .

Clausius deduced his proof of the Second Law of Thermodynamics from the hypothesis that "*heat will not pass spontaneously from a colder to a hotter body.*" But this hypothesis must be supplemented by a clarifying explanation. For it is meant to express not only that heat will not pass directly from a colder into a warmer body, but also that it is impossible to transmit, by any means, heat from a colder into a hotter body without there remaining in nature some change to serve as compensation.

In my endeavor to clarify this point as fully as possible, I discovered a way to express this hypothesis in a form which I considered to be simpler and more convenient, namely: "*The process of heat conduction cannot be completely reversed by any means.*" This expresses the same idea as the wording of Clausius, but without requiring an additional clarifying explanation. A process which in no manner can be completely reversed I called a "*natural*" one. The term for it in universal use today, is: "*Irreversible.*"

9

Raising and Lowering Weights Reversibly

Heat Engines, Heat Pumps, and Refrigerators

Complete conversion of thermal energy to work is virtually impossible; it is, as mentioned, about as likely as the transcription of Shakespeare's complete works by a tribe of wild monkeys punching randomly on a set of typewriters. In all likelihood no one will ever witness either event; moreover, as once noted, were someone to see thermal energy completely converted to work, or monkey-business turn out Shakespeare, he probably would not believe it, for the thermodynamic probability that any such process will occur—more specifically, the number of internal arrangements accessible to the universe in its final state compared to the number of internal arrangements accessible to the universe in its initial state (on this, more later in Chapter 21)—is

$$10^{\frac{\Delta S_{\text{total}}}{2.303\,R}\cdot N_0} \text{ to } 1.$$

Here R is the gas constant, 1.987 cal/deg-mole, and N_0 is Avogadro's number, 6×10^{23}. Because of the great size of Avogadro's number,* the value of the expression on the left is very large, much larger than 1 (many more internal arrangements for the universe in its final state than in its initial state) if ΔS_{total} is significantly greater than zero, but very small, much less than 1 (many fewer internal arrangements for the universe in its final state than in its initial state) if ΔS_{total} is significantly less than zero, as it would be if any significant quantity of thermal energy were converted completely to work.

For example, consider a process that has as its net effect removal from a 400°K thermal reservoir of 1200 calories and the raising of a weight by an

* Perhaps the important point is not the bigness of Avogadro's number but the bigness of Avogadro (see Schrödinger, *What Is Life*, Anchor Books, 1956). Lilliputians 10^{23} times smaller than Avogadro or Gulliver would use a much smaller value for N_0 and would probably witness on occasion the complete conversion of random thermal motion to work.

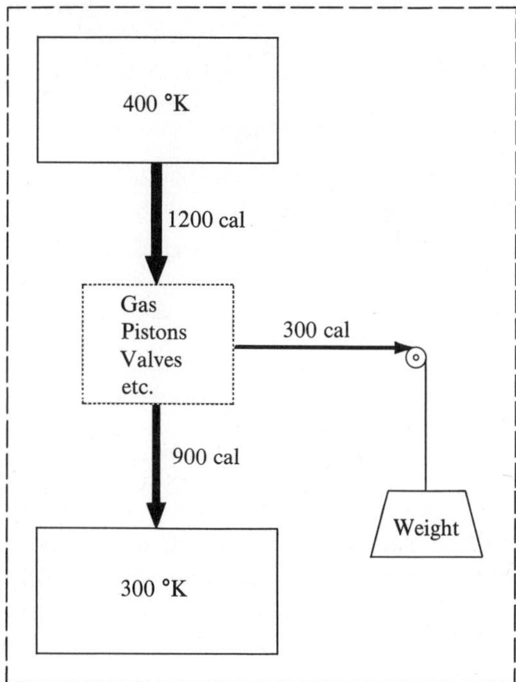

Fig. 9.1 Heat engine operating reversibly between the temperatures 400°K and 300°K. For the changes shown $\Delta S_{\text{total}} = 0$.

equivalent amount. In this process the entropy of the weight does not change; the entropy of the thermal reservoir, however, changes by -1200 cal/400°K $= -3$ cal/deg; therefore the entropy change for the two taken together (the universe of this event) is -3 cal/deg. The odds for this unearthly process are very poor, only $10^{-400,000,000,000,000,000,000,000}$ to 1, or one chance in $10^{400,000,000,000,000,000,000,000}$.

On the other hand, nothing on earth prohibits a system composed of three objects, a thermal reservoir at 400°K, another one at 300°K, and a weight, from being carried through a process whose net effect is the following: thermal energy of 400°K reservoir down by 1200 calories, entropy down by 3 cal/deg; thermal energy of 300°K reservoir up by 900 calories, entropy up by 3 cal/deg; potential energy of weight up by 300 calories, entropy unchanged. For this process, illustrated in Fig. 9.1 and summarized in tabular form below,

$$(E_{\text{total}})_{\substack{\text{initial}\\\text{state}}} = (E_{\text{total}})_{\substack{\text{final}\\\text{state}}}$$

and

$$(S_{\text{total}})_{\substack{\text{initial}\\\text{state}}} = (S_{\text{total}})_{\substack{\text{final}\\\text{state}}},$$

since

$$\Delta S_{\text{total}} = (S_{\text{total}})_{\substack{\text{final}\\ \text{state}}} - (S_{\text{total}})_{\substack{\text{initial}\\ \text{state}}} = \Delta S_{\substack{400^\circ\text{K}\\ \text{res.}}} + \Delta S_{\substack{300^\circ\text{K}\\ \text{res.}}} + \Delta S_{\text{Wt.}}$$

$$= (-3\ \text{cal/deg}) + (+3\ \text{cal/deg}) + 0 = 0.$$

From the point of view of the First and Second Laws, the initial and final states of this process are equivalent.

	Change in E (cal)	*Change in S* (cal/deg)
400°K thermal reservoir	−1200	−3
300°K thermal reservoir	+ 900	+3
Weight	+ 300	0
Three taken together (the universe)	0	0

One way to carry out such a process is illustrated in Fig. 9.2. By allowing a gas to expand at high temperatures and by compressing it at lower temperatures, the average pressure during expansion—a stage during which the gas does work—is made greater than during compression—where work is done

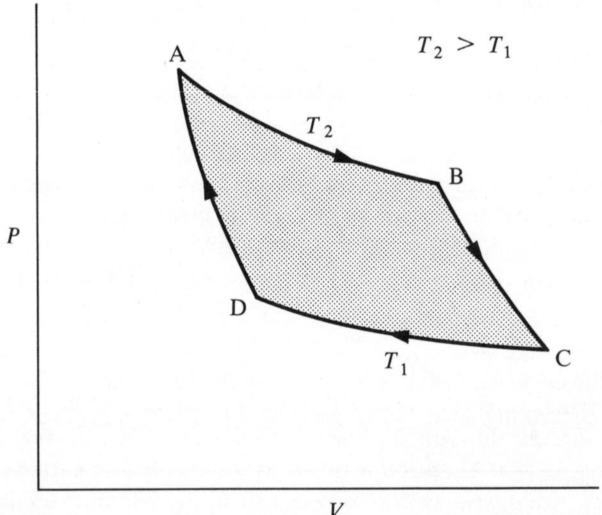

Fig. 9.2 Indicator diagram giving the pressure and volume for the working fluid of a heat engine that operates between the two temperatures T_2 and T_1 (schematic). $A \to B$ is an isothermal expansion at temperature T_2. $B \to C$ is an adiabatic expansion from T_2 to T_1; in this step there is no change in the thermal energy of any surrounding object. $C \to D$ is an isothermal compression at T_1. $D \to A$ is an adiabatic compression from T_1 to T_2. The area *ABCD* is the net work done by the fluid and its thermal surroundings in one complete cycle.

on the gas. The net effect of the cycle is the performance of useful work; at the same time, some thermal energy is necessarily transferred from a hot object (in a steam engine, the boiler) to a colder object (the condenser). The cycle might be begun, for example, at A (Fig. 9.2) with the gas expanding while in thermal contact with the hotter of two thermal reservoirs. During this expansion, the temperature of the gas does not change as the gas does mechanical work at the expense of some thermal energy of the hot reservoir, from which the gas absorbs energy as it expands. At B the gas is allowed to expand further, but without the absorption of thermal energy from any surrounding object. During this expansion, the gas continues to do work, but its temperature now drops.* When the temperature of the colder thermal reservoir has been reached, point C, the gas is placed in contact with this body and compressed. During this compression, the temperature of the gas remains constant as work is done on it and as thermal energy flows from it to the colder thermal reservoir. At D the gas is further compressed, but now without the loss of thermal energy to any surrounding object. During this compression, work continues to be done on the gas, and its temperature increases† until finally the temperature and pressure (and volume and energy and entropy) of the gas are back to those values characteristic of the starting point A. For the complete cycle, $\Delta T_{gas} = \Delta P_{gas} = \Delta V_{gas} = \Delta E_{gas} = \Delta S_{gas} = 0$.

To these comments concerning the operation of a heat engine should be added the remark that the interest a classical thermodynamicist may have in the intimate details of heat engines is generally academic. It is similar to the interest a gambler may have in footballs and football players: his chief interest in football is the final score. Although it is recognized that such items as a working fluid, pistons, valves, etc., are necessary for the proper functioning of a heat engine, it is known that with care friction and other kinds of wear and tear on moving parts can be diminished to the point where these parts suffer in one complete cycle virtually no change at all in any of their thermodynamic properties, T, P, V, M, E, or S. From the thermodynamic point of view, the interesting changes that occur in the universe during the functioning of a heat engine are the changes that occur in the two thermal reservoirs and the weight.

It is interesting to consider the changes that would occur in the universe if the four-step cycle‡ depicted in Fig. 9.2 were run through counterclockwise. This would correspond to running a heat engine backward. The working fluid would be allowed to expand at relatively low temperatures and pressures (steps A to D and D to C) and would be compressed at higher temperatures

* The gas molecules rebound from the retreating piston with diminished kinetic energy.
† The gas molecules rebound from the advancing piston with enhanced kinetic energy.
‡ This cycle is known as Carnot's cycle.

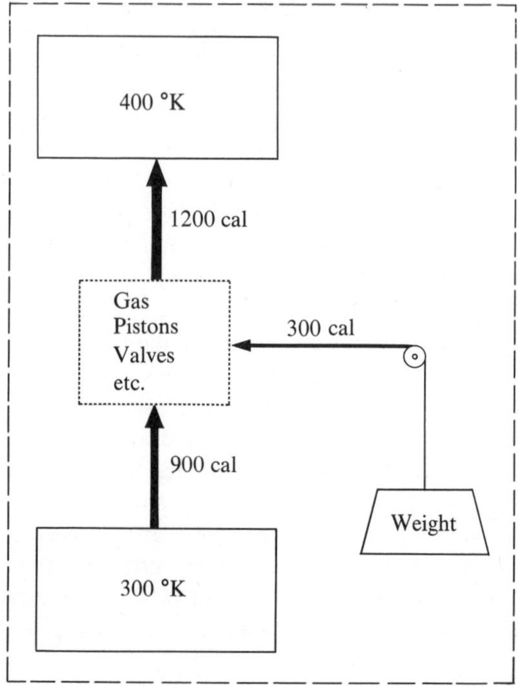

Fig. 9.3 A reversible heat engine run in reverse: a heat pump or a refrigerator, depending upon one's point of view.

and pressures (steps C to B and B to A). The net effect of the cycle would be removal of thermal energy from the colder reservoir (step D to C), deliverance of thermal energy to the hotter reservoir (step B to A), and the consumption of work (weight lower at the end, rather than higher). Reversing the directions of the three arrows in Fig. 9.1 serves to illustrate these effects in a particular instance (Fig. 9.3). For the expenditure of 300 calories of work by the weight, 900 calories of thermal energy can be removed from a 300°K thermal reservoir and the sum, 300 cal + 900 cal = 1200 calories, delivered to a 400°K reservoir. For this process, as for the original process,

$$\Delta E_{\text{total}} = \Delta E_{300^\circ\text{K}} + \Delta E_{400^\circ\text{K}} + E_{\text{Wt.}} = 0$$

and

$$\Delta S_{\text{total}} = \Delta S_{300^\circ\text{K}} + \Delta S_{400^\circ\text{K}} + \Delta S_{\text{Wt.}} = 0.$$

A heat engine run in reverse (Fig. 9.3) may be viewed from two standpoints. From the standpoint of the high-temperature thermal reservoir it functions as a heat pump. Work is done to discharge to a hot body (a house, for example) thermal energy removed from a colder one (the ground, perhaps).

From the standpoint of the low-temperature reservoir it functions as a refrigerator. Work is done to remove from a cold body (the freezing compartment) energy that is discharged eventually to a warmer body (the kitchen). In both devices, the spontaneous character of a weight falling downward is harnessed to make thermal energy migrate up a temperature gradient. By contrast, in a heat engine functioning in the normal fashion (Fig. 9.1), the spontaneous flow of thermal energy down a temperature gradient is harnessed to make a weight move up a potential energy gradient.

The two natural processes, thermal energy flow down a temperature gradient and a weight falling from a higher to a lower altitude with the production of thermal energy, are by themselves processes for which ΔS_{total} is greater than zero. Either process can be harnessed to an intrinsically unnatural process, one that by itself would have ΔS_{total} less than zero (this might be, as above, the reverse of one of the natural processes just mentioned), to give a combined operation that is possible provided the corresponding value of ΔS_{total} is equal to or greater than zero. If ΔS_{total} for the combined operation equals zero (Fig. 9.1 or Fig. 9.3), the operation is reversible.

Conversely, if a process is known to be reversible, it can be shown that ΔS_{total} for the process must have the value zero. For suppose the process consists of passage from State 1 to State 2; this implies that $S_2 \geqslant S_1$. But if the process is reversible, the change from State 2 back to State 1 can also occur; this implies that $S_1 \geqslant S_2$. Hence $S_1 = S_2$. Therefore, for any reversible process

$$\Delta S_{\text{total}} = S_2 - S_1 = 0.$$

Postscript

The reader may (and perhaps should) react to this last statement with a feeling of ambivalence. It seems to say so much. But, in fact, what does it say about any particular process? What does it say, for example, about the freezing of water, or about the reaction of hydrogen with oxygen? Are these processes reversible?

Under certain conditions they are. Under other conditions they are not. Under some conditions they are not even possible.

How does one know this?

From experiment.

Of what practical value, then, is thermodynamics?

Its practical value lies in its method of bookkeeping events—as exemplified by the two statements: For all processes, $\Delta E_{\text{total}} = 0$; For all reversible processes, $\Delta S_{\text{total}} = 0$. These statements have revealed many indirect, useful, and often unsuspected avenues of attack on problems. They make it possible, for example, to combine low-temperature heat capacity data with data on heats of combustion and data on the color of substances in the various spectral regions to calculate the value of an equilibrium constant for a chemical reaction and, hence, to predict whether under certain specified conditions a reaction will be reversible, irreversible, or impossible, and, if the reaction is possible, what the maximum yield of a desired product will be.

Prior to the discovery and application of the Second Law predictions of this type were seldom correctly made. In the nineteenth century, for example, the urgent needs of the Industrial Revolution led to the construction of blast furnaces of ridiculous heights in vain efforts to increase efficiency in the production of pig iron. Increasing the contact time between unspent blast furnace gases and iron ore does not yield more iron if the ore and the furnace gases have previously come to chemical equilibrium.

Problems for Chapter 9

1. What are the odds favoring a process for which $\Delta S_{\text{total}} = 0$?
2. What are the odds favoring a reversible process?
3. Show that the probability that a weight might spontaneously alter its potential energy by an amount Δ(P.E.) is

$$10^{-\frac{\Delta(\text{P.E.})}{2.303RT} \cdot N_0} \text{ to } 1.$$

4. What is the probability that a 1 gram weight resting on a table might spontaneously spring 1 centimeter into the air? Take the temperature of the table to be 300°K and the acceleration of gravity to be 980 cm/sec^2.
5. The entropy of a system is related to Ω, the number of distinct quantum-mechanical configurations readily accessible to the system, by the formula

$$S = (R/N_0) \ln \Omega.$$

Show that

$$\frac{(\Omega_{\text{total}})_{\substack{\text{final}\\ \text{state}}}}{(\Omega_{\text{total}})_{\substack{\text{initial}\\ \text{state}}}} = 10^{\frac{\Delta S_{\text{total}}}{2.303\,R} \cdot N_0}.$$

6. The following questions refer to Fig. 9.2.
   (a) What happens to the entropy of the gas in the isothermal expansion from A to B? In the isothermal compression from C to D?
   (b) What happens to the entropy of the gas in the expansion from B to C? In the compression from D to A?
   (c) Plot the temperature of the gas against its entropy for the cycle $ABCDA$.

7. The efficiency of a heat engine is defined as the ratio of the work done by the engine to the energy removed from the high-temperature reservoir of thermal energy. Derive a formula for the efficiency of a reversible heat engine.
8. How might the theoretical efficiency of a steam engine be improved?
9. How much energy can be pumped from a lake bottom whose winter temperature is 4°C into a house whose interior temperature is 27°C by the expenditure of 100 calories of mechanical or electrical energy?
10. Show that if thermal energy could be induced to flow spontaneously from a cold object to a warmer one, thermal energy could be converted completely to work.
11. State the First Law for a system that absorbs from its surroundings Q calories of thermal energy under circumstances where, simultaneously, there appear W calories of useful mechanical work.

Sadi Carnot

In the daily events of life, energy is never created and it is never destroyed. It is only transformed. Through gentle metabolic processes, a day laborer gradually transforms the chemical energy of the food he eats and the oxygen he breathes into heat, sound, and useful work. Through more violent processes, a jet engine during take-off rapidly transforms chemical energy into heat, sound, kinetic energy, and (hopefully) potential energy. In a few seconds a jet engine transforms more energy than can be transformed by a day laborer in many days. Jets, to use a technical term, are more *powerful* than day laborers.

Civilizations are often characterized by the power of their energy converters. The shorter the time required to transport a cargo across a continent, or to flatten a city, the more advanced the civilization is said to be.

Man's first energy converter was man. Later draft animals were used; then steam engines. Today (1965) man's principal energy converters are internal combustion engines, jet engines, and rockets.

In ancient times, societies that domesticated draft animals enjoyed two distinct advantages over simpler, food gathering societies. Animals such as horses and oxen can convert directly to useful mechanical work the energy of grasses, which man himself cannot, and these animals are more powerful than men. Thus, when time was precious, as often it was during the spring planting season, draft animals, though usually less efficient than man in converting plant energy to mechanical energy, proved to be worth their keep.

By the nineteenth century, however, a more powerful and more versatile energy converter was being developed, and draft animals were becoming obsolete. The transition from horses and oxen, which had been used by mankind for thousands of years, to the newer energy converter did not occur quickly, however, and the nation where the newer, more powerful converter was first widely used enjoyed for the better part of a century a signal advantage over her sovereign rivals. That nation was England. Her advantage was the steam engine, the most perfect device then known to man for rapidly converting to useful work anything that would burn. With the steam engine, England could keep dry her deepening coal mines from which came the coal to make the coke to stoke the furnaces of the converters that made the iron from which were fashioned the hulls of the ships of the British navy and the tools of the Industrial Revolution.

Improvements in the steam engine came slowly at first, by trial and error and occasional success, for no one understood what factors determined the motive-power of heat, until this problem was examined in a deeply fundamental and strikingly original manner by a young French military engineer, Sadi Carnot.

Around 1820, when he was twenty-three or twenty-four years old, Carnot published privately a brochure titled *Reflections on the Motive-Power of Heat, and on Machines Fitted to Develop that Power.* In this brochure, neglected until the Englishman William Thomson recognized its full merits in 1848, Carnot showed that the maximum efficiency of a steam engine is determined by the temperature of its boiler and the temperature of its condenser.

Below is a brief digest of Carnot's brochure. This brochure turned the art of making heat engines into a science and laid the foundations for the Second Law of Thermodynamics.*

Everyone knows that heat can produce motion. That it possesses vast motive-power no one can doubt, in these days when the steam-engine is everywhere so well known. The study of these engines is of the greatest interest, their importance is enormous, their use is continually increasing, and they seem destined to produce a great revolution in the civilized world.

The most signal service that the steam-engine has rendered to England is undoubtedly the revival of the working of the coal-mines, which had declined, and threatened to cease entirely, in consequence of the continually increasing difficulty of drainage, and of raising the coal. We should rank second the benefit to iron manufacture. To take away today from England her steam-engines would be to take away at the same time her coal and iron. It would be to dry up all her sources of wealth, to ruin all on which her prosperity depends, in short, to annihilate that colossal power.

* There has been some discussion as to what Carnot meant by the term "caloric." Early writers supposed that he meant "heat." In a footnote Carnot says that he employs the two expressions indifferently. In certain passages the term we would use today is "entropy."

If the honor of the discovery of the steam engine belongs to the nation in which it has acquired its growth and all its developments, this honor cannot be here refused to England. But notwithstanding the work of all kinds done by steam-engines, notwithstanding the satisfactory condition to which they have been brought to-day, their theory is very little understood, and the attempts to improve them are still directed almost by chance.

The question has often been raised whether the motive power of heat is unbounded, whether the possible improvements in steam-engines have an assignable limit,—a limit which the nature of things will not allow to be passed by any means whatever; or whether, on the contrary, these improvements may be carried on indefinitely. We propose now to submit these questions to a deliberate examination.

The phenomenon of the production of motion by heat has not been considered from a sufficiently general point of view. It is necessary to establish principles applicable not only to steam-engines but to all imaginable heat-engines, whatever the working substance and whatever the method by which it is operated.

The production of motion in steam-engines is always accompanied by a circumstance on which we should fix our attention. This circumstance is the re-establishing of equilibrium in the caloric; that is, its passage from a body in which the temperature is more or less elevated, to another in which it is lower. The steam is here only a means of transporting the caloric.

The production of motive power is then due in steam-engines not to an actual consumption of caloric, but *to its transportation from a warm body to a cold body.* According to this principle, the production of heat alone is not sufficient to give birth to the impelling power: it is necessary that there should also be cold; without it, the heat would be useless.

Wherever there exists a difference of temperature, it is possible to have also the production of impelling power. Steam is a means of realizing this power, but it is not the only one. All substances in nature can be employed for this purpose, all are susceptible of changes of volume, of successive contractions and dilatations, through the alternation of heat and cold. All are capable of overcoming in their changes of volume certain resistances, and of thus developing the impelling power. A solid body—a metallic bar for example—alternately heated and cooled increases and diminishes in length, and can move bodies fastened to its ends.

It is natural to ask here this curious and important question: Is the motive power of heat invariable in quantity, or does it vary with the agent employed to realize it as the intermediary substance, selected as the subject of action of the heat?

We take, for example, one body A kept at a temperature of 100° and another body B kept at a temperature of 0°, and ask what quantity of motive power can be produced by the passage of a given portion of caloric (for example, as much as is necessary to melt a kilogram of ice) from the first of these bodies to the second. We inquire whether this quantity of motive power is necessarily limited, whether it varies with the substance employed to realize it, whether the vapor of water offers in this respect more or less advantage than the vapor of alcohol, of mercury, a permanent gas, or any other substance.

Carnot began his analysis by noting that a heat engine can be run in reverse.

> Whenever there exists a difference of temperature, motive-power can be produced. Reciprocally, whenever we can consume this power, it is possible to produce a difference of temperature.

This led Carnot to the idea of a reversible ("Carnot") cycle, which he twice described, first in general terms and later in specific detail. Between these two descriptions Carnot presented this interesting observation:

> By our first operations there would have been at the same time production of motive power and transfer of caloric from the body A to the body B. By the inverse operations there is at the same time expenditure of motive power and return of caloric from the body B to the body A. But if we have acted in each case on the same quantity of vapor, if there is produced no loss either of motive power or caloric, the quantity of motive power produced in the first place will be equal to that which would have been expended in the second, and the quantity of caloric passed in the first case from body A to the body B would be equal to the quantity which passes back again in the second from the body B to the body A; so that an indefinite number of alternative operations of this sort could be carried on without in the end having either produced motive power or transferred caloric from one body to the other.

Following directly upon this observation is one of the most pregnant observations in the history of thermodynamics.

> Now if there existed any means of using heat preferable to those which we have employed, that is, if it were possible by any method whatever to make the caloric produce a quantity of motive power greater than we have made it produce by our first series of operations, it would suffice to divert a portion of this power in order by the method just indicated to make the caloric of the body B return to the body A from the refrigerator to the furnace, to restore the initial conditions, and thus to be ready to commence again an operation precisely similar to the former, and so on: this would be not only perpetual motion, but an unlimited creation of motive power without consumption either of caloric or of any other agent whatever. Such a creation is entirely contrary to ideas now accepted, to the laws of mechanics and of sound physics. It is inadmissible. We should then conclude that *the maximum of motive power resulting from the employment of steam is also the maximum of motive power realizable by any means whatever.*

This conclusion is sufficient, William Thomson showed, to define a temperature scale that is independent of the specific properties of substances.

Thomson called it the *absolute temperature* scale. It is, we have seen, the key to the uses of the Second Law.

Carnot also established the following conclusions:

The necessary condition that the motive power of a heat engine be a maximum is *that in the bodies employed to realize the motive power of heat there should not occur any change of temperature which may not be due to a change of volume.* Reciprocally, every time that this condition is fulfilled the maximum will be attained. This principle should never be lost sight of in the construction of heat-engines; it is its fundamental basis. If it cannot be strictly observed, it should at least be departed from as little as possible.

The fall of caloric produces more motive power at inferior than at superior temperatures. Thus a given quantity of heat will develop more motive power in passing from a body kept at 1 degree to another maintained at zero, than if these two bodies were at the temperature of 101° and 100°.

When a gas varies in volume without change of temperature, the quantities of heat absorbed or liberated by this gas are in arithmetical progression, if the increments or decrements of volume are found to be in geometrical progression.

The difference between specific heat under constant pressure and specific heat under constant volume is the same for all gases.

Sadi Carnot was born in 1796. His brother, Hyppolyte, wrote that he was of delicate constitution, but that he managed to increase his strength "by means of varied and judicious bodily exercises." This note appears in an extract from Carnot's unpublished writings: "Vary the mental and bodily exercises with dancing, horsemanship, swimming, fencing with sword and with sabre, shooting with gun and pistol, skating, the sling, stilts, tennis, bowls; hop on one foot, cross the arms, jump high and far, turn on one foot propped against the wall, exercise in shirt in the evening to get up a perspiration before going to bed; turning, joinery, gardening, reading while walking, declamation, singing, violin, versification, musical composition; eight hours of sleep; a walk on awakening, before and after eating; great sobriety; eat slowly, little, and often; avoid idleness and useless meditation." On one occasion, Hyppolyte wrote, Carnot was out walking when a horseman "who was evidently intoxicated, passed along the street on the gallop, brandishing his sabre and striking down the passers-by. Sadi darted forward, cleverly avoided the weapon of the soldier, seized him by the leg, threw him to the earth and laid him in the gutter, then continued on his way to escape from the cheers of the crowd, amazed at this daring deed." Not long thereafter Sadi Carnot died of cholera following an attack of scarlet fever, at the age of thirty-six.

10

Melting and Freezing

Why does hot ice melt and cold water freeze?
Because the entropy of the universe is thereby increased.

It is known that ice melts reversibly at 0°C and 1 atm.* Under these conditions the change

$$H_2O(\text{solid}) + 1440 \text{ cal/mole} = H_2O(\text{liq})$$

can occur in either direction. The fact that the change is reversible at 0°C means that at this temperature the universe composed of the ice-water and its thermal surroundings (also at 0°C) suffers no net change in entropy when ice melts or water freezes, although of course various parts of the universe do individually suffer entropy changes. (This happens, also, in the reversible operation of a heat engine.) But the total entropy within the universe after the reversible change has occurred—whether this be melting or freezing—is the same as before. Briefly stated,

$$(S_{\text{total}})_{\substack{\text{final}\\\text{state}}} = (S_{\text{total}})_{\substack{\text{initial}\\\text{state}}},$$

or

$$(S_{\text{total}})_{\substack{\text{final}\\\text{state}}} - (S_{\text{total}})_{\substack{\text{initial}\\\text{state}}} \equiv \Delta S_{\text{total}} = 0.$$

In both melting and freezing, the two parts of the universe directly involved in the process always undergo opposing changes in energy and opposing changes in entropy. In melting, for example, the energy and the entropy of the substance that is melting increase; simultaneously the energy and entropy of the material surrounding the melting substance decrease. The First Law states that whatever the individual energy changes in the various parts of the universe, they must together add up exactly to zero at all temperatures. The corresponding entropy changes, however, add up to zero at only one temperature—if the substance in both phases is pure. At this temperature, the normal melting temperature, or the normal freezing temperature, $\Delta E_{\text{total}} = 0$ and $\Delta S_{\text{total}} = 0$. At other temperatures ΔE_{total} still vanishes, but ΔS_{total}

* A pressure of 1 atm. will henceforth be assumed in this chapter.

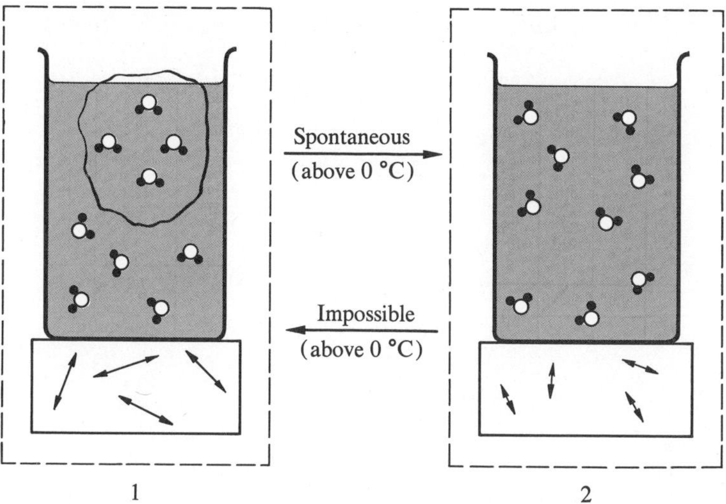

Fig. 10.1 Changes in disorder in the universe when ice melts (State 1 → State 2), or when liquid water freezes (State 2 → State 1).

does not. These facts may be illustrated by considering as a special case the system illustrated in Fig. 10.1.

Because the entropy of liquid water, mole for mole, is larger than the entropy of solid water at the same temperature, the entropy of water increases when it changes from the solid to the liquid state; at the same time, because melting is an endothermic process, the entropy of whatever substance surrounds the melting water decreases. The entropy lost by the surroundings can be calculated from the relation $\Delta S = \Delta E/T$ if the melting temperature and the heat of fusion of ice at this temperature are known. The entropy gained by the water can then be calculated if melting occurred at 0°C, for at this temperature melting is reversible and $\Delta S_{\text{total}} = 0$. Listed below are the separate steps leading up to this calculation.

As in Fig. 10.1, σ will be used to represent the melting substance and its surroundings will be signified by the symbol θ. By the additivity of S,

$$S_{\text{total}} = S_\sigma + S_\theta.$$

This equation is valid for both initial and final states. For the melting of ice at 0°C (a reversible change),

$$(S_{\text{total}})_{\substack{\text{final}\\\text{state}}} = (S_{\text{total}})_{\substack{\text{initial}\\\text{state}}}.$$

Hence, at this temperature,

$$(S_\sigma)_{\substack{\text{final}\\\text{state}}} + (S_\theta)_{\substack{\text{final}\\\text{state}}} = (S_\sigma)_{\substack{\text{initial}\\\text{state}}} + (S_\theta)_{\substack{\text{initial}\\\text{state}}}.$$

If the terms for σ and θ are brought together, this may be written

$$[(S_\sigma)_{\substack{\text{final}\\\text{state}}} - (S_\sigma)_{\substack{\text{initial}\\\text{state}}}] = -[(S_\theta)_{\substack{\text{final}\\\text{state}}} - (S_\theta)_{\substack{\text{initial}\\\text{state}}}],$$

which in Δ-notation becomes

$$\Delta S_\sigma = -\Delta S_\theta.$$

Alternatively, this might be written

$$\Delta S_\sigma + \Delta S_\theta = \Delta S_{\text{total}} = 0.$$

For θ, which suffers only thermal changes,

$$\Delta S = \frac{\Delta E_\theta}{T}.$$

In the reversible melting of 1 mole of ice, $\Delta E_\theta = -1440$ calories and $T = 273.15°\text{K}$. Therefore, for this change

$$\Delta S_\theta = \frac{-1440 \text{ cal/mole}}{273.15°\text{K}}$$
$$= -5.27 \text{ cal/deg-mole}$$

and, therefore,

$$\Delta S_\sigma = +5.27 \text{ cal/deg-mole}.$$

The number $+5.27$ cal/deg-mole represents the difference between the molar entropies of liquid and solid water at 0°C. To see this analytically, let these entropies be represented by the symbols $S^{\text{liq}}_{\text{H}_2\text{O}}$ and $S^{\text{solid}}_{\text{H}_2\text{O}}$ and let the symbols $n^{\text{liq}}_{\text{H}_2\text{O}}$ and $n^{\text{solid}}_{\text{H}_2\text{O}}$ denote the number of moles of liquid and solid water in the system initially. Then, by the additive property of S,

$$(S_\sigma)_{\substack{\text{initial}\\\text{state}}} = n^{\text{liq}}_{\text{H}_2\text{O}} S^{\text{liq}}_{\text{H}_2\text{O}} + n^{\text{solid}}_{\text{H}_2\text{O}} S^{\text{solid}}_{\text{H}_2\text{O}}. \quad (*)$$

Melting of a mole of ice increases by 1 the number of moles of liquid in the system and decreases by 1 the number of moles of solid. Therefore

$$(S_\sigma)_{\substack{\text{final}\\\text{state}}} = (n^{\text{liq}}_{\text{H}_2\text{O}} + 1) S^{\text{liq}}_{\text{H}_2\text{O}} + (n^{\text{solid}}_{\text{H}_2\text{O}} - 1) S^{\text{solid}}_{\text{H}_2\text{O}},$$

and, therefore,

$$\Delta S_\sigma \equiv (S_\sigma)_{\substack{\text{final}\\\text{state}}} - (S_\sigma)_{\substack{\text{initial}\\\text{state}}}$$
$$= S^{\text{liq}}_{\text{H}_2\text{O}} - S^{\text{solid}}_{\text{H}_2\text{O}}.$$

* In writing this expression, the valid assumption is made that the individual molar entropies $S^{\text{liq}}_{\text{H}_2\text{O}}$ and $S^{\text{solid}}_{\text{H}_2\text{O}}$ do not change when liquid and solid water are placed in contact with each other at 0°C. In some cases, when liquid water and alcohol are mixed, for example, $S^{\text{liq}}_{\text{H}_2\text{O}}$ does change (Chapter 25).

Reference to the previously derived value for ΔS_σ shows that at 0°C

$$S_{H_2O}^{\text{liq}} - S_{H_2O}^{\text{solid}} = +5.27 \text{ cal/deg-mole.}$$

At temperatures below 0°C, the molar entropies of liquid and solid water are less than their zero-degree values; however, their difference, which determines ΔS_σ (see above), like the difference between their molar energies, which determines the heat of fusion, remains relatively constant compared with the variation with temperature of ΔS_θ. For ΔS_θ depends directly on the absolute temperature through the factor T in the denominator of the expression $\Delta E_\theta/T$. Thus, while the values of ΔS_σ and ΔE_σ at -10°C differ little from their zero-degree values ($+5.27$ cal/deg-mole and 1440 cal/mole, respectively), the value of ΔS_θ at -10°C is calculated to be $-$(1440 cal/mole)/263°K $= -5.48$ cal/deg-mole (for convenience it is assumed here that the heat of fusion of ice is independent of temperature over the interval 0–(-10°C)). This figure for ΔS_θ is 0.2 cal/deg-mole less than its zero-degree value and the sum $\Delta S_\sigma + \Delta S_\theta$ now is a negative number. The implication is that ice will not melt to the pure liquid at -10°C. On the other hand, at $+10$°C, the value estimated for ΔS_θ is -5.09 cal/deg-mole and the sum $\Delta S_\sigma + \Delta S_\theta$ is now positive. Ice this hot should melt. These results are summarized in Table 10.1.

TABLE 10.1

Entropy Changes When Ice Melts

(Entropy values in cal/deg-mole)

H_2O(solid) + 1440 cal → H_2O(liq)

Temperature (°C)	ΔS_σ $(S_{H_2O}^{\text{liq}} - S_{H_2O}^{\text{solid}})$	$\Delta S_\theta = \Delta E_\theta/T$ $\left(\frac{\Delta E_\theta}{T} = \frac{-1440}{T}\right)$	ΔS_{total} $(\Delta S_\sigma + \Delta S_\theta)$	*Comments*
-10	$+5.3$	-5.5	-0.2	Won't melt
0	$+5.3$	-5.3	0.0	Could melt
$+10$	$+5.3$	-5.1	$+0.2$	Should melt

From the broad point of view (Table 10.1, column 4), the melting of hot ice and the freezing of cold water occur for identical reasons. Both processes increase the entropy of the universe.* The origins of this increase, however, differ. In melting, a favorable entropy change in the chemical system σ overwhelms an unfavorable entropy change in the surroundings θ. In freezing, a favorable entropy change in θ overwhelms an unfavorable entropy change in σ. Thus when hot ice melts, the entropy of the universe increases

* The words "hot ice" here and elsewhere stand for "ice in hot water." While liquids often can be supercooled, solids generally cannot be superheated.

because the increase in entropy of the water itself, as it changes from the low energy, low entropy solid to the higher energy, higher entropy liquid, more than offsets, at high temperatures, the decrease in entropy of the water's thermal surroundings, which lose thermal energy when ice melts. On the other hand, when cold water freezes, the entropy of the universe increases because at low temperatures the energy released in this exothermic process produces in the surroundings θ an increase in entropy that is sufficient to offset the decrease in entropy of water when water crystallizes.

In summary, at high temperatures (large T), the absolute value of ΔS_θ is small,

$$\Delta S_\theta = \frac{\Delta E_\theta}{T} \rightarrow 0 \qquad (T \text{ large})$$

and the sign of ΔS_{total} depends on the sign of ΔS_σ .

$$\Delta S_{\text{total}} = \Delta S_\theta + \Delta S_\sigma \rightarrow \Delta S_\sigma \ . \qquad (T \text{ large})$$

For ΔS_{total} to be positive, ΔS_σ must be positive. For ΔS_σ to be positive, the entropy of the products of the reaction must be larger than the entropy of the reactants.

$$\Delta S_\sigma (= S_{\text{products}} - S_{\text{reactants}}) > 0 \Rightarrow S_{\text{products}} > S_{\text{reactants}} \ .$$

At high temperatures, therefore, liquid water is more stable than solid water.

On the other hand, at low temperatures (small T), the absolute value of ΔS_θ is large and the sign of ΔS_{total} depends on the sign of ΔS_θ rather than on the sign of ΔS_σ. For ΔS_{total} to be positive under these conditions, ΔS_θ must be positive. For ΔS_θ to be positive, ΔE_θ must be positive (since T is always positive). And for ΔE_θ to be positive, ΔE_σ must be negative, since by the First Law

$$\Delta E_\theta = -\Delta E_\sigma \ . \qquad (*)$$

For ΔE_σ to be negative, the energy of the products of the reaction must be less than the energy of the reactants.

$$\Delta E_\sigma (= E_{\text{products}} - E_{\text{reactants}}) < 0 \Rightarrow E_{\text{products}} < E_{\text{reactants}} \ .$$

At low temperatures, therefore, solid water is more stable than liquid water.

The rule that the chemical stability of a particular arrangement of atoms

* Strictly speaking ΔE_σ should be replaced by the expression $\Delta(E + PV)_\sigma$, to allow for a change in the energy of the atmosphere during melting. This matter is discussed in more detail in a later chapter. For most practical applications at pressures of several atmospheres or less the PV correction term is negligible.

and molecules depends at low temperatures on the energy of the arrangement and at high temperatures on its entropy is very general.*

At room temperature, for example, the entropy change in σ alone is negative for the reaction

$$H_2(\text{gas}) + (1/2)O_2(\text{gas}) = H_2O(\text{liq}).$$

When $P_{H_2} = P_{O_2} = 1$ atm., for example,

$$\begin{aligned}\Delta S_\sigma &= S_{\text{products}} - S_{\text{reactants}} \\ &= S^{\text{liq}}_{H_2O} - (S^{\text{gas}}_{H_2} + 1/2\, S^{\text{gas}}_{O_2}) \\ &= 16.7 \text{ e.u.} - (31.2 \text{ e.u.} + 1/2 \cdot 49.0 \text{ e.u.}) \\ &= -39.0 \text{ cal/deg-mole } H_2O(\text{liq}) \text{ formed.}\end{aligned}$$

Despite this unfavorable entropy change in σ, the reaction occurs spontaneously at room temperature because the molar energy of liquid water is very much less than the energy of the corresponding hydrogen and oxygen. Briefly stated, the reaction is very exothermic. Per mole of liquid water formed, 68 kilocalories of energy are released to the surroundings. At 300°K, ΔS_θ is therefore the overwhelmingly positive number +68,000 cal/300°K = +227 e.u. It is easily seen, however, that at very high temperatures —10,000°K, for example—ΔS_θ will play a less important part in determining the sign of ΔS_{total}.

In several respects the explosion of hydrogen and oxygen at room temperature is similar to the freezing of cold water. Both reactions are spontaneous at low temperatures, but not at high temperatures; both reactions are exothermic; both reactions increase the entropy of the thermal surroundings and diminish the entropy of the reactive system; and both reactions (being low-temperature reactions) have a total entropy change whose sign is the same as the sign of the term $-(E_{\text{products}} - E_{\text{reactants}})$. This sign is plus if, and only if, E_{products} is less than $E_{\text{reactants}}$.

In analogous fashion, the thermal dissociation of very hot water vapor may be compared with the melting of warm ice. Both of these reactions are spontaneous at high temperatures, but not at low temperatures; both reactions are endothermic; both reactions decrease the entropy of the surroundings and increase the entropy of the reactive system; and both reactions (being essentially high-temperature reactions) have a total entropy

* At high pressures, the volume, too, must be considered. The term "chemical stability" means stability with respect to some other arrangement of the same atoms and molecules. The chemical stability of liquid water, for example, may be compared with the stability of ice, or steam, or gaseous hydrogen and oxygen, but not with the stability of graphite, laughing gas, or gold. To say that a reaction is spontaneous is the same thing as saying that the products of the reaction are more stable than the reactants.

change whose sign is the same as the sign of the entropy term ($S_{\text{products}} - S_{\text{reactants}}$). This sign is plus if, and only if, S_{products} is greater than $S_{\text{reactants}}$.

The association of chemical stability at low temperatures with low energy and at high temperatures with high entropy can be neatly expressed by introducing into thermodynamics a function called free energy.

Problems for Chapter 10

1. Show that the normal melting point of a substance is directly proportional to its heat of fusion and inversely proportional to its entropy of fusion.
2. Can you rationalize the melting points of the following isomeric substances?

Substance	*Melting Point* (°C)	*Boiling Point* (°C)
n-octane	− 57	125
2,2,3-trimethylpentane	−109	114
2,2,4-trimethylpentane	−107	99
2,2,3,3-tetramethylbutane	+101	106

3. Calculate the difference between the entropy of liquid and solid ethylpentachlorobenzene at this substance's normal melting point, 56°C. The heat of fusion at 56°C is 5.4 kcal/mole.
4. Calculate the difference between the entropy of liquid and solid pentachlorotoluene at this substance's normal melting point, 224°C. Assume that its heat of fusion is the same as that of ethylpentachlorobenzene.
5. Why does molecular hydrogen form spontaneously from hydrogen atoms at low temperatures and dissociate spontaneously into hydrogen atoms at high temperatures?
6. Why is the value of ΔS_σ positive for the following reactions?

$$H_2(\text{gas}) = 2H(\text{gas})$$

$$H_2O(\text{gas}) = H_2(\text{gas}) + (1/2)O_2(\text{gas})$$

7. Why are attempts to form the elusive fragment BH_3 by heating elemental boron in an atmosphere of hydrogen unlikely to succeed?
8. What would happen to the values of ΔS_σ and ΔS_{total} for the reaction

$$H_2(\text{gas}) = 2H(\text{gas})$$

if the total pressure on a system at equilibrium with respect to this reaction were suddenly diminished? Assume that the temperature does not change.

9. At 25°C the molar entropies of liquid and gaseous water at one atmosphere are 16.716 and 45.106 cal/deg-mole, respectively, and the heat of vaporization is 10.520 kcal/mole. Can the following change occur at 25°C?

$$H_2O(\text{liq}) = H_2O(\text{gas}, P = 1 \text{ atm.})$$

10. Can the following change occur at 25°C?

$$H_2O(\text{liq}) = H_2O(\text{gas}, P = 0.01 \text{ atm.})$$

11. Under what conditions is $\Delta S_{\text{total}} = 0$ for the evaporation of pure water at 25°C?
12. How large must the partial pressure of water vapor be at 25°C before condensation to the pure liquid can occur?
13. What is the thermodynamic significance of the term "vapor pressure"?
14. Describe what is meant by the term "normal boiling point."
15. Estimate the normal boiling point of water from the data given in problem 9.

11

Free Expansion of an Ideal Gas

$$E(P_2, V_2, T, n) = E(P_1, V_1, T, n)$$

According to the laws of Boyle and Charles the pressure of an ideal gas depends upon its volume and its temperature. One might ask, do the energy and entropy of an ideal gas depend upon its volume and its temperature? There are four questions here. One of these is considered in this chapter. *Does the internal energy of an ideal gas depend upon its volume?* The answer is that it does not. Given this fact, it can be shown (Chapter 12) that the entropy of an ideal gas contains the term $R \ln V$, or its physical equivalent, $-R \ln P$. Together with entropy's other properties—its additivity and the property that at equilibrium $\Delta S_{\text{total}} = 0$—this logarithmic term yields the characteristic equilibrium constant expressions of chemistry.

About a hundred years ago James Clerk Maxwell, the Scot, discoverer of Maxwell's equations of electromagnetic theory, invented a much-discussed demon, called Maxwell's demon. Maxwell proposed to station this demon, whom he endowed with a modest intelligence, at a small trap door in a partition between two gases with instructions to let fast-moving molecules, and only fast-moving molecules, pass through the door one way and only slow-moving molecules through the other way. In time all the fast-moving molecules would be trapped on one side and all the slow-moving molecules on the other. To observers such as ourselves it would appear that a temperature difference had been created spontaneously where previously none existed. This would be very useful. The hot side could serve as a heater and the cold side as a refrigerator for the duration of the demon's life without the consumption of any outside power. Of course this would violate the Second Law.* A more practical method of sorting out "hot" molecules from "cold" ones is illustrated in Fig. 11.1.

* The problems encountered with Maxwell's demon and the related problem of the ratchet and pawl machine are discussed in an entertaining fashion by Feynman in *The Feynman Lectures on Physics*, Addison-Wesley, 1963, Vol. I, Chapter 46.

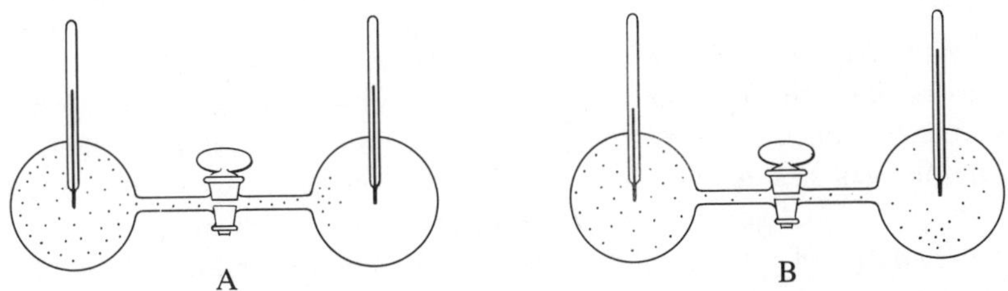

Fig. 11.1 (A) Free, irreversible expansion of a gas through a narrow orifice. (B) State of the previous system when pressure—but not temperature—equilibrium is first attained.

A flask containing a gas is connected by a narrow orifice to an evacuated flask. When the stopcock between the two is opened, molecules diffuse down the orifice from the high-pressure side to the low-pressure side (Fig. 11.1A). Since fast-moving molecules diffuse more rapidly than slower-moving ones, the gas as it expands becomes warmer on the right side and cooler on the left (Fig. 11.1B).

The apparatus produces in this way a transient temperature difference. In time, of course, both the temperature and the pressure become uniform throughout the system. It is natural to ask, how does the final temperature of the expanded gas compare with its initial temperature? For gases that obey closely the equation of state $PV = nRT$, there is no change in temperature. The initial and final temperatures are the same. The apparatus illustrated in Fig. 11.1 is not, however, a good apparatus with which to demonstrate this fact. By its very nature, the experiment with this apparatus takes time, a time during which undetermined quantities of thermal energy might be exchanged between the flasks and their thermal surroundings. A better method for examining these effects is illustrated in Fig. 11.2.

The orifice joining the two flasks is widened and the unit together with a thermometer is immersed in a water bath (Fig. 11.2). Now when the stopcock

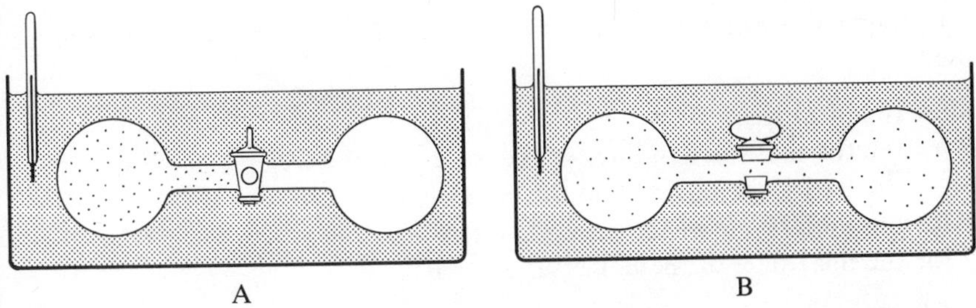

Fig. 11.2 Apparatus for determining how the internal energy of a gas varies with its volume. Note that the thermometer reading in **B** is the same as in **A**.

is opened the temperature and the pressure in the system rapidly become uniform. It is found, moreover, that for ideal gases, i.e. for gases that obey closely the equation of state $PV = nRT$, *the temperature of the surrounding water does not change*. This observation implies that the surrounding water has neither gained nor lost thermal energy. By the First Law, then, the expanding gas has neither lost nor gained thermal energy. *The internal energy of an ideal gas is independent of its volume.* Put another way, the internal energy of n moles of an ideal gas at some temperature T and some initial pressure and volume P_1, V_1 is equal to the internal energy of the same number of moles of gas at some different pressure and volume, P_2, V_2, if the temperature is the same. In symbols,

$$E(P_2, V_2, T, n) = E(P_1, V_1, T, n). \qquad (*)$$

The statement that the internal energy of an ideal gas does not depend upon its volume is equivalent to the statement that the internal energy of an ideal gas depends only upon its temperature. The internal energy of an ideal gas changes when, and only when, the gas temperature changes.

But how about the entropy of the gas? Is this, like the energy, unchanged by the free expansions illustrated in Figs. 11.1 and 11.2? Certainly the entropy of the universe—the gas plus its thermal surroundings—is changed. For the free expansion from State A to State B (Figs. 11.1 and 11.2) is highly irreversible. Never does the system through self-compression of the gas change back spontaneously from State B to State A. The entropy of the universe must therefore be greater in State B than in State A. But where is it greater? Is the entropy of the surrounding water (Fig. 11.2) greater? No, neither the volume nor the temperature nor the internal energy nor the entropy of the water has changed.

The conclusion seems inescapable. The entropy of the gas must be greater.

Problems for Chapter 11

The following problems refer to Fig. 11.2. These four subscripts will be used:

σ for the ideal gas.

θ for the thermal surroundings of the ideal gas (the water, the thermometer, the flasks, the orifice, and the stopcock).

1 for the initial state of the universe (State A).

2 for the final state of the universe (State B).

* The variables within the parentheses cannot all be varied independently. Given three of them, the fourth can be calculated by the equation $PV = nRT$.

1. Express in words what these symbols represent: T_θ, T_σ, $(T_\theta)_1$, $(T_\theta)_2$.
2. What relation, if any, exists between T_θ and T_σ?
3. What relation, if any, is observed to exist between $(T_\theta)_1$ and $(T_\theta)_2$?
4. Write the definition of the symbols ΔT_θ and ΔV_σ.
5. What can be said about the values of ΔT_θ, ΔT_σ, ΔV_θ and ΔV_σ?
6. The energy of a simple, chemically stable system depends upon two variables; these we may take to be its temperature and its volume. Another way to express this fact is as follows. The energy change in a system produced by a change in the system's temperature and by a concurrent change in the system's volume can be expressed as an average rate of change of the system's energy as its temperature (but not its volume) changes times the temperature change, plus an average rate of change of the system's energy as its volume (but not its temperature) changes times the volume change; in symbols,

$$\Delta E = \begin{pmatrix}\text{rate of change of} \\ E \text{ as } T \text{ changes}\end{pmatrix}_{V\ \text{constant}} \times \Delta T$$
$$+ \begin{pmatrix}\text{rate of change of} \\ E \text{ as } V \text{ changes}\end{pmatrix}_{T\ \text{constant}} \times \Delta V.$$

   If for some part of the universe it is known that $\Delta T = \Delta V = 0$, what can be concluded? To what part of the universe (σ or θ) does this conclusion apply?
7. What relation, if any, exists between E_{total}, E_θ, and E_σ? Does this relation apply to both States 1 and 2?
8. Write down the definition of $\Delta(E_{\text{total}})$.
9. What, if anything, is known about $\Delta(E_{\text{total}})$?
10. What relation, if any, exists between ΔE_θ and ΔE_σ?
11. What, if anything, can be said about ΔE_σ?
12. If for some part of the universe it is known that $\Delta E = \Delta T = 0$ and that $\Delta V \neq 0$, what can be concluded? To what part of the universe (σ or θ) does this conclusion apply?

The following questions do not necessarily refer to Fig. 11.2.

13. Is the free expansion of an ideal gas adiabatic?
14. Is the energy of an ideal gas independent of its pressure?
15. Is the entropy of an ideal gas independent of its pressure?
16. To "expand a gas at constant temperature" means to cause the volume of the gas to increase without any net change in the temperature of the gas. Is this possible for a non-ideal gas?
17. Can an ideal gas be expanded at (a) constant pressure, (b) constant energy, (c) constant entropy, and (e) constant volume? And if so, how?

18. If it were possible to expand a gas adiabatically and reversibly, what, if anything, could be said about its change in entropy?
19. What is the driving force behind the production of the temperature difference illustrated in Fig. 11.1?
20. A gas that has undergone a free expansion could, of course, be mechanically compressed back to its original volume. Or, if the gas is not too low boiling, it could be condensed back into the original flask with liquid nitrogen or some other coolant and then allowed to warm up to its original temperature. Do either of these operations represent a reversal of the original event?

Obituary
Maxwell's Demon (1871—c. 1949)

The paradox posed by Maxwell's demon bothered generations of physicists. In 1912 Smoluchowski noted that Brownian agitation of the trap door, which would result in a random opening and closing of the door, would render ineffective the long range operation of any automatic device, such as a spring valve or a ratchet and pawl. In 1939 Slater suggested that the uncertainty principle might play a role in the problem. Later it was shown that this would not be the case for heavy atoms at low pressures. Not until 1944–51, however, did two physicists, Demers and Brillouin, call attention to the fact that in an isolated enclosure in internal thermal equilibrium *it would be impossible for the demon to see the individual molecules.* To make the molecules visible against the background black-body radiation, the demon would have to use a torch. Happily, as Demers and Brillouin showed, the entropy produced in the irreversible operation of the torch would always exceed the entropy destroyed by the demon's sorting procedure. A real demon could not produce a violation of the Second Law.*

* For leading references, see Leon Brillouin, *Science and Information Theory*, New York: Academic Press Inc., 1956, Chapter 13.

12

Reversible Isothermal Expansion of an Ideal Gas

$$S = S^0 - R \ln P$$

A question of central importance to the thermodynamic theory of chemical equilibrium is this: How much does the entropy of an ideal gas increase during an isothermal expansion?

This question can be answered by asking another: What if the isothermal expansion were reversible?

If the expansion were reversible, the entropy of the universe would not change; for if the entropy of the universe did change, there would be only one way for it to go: it would have to increase. But then the expansion would not be reversible, for the entropy of the universe never decreases. To be reversible, ΔS_{total} must vanish. This means that during a reversible expansion whatever change occurs in the entropy of a gas must in some way be balanced by a change equal in magnitude but opposite in sign in the entropy of some other part of the universe. That other part of the universe is, in fact, the thermal surroundings of the gas. If the entropy change of the thermal surroundings during a reversible expansion can be determined (from its change in thermal energy and its absolute temperature), the entropy change of the gas will be known.

That, briefly, is the strategy of thermodynamics. Make the desired change in a chemical system part of a larger, reversible event; determine for this larger event the energy change of the thermal surroundings of the chemical system; from this energy change and the absolute temperature, calculate for the event the entropy change of the thermal surroundings; that, with its sign changed, is the entropy change of the chemical system.

Statement of the strategy is simple enough. Its execution may be another matter. Thermodynamics tells us in what way the quantities of a reversible event are related to each other. But, characteristically, it does not tell us how to measure these quantities or, even more importantly, how to make a particular, desired change part of a larger, reversible event. How, for example, does one expand a gas isothermally and reversibly? One way, which has often been described in textbooks, is the following.

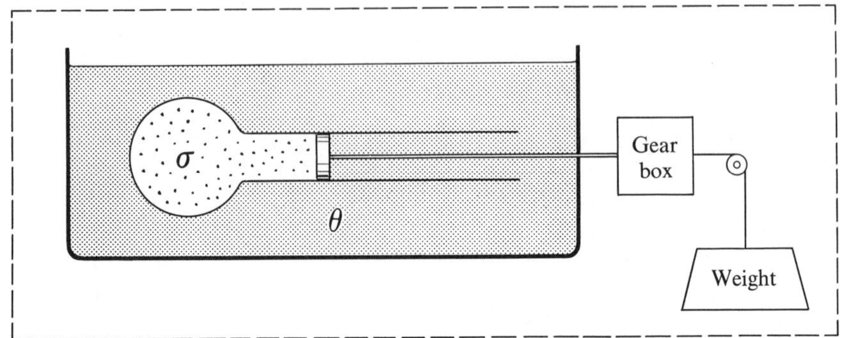

Fig. 12.1 Apparatus for expanding (or compressing) a gas isothermally and reversibly.

Lengthen the orifice in Fig. 11.2 and place in it a frictionless piston. Couple this piston mechanically to a weight, which the gas as it expands raises as rapidly as the First and Second Laws allow. For convenience in discussion designate the three thermodynamically significant parts of the apparatus, the gas, the thermostat, and the weight, as σ, θ, and Wt., respectively (Fig. 12.1). Now ask, what happens to the energies and entropies of these three parts of the universe as the gas expands reversibly ($\Delta S_{\text{total}} = 0$) and isothermally ($\Delta T_\sigma = 0$) from V_1, P_1 to V_2, P_2 ($V_2 > V_1$; $P_2 < P_1$)?

The energy of the gas does not change; that is, $\Delta E_\sigma = 0$; because for an ideal gas E_σ depends only upon T_σ (Chapter 11), which by supposition does not change. But as the gas expands, work is done on the weight and the potential energy of the weight increases (Fig. 12.1). The First Law states, however, that for any process $\Delta E_{\text{total}} = 0$; that is, for the present process

$$\Delta E_\sigma + \Delta E_\theta + \Delta E_{\text{Wt.}} = 0.$$

Therefore,

$$\Delta E_\theta = -\Delta E_{\text{Wt.}}.$$

In the reversible isothermal expansion of an ideal gas the energy of the thermostat decreases by exactly the amount by which the energy of the weight increases.

As for the entropy, we know that, as the thermal energy of the thermostat decreases ($\Delta E_\theta < 0$), the entropy of the thermostat must also decrease. We know, also, that the entropy of the weight does not change: $\Delta S_{\text{Wt.}} = 0$. The Second Law states, however, that for any reversible process $\Delta S_{\text{total}} = 0$; that is, for the present process

$$\Delta S_\sigma + \Delta S_\theta + \Delta S_{\text{Wt.}} = 0.$$

Therefore,

$$\Delta S_\sigma = -\Delta S_\theta.$$

In the reversible isothermal expansion of a gas the entropy of the gas increases by exactly the amount by which the entropy of the thermostat decreases. Since by the definition of T

$$\Delta S_\theta = \frac{\Delta E_\theta}{T},$$

it follows that in the reversible isothermal expansion of any gas

$$\Delta S_\sigma = -\frac{\Delta E_\theta}{T}.$$

In the reversible isothermal expansion of an ideal gas ($\Delta E_\theta = -\Delta E_{\text{Wt.}}$),

$$\Delta S_\sigma = +\frac{\Delta E_{\text{Wt.}}}{T}.$$

The last equation expresses quantitatively the qualitative conclusion of the previous chapter, namely that the entropy of an ideal gas increases during an isothermal expansion. For during expansion the weight rises and $\Delta E_{\text{Wt.}}$ is positive; by the last equation, ΔS_σ is positive, too. During a compression the weight falls and ΔS_σ, which has the same sign as $\Delta E_{\text{Wt.}}$, is negative.

It will be shown later (Chapter 15) that in a reversible expansion of an ideal gas the change in the potential energy of the weight can be expressed in terms of the initial and final volumes of the expanding gas, the number of moles of gas, and the absolute temperature of the gas, according to the equation

$$\Delta E_{\text{Wt.}} = nRT \ln (V_2/V_1).$$

$\Delta E_{\text{Wt.}}$ is positive when, as in an expansion, $V_2 > V_1$. Also, for a given expansion ratio (V_2/V_1) the work done on the weight is directly proportional to n and T; the hotter the gas and the more there is of it, the greater the work term. The presence of the gas constant R makes the equation dimensionally homogeneous.*

Dividing this expression for $\Delta E_{\text{Wt.}}$ by T, one finds that the difference between the entropy of n moles of an ideal gas at some temperature T and volume V_2 and its entropy at the same temperature but some other volume V_1 can be expressed as follows:

$$\Delta S_\sigma = nR \ln (V_2/V_1).$$

If S_2 and S_1 represent, respectively, the entropies of one mole of an ideal gas at conditions P_2, V_2, T, n and P_1, V_1, T, n, then

$$\mathsf{S}_2 - \mathsf{S}_1 = R \ln (V_2/V_1).$$

* The factor $\ln(V_2/V_1)$ is dimensionless. The product nRT has the units (moles)(cal/deg-mole)(deg) = cal.

For an ideal gas, $(V_2/V_1) = (P_1/P_2)$. Therefore, one may also write that for an ideal gas

$$\mathsf{S}_2 - \mathsf{S}_1 = R \ln (P_1/P_2),$$

or

$$\mathsf{S}_2 = \mathsf{S}_1 - R \ln (P_2/P_1).$$

If $P_2 < P_1$, $-R \ln (P_2/P_1)$ is positive and $\mathsf{S}_2 > \mathsf{S}_1$.

It will be noticed that the equations of this chapter express only entropy differences. It is frequently useful to select as a reference for the molar entropy of an ideal gas its molar entropy at one atmosphere, symbolized S^0. The molar entropy of an ideal gas at some other pressure can then be expressed as follows:

$$\mathsf{S} = \mathsf{S}^0 - R \ln P.$$

S stands for the molar entropy of the gas when its actual pressure (or partial pressure) is P atm. When $P = 1$ atm., $\mathsf{S} = \mathsf{S}^0$.

Two concluding remarks. Our first concluding remark concerns the significance in thermodynamics of initial and final states. When a defined chemical system passes from a given initial state to a given final state, the changes that occur in the chemical system's thermodynamic properties are completely independent of the manner in which the chemical system is coupled to its surroundings. What happens in the surroundings will, of course, determine whether or not the over-all event is reversible. However, so far as changes in the thermodynamic properties of the particular chemical system under consideration are concerned, only two things count: its initial state and its final state. Once these have been specified, all changes in the thermodynamic properties of the system are determined. When an ideal gas expands from volume V_1 to volume V_2 without any net change in temperature or mass, the change in entropy of the gas—although not of its surroundings—is the same ($nR \ln V_2/V_1$) whether (Fig. 12.1) or not (Fig. 11.1) elsewhere in the universe the potential energy of a weight increases at the expense of the thermal energy of a thermostat.

The second remark concerns the condition of reversibility implied in some of the previous equations. Take the two equations $\Delta S_\sigma = -\Delta S_\theta$ and $\Delta S_\sigma = -\Delta E_\theta/T$. These equations are valid only if ΔS_{total} is zero. They should be applied only to events in which the total entropy change of the universe does not change; i.e. to reversible events. As a reminder of this, the second equation could be written

$$\Delta S_\sigma \overset{\substack{\text{reversible}\\ \text{event}}}{=} -\frac{\Delta E_\theta}{T};$$

or, what is the same thing,

$$\Delta S_\sigma \overset{\substack{S_{\text{total}} \\ \text{constant}}}{=} -\frac{\Delta E_\theta}{T}.$$

Problems for Chapter 12

1. What is the entropy change of one mole of an ideal gas that expands isothermally at 298°K from (a) one liter to ten liters? (b) From ten liters to one hundred liters? (c) What would the corresponding changes be for ten moles of gas? (d) What would the changes in part (a) be if the gas were expanded isothermally at 398°K?
2. What substitutions for S_1, S_2, P_1, and P_2 have been made in going from the equation $\mathsf{S}_2 = \mathsf{S}_1 - R \ln (P_2/P_1)$ to the equation $\mathsf{S} = \mathsf{S}^0 - R \ln P$?
3. In terms of S^0 and P, what is the entropy of three moles of an ideal gas?
4. The derivation of the equation $\mathsf{S}_2 = \mathsf{S}_1 + R \ln (V_2/V_1)$ was based upon an analysis of changes that occur during a reversible event. Does this mean that the equation should only be applied to ideal gases that have participated in reversible events?
5. What is the entropy change in the universe when one mole of an ideal gas expands freely into an evacuated flask whose volume equals the initial volume of the gas?
6. Give an example where the equation $\Delta S_\sigma = -\Delta E_\theta/T$ is not valid.
7. What is the relation between ΔS_σ and the quantity $-\Delta E_\theta/T$ for an irreversible event? Give an illustrative example.
8. Compare the changes that occur in the energies and entropies of an ideal gas and its surroundings when the gas expands freely with those that occur in the same parts of the universe when the gas expands reversibly and isothermally.
9. What would happen to the entropy of the gas in Fig. 12.1 if the thermostat were not present during the reversible expansion? What might such an expansion be called?
10. Two ways of indicating explicitly the reversibility condition on the equation $\Delta S_\sigma = -\Delta E_\theta/T$ were given in the text. Suggest another way this might be done.
11. Of the first seven separately listed equations in this chapter, which apply to the reversible isothermal expansion of ideal gases? Which apply to the reversible isothermal expansion of any gas? Which apply to any reversible process? Which apply to all processes?

12. Consider as a single event this two-step sequence: (i) free expansion of an ideal gas; (ii) isothermal compression of the gas back to its initial volume. What is the net effect of this event?
13. In the isothermal expansion of an ideal gas (Fig. 12.1), thermal energy is converted completely to work. How does this process escape violating the Second Law, which states that any process whose net effect is equivalent to the raising of a weight at the expense of the thermal energy of some object is impossible?
14. Does complete specification of the changes that occur within a chemical system determine what happens to the surroundings of the chemical system?
15. It has been said that the changes that occur in the thermodynamic properties of a chemical system are completely determined once the initial and final states of the chemical system have been specified. Does a similar statement hold for a universe? Are the changes that occur within the universe of an event completely determined once the initial and final states of the universe have been specified?
16. This problem is a verification of the equation

$$\Delta E_{\text{Wt.}} = nRT \ln (V_2/V_1)$$

for expansion ratios (V_2/V_1) close to unity.

Let n moles of an ideal gas at temperature T be confined in a cylinder to a volume V_1 by a weight resting upon a piston of cross-sectional area A (Fig. 12.2), the mass m of the piston and weight being such that the pressure exerted by them on the gas is very slightly less than the internal pressure of the gas as given by the equation $PV = nRT$, and suppose

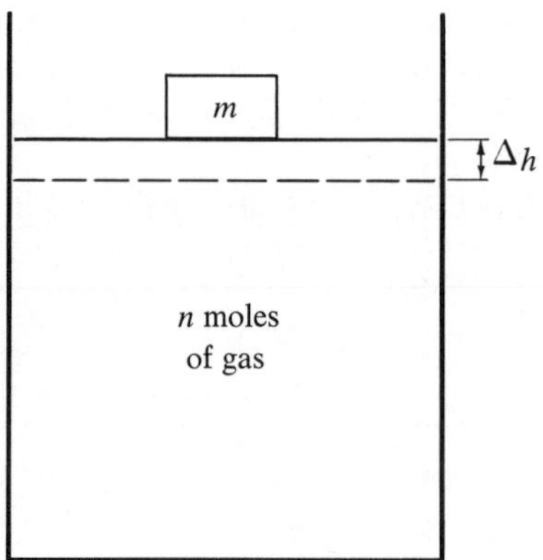

Fig. 12.2

that the gas expands slightly from V_1 to V_2, lifting the weight by the amount Δh. Let $\Delta V = V_2 - V_1$. In terms of ΔV and V_1, the expansion ratio (V_2/V_1) can be written as 1 plus ——— (a). Now for small x, $\ln(1 + x)$ is approximately equal to ——— (b). (Examination of the natural logarithms of a few numbers close to one quickly reveals the nature of this approximation.*) Thus, when ΔV is small compared to ——— (c), $\ln(V_2/V_1)$ is approximately equal to ——— (d) and the quantity $nRT \ln(V_2/V_1)$ is approximately equal to ——— (e). Since the initial pressure of the gas is equal to ——— (f), this expression can be simplified to the product of two factors ——— (g). Now, in terms of the cross-sectional area A and Δh, ΔV is equal to ——— (h), and in terms of the mass m and the cross-sectional area A, the initial pressure on the gas—which by supposition is nearly equal to the pressure defined by the equation $PV = nRT$—is equal to ——— (i). Hence, for small expansions, $nRT \ln(V_2/V_1)$ is approximately equal to ——— (j), which is just the change in the ——— (k) of the ——— (l). Q.E.D.

* See also Chapter 15.

13

Chemical Equilibrium (I)

$H_2 = 2H$

Why is molecular hydrogen stable at room temperature?

This may seem like a foolish question to ask. Everyone knows that hydrogen is stable at room temperature. Still, consider the fact that the standard entropy of two moles of atomic hydrogen is greater than the standard entropy of one mole of molecular hydrogen. From this fact alone, one might suppose (incorrectly) that the reaction $H_2 = 2H$ is spontaneous at room temperature. Sometimes, "a little learning is a dangerous thing."*

"Hydrogen hasn't enough energy to dissociate at room temperature," it is sometimes said. But doesn't this miss the point? True, the kinetic energy associated with the translational motion of molecular hydrogen at 300°K is only $(3/2)RT = 900$ cal/mole,† whereas rupture of the H-H bond requires over 100,000 cal/mole. Does this mean, however, that for hydrogen to dissociate its temperature must be sufficient to make its kinetic energy as great as its bond dissociation energy? Were that so, hydrogen would not dissociate below 34,000°K. In fact, hydrogen is appreciably dissociated at 5000°K. At this temperature its average kinetic energy is less than one-sixth its bond dissociation energy. Where, then, does hydrogen at 5000°K obtain the energy to dissociate? The only possibility seems to be: From its thermal surroundings. Briefly stated,

$$H_2 + \text{Thermal energy from surroundings} = 2H.$$

The question, then, is not, why does hydrogen fail to dissociate at room temperature, even though the product of dissociation has a greater entropy than the reactant? Nor is it, why does hydrogen dissociate spontaneously

* "... Drink deep, or taste not the Pierian spring:
There shallow draughts intoxicate the brain,
And drinking largely sobers us again."
POPE, *An Essay on Criticism*

† $R = 0.08205$ liter-atm./deg-mole $= 1.987$ cal/deg-mole.

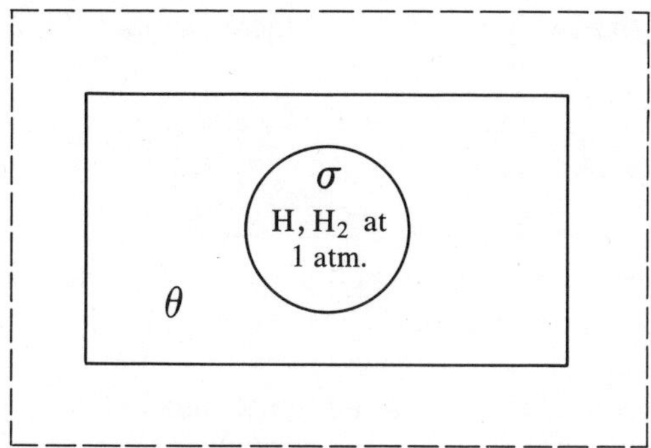

Fig. 13.1 The universe of a reaction that occurs at constant volume without the production or consumption of useful work. σ is the reaction mixture, θ its thermal surroundings.

at 5000°K, even though its thermal energy at that temperature is less than its bond dissociation energy?

The question is, why is thermal energy absorbed spontaneously by molecular hydrogen at high temperatures but not at low temperatures?

Ice, one may recall, exhibits a similar behavior. At high temperatures ($T > 273$°K) ice absorbs thermal energy from its surroundings and melts; at low temperature it does not.

The thermal behavior of hydrogen can be explained as follows. Suppose one had a mixture of atomic and molecular hydrogen, each at a partial pressure of 1 atm., in a flask surrounded by a thermal reservoir (Fig. 13.1). At 25°C the molecular hydrogen would have an entropy of 31.2 cal/deg-mole (28.1 from translation, 3.1 from rotation); atomic hydrogen's entropy under these conditions would be 27.4 cal/deg-mole. Now ask, what would happen to the entropy of the universe if some of the molecular hydrogen were to dissociate at the expense of the thermal energy of its surroundings? The entropy of the flask's contents would increase at the rate of 2(27.4 cal/deg-mole) − 31.2 cal/deg-mole − R = +21.6 cal/deg per mole of H_2 consumed.* Simultaneously the entropy of the surroundings would change at the rate of −104,000 cal/298°K or −349 cal per deg per mole of H_2 consumed. Clearly hydrogen should not dissociate under these conditions. If it did, the entropy of the universe would change at the rate of −327 cal per deg per mole of H_2 consumed. This is improbable. It is more probable that some of the atomic hydrogen would recombine to form molecular hydrogen. Thus, at 25°C, occurrence of the reaction

$$2H = H_2 + 104 \text{ kcal}$$

* The term R is included to allow for the fact that when the reaction $H_2 = 2H$ occurs at constant volume the total pressure in σ increases. The origin of this term is discussed more fully in Chapter 15.

in a large flask of fixed volume produced these changes in entropy when the partial pressures P_{H_2} and P_H are 1 atm.

$$\Delta S_\sigma \quad = - \ 21.6 \text{ cal/deg}$$
$$\Delta S_\theta \quad = +349. \text{ cal/deg}$$
$$\Delta S_{\text{total}} = +327. \text{ cal/deg}$$

At 10,000°K, however, ΔS_θ would be numerically smaller, not larger, than ΔS_σ. At this temperature some of the hydrogen molecules in the flask would dissociate.

To summarize, cold hydrogen atoms unite and hot hydrogen molecules disintegrate because in both cases the entropy of the universe is thereby increased. The thermodynamic reasons for this behavior are these.

At low temperatures (small T), the absolute magnitude of the ratio $\Delta E_\theta/T$ is large. At low temperatures, therefore, the sign of ΔS_{total} is likely to be determined by the sign of $\Delta E_\theta/T$, i.e. by the sign of ΔE_θ, which for exothermic reactions, such as $2H = H_2$, is positive. Hence at low temperatures exothermic reactions are likely to be spontaneous, despite the fact that the products (H_2) may have less entropy than the reactants (2H), because any decrease in entropy of the reaction mixture is more than balanced by a large increase in entropy of the thermal surroundings.

On the other hand, at high temperatures (large T), the ratio $\Delta E_\theta/T$ approaches the value zero. At high temperatures, therefore, the sign of ΔS_{total} is likely to be determined by the sign of ΔS_σ, which for dissociative reactions, such as $H_2 = 2H$, is positive. Hence at high temperatures dissociative reactions are likely to be spontaneous, despite the fact that dissociative reactions are generally endothermic reactions and therefore cause the entropy of the thermal surroundings to decrease, for any decrease in the thermal entropy of the surroundings is more than balanced by an increase in entropy of the reaction mixture.

Of course, the magnitude and even the sign of ΔS_σ will depend upon the partial pressures of the reactants and the products; for the entropy of a chemical depends to some extent upon its partial pressure (Chapter 12). In principle it is possible to so diminish the partial pressure of the products of a reaction as to make the entropies of the reaction products so large as to render the reaction, however endothermic (and however low the temperature), spontaneous.

For example, if the partial pressure of atomic hydrogen were 10^{-50} atm., the entropy of atomic hydrogen would be greater than its standard 1-atm. value by the amount $-R \ln (10^{-50}/1) = 230$ cal/deg-mole. This is sufficient to make the reaction $H_2 = 2H$ spontaneous at 25°C—unless the partial pressure of H_2 is also very low. (When P_{H_2} is small, S_{H_2} is large and the reaction $2H = H_2$, not $H_2 = 2H$, tends to be favored.)

Of particular interest are the values of the partial pressures of the reactants and the products of a reaction that make ΔS_{total} for the reaction vanish. These values can be found as follows.

Consider again the reaction $H_2 = 2H$. When this reaction occurs in a flask of fixed volume, the contributions H_2 and H make to the entropy of the chemical system at 25°C can be expressed as follows (all entropies are in cal/deg-mole, or in cal/deg per mole of H_2 consumed).

$$S_{H_2} = 29.2 - R \ln P_{H_2} \qquad (*)$$
$$S_H = 25.4 - R \ln P_H$$

Now, in general,

$$\Delta S_\sigma = S_{\text{products}} - S_{\text{reactants}} \cdot$$

For the reaction $H_2 = 2H$,

$$\begin{aligned} \Delta S_\sigma &= 2S_H - S_{H_2} \\ &= 2(25.4 - R \ln P_H) - (29.2 - R \ln P_{H_2}) \\ &= 21.6 - R \ln (P_H{}^2/P_{H_2}). \end{aligned}$$

When $P_H = P_{H_2} = 1$ atm., ΔS_σ for the reaction $H_2 = 2H$ is $+21.6$. On the other hand, if $P_{H_2} = 1$ atm. and $P_H = 10^{-50}$ atm., $\Delta S_\sigma = 21.6 - 4.6 \log (10^{-100}) = +481.6$. The value of ΔS_σ that makes ΔS_{total} vanish is the value of ΔS_σ that makes the sum $\Delta S_\sigma + \Delta S_\theta$ vanish.

Now, in general,

$$\Delta S_\theta = \Delta E_\theta / T.$$

For the reaction $H_2 = 2H$ in a constant-volume container at 25°C,

$$\begin{aligned} \Delta S_\theta &= -104{,}000/298 \\ &= -349. \end{aligned}$$

Therefore, ΔS_{total} for the change $H_2 + 104$ kcal $= 2H$ will vanish at 25°C if the values of P_H and P_{H_2} are such that

$$[21.6 - R \ln (P_H{}^2/P_{H_2})] + [-349] = 0.$$

That is to say, for the reaction $H_2 = 2H$, $\Delta S_{\text{total}} = 0$ at 25°C whenever

$$\begin{aligned} R \ln (P_H{}^2/P_{H_2}) &= -349 + 21.6 \\ &= -327; \end{aligned}$$

* If the pressure, rather than the volume, were to remain fixed during the chemical reaction, the numerical constants in these two equations would be $29.2 + R = 31.2$ and $25.4 + R = 27.4$, respectively. See Chapter 15.

that is, whenever

$$\frac{P_{\text{H}}^{2}}{P_{\text{H}_2}} = e^{-327/R}$$

$$= 10^{-327/2.3R} = 10^{-327/4.6} \qquad (*)$$

$$= 10^{-71}.$$

A system is said to be at equilibrium with respect to a specified change if the value of ΔS_{total} for the change is zero. The last set of equations states that when a system is at equilibrium with respect to the change $H_2 = 2H$, the square of the partial pressure of the atomic hydrogen is proportional to the partial pressure of the molecular hydrogen. At 25°C the value of the proportionality constant is 10^{-71}. This proportionality constant is called the "equilibrium constant" for the reaction and is given the symbol K. For the reaction $H_2 = 2H$ at 25°C

$$K = 10^{-71}$$

$$= \frac{P_{\text{H}}^{2}}{P_{\text{H}_2}} \qquad (\text{when } \Delta S_{\text{total}} = 0).$$

Since $10^{-71} = e^{-327/R}$, one could also write that

$$R \ln K = -327$$

$$= -349 + 21.6$$

$$= -104{,}000/298 + (\Delta S^0{}_\sigma)_V \qquad (\dagger)$$

$$= (\Delta E_\theta)_V/T + (\Delta S^0{}_\sigma)_V$$

$$= -(\Delta E_\sigma)_V/T + (\Delta S^0{}_\sigma)_V. \qquad (\ddagger)$$

The last equation shows that the equilibrium constant K of a reaction depends upon two characteristics of the reaction: ΔE_σ and $\Delta S^0{}_\sigma$. Regarding the former, K tends to be large (at least at low temperatures) if ΔE_σ is negative; i.e. if the energy of the products is less than the energy of the reactants. Regarding the latter, K tends to be large (at least at high temperatures) if $\Delta S^0{}_\sigma$ is positive; i.e. if the standard entropy of the products is greater than the standard entropy of the reactants.

* $e = 10^{1/2.3}$. Therefore $e^{-327/R} = (10^{1/2.3})^{-327/R} = 10^{-327/2.3R}$.

† The subscript V indicates that the quantity within the parentheses refers to a change that occurs at constant volume. The superscript zero indicates that it is the partial-pressure-1-atm. values that are involved. This superscript has not been placed on ΔE_σ since the energy of an ideal gas (e.g. H_2 or H) is independent of its partial pressure. (For the same reason, the subscript V on (ΔE_σ) is not necessary in the present case.)

‡ By the First Law, $\Delta E_\theta = -\Delta E_\sigma$.

It is seen, also, that the difference between the energy of the reactants and the products, while very important at low temperatures (small T), becomes progressively less important as T increases.

This last statement is valid if ΔE_σ and $\Delta S^0{}_\sigma$ do not vary greatly with T. Usually this is the case.* Moreover, what changes do occur in ΔE_σ and $\Delta S^0{}_\sigma$ always occur in the same direction; if ΔE_σ increases with T, $\Delta S^0{}_\sigma$ increases, too. (This behavior is directly related to the fact that the energy and entropy of a substance both increase as T increases and both decrease as T decreases, the change in one always being in the same direction as the change in the other.) Thus, in their effects on K (or $\ln K$), the variation in ΔE_σ and $\Delta S^0{}_\sigma$ with T tend to cancel. Suppose, for example, that with a change in T, ΔE_σ increases (becomes more positive); this tends to decrease K. Simultaneously, there is an increase in $\Delta S^0{}_\sigma$ that tends to increase K. The net change in K is close to what would have been the change in K had ΔE_σ and $\Delta S^0{}_\sigma$ not changed at all. This fact is the basis of a very useful approximation.

Suppose, for example, one wishes to know approximately the value of the equilibrium constant for the reaction $H_2 = 2H$ at 10,000°K. To a first approximation, one may use in place of the true values of $(\Delta E_\sigma)_V$ and $(\Delta S^0{}_\sigma)_V$ at 10,000°K their values at some lower temperature, say 298°K.† In this way one estimates that

$$R \ln K(T = 10{,}000°\text{K}) \approx -104{,}000/10{,}000 + 21.6 = 11.2$$

$$K \approx 10^{11.2/4.6} = 10^{2.4}.$$

Evidently at 10,000°K most of the hydrogen is dissociated. If, for example, $P_{\text{H}} = 1$ atm., $P_{\text{H}_2} \approx 10^{-2.4}$ atm.

Or suppose one wishes to know roughly the temperature at which the equilibrium constant for the reaction $H_2 = 2H$ is unity ($R \ln K = 0$). At this temperature

$$-(\Delta E_\sigma)_V/T + (\Delta S^0{}_\sigma)_V = 0.$$

Hence,

$$T = \frac{(\Delta E_\sigma)_V}{(\Delta S^0{}_\sigma)_V}.$$

Using for $(\Delta E_\sigma)_V$ and $(\Delta S^0{}_\sigma)_V$ their values at 298°K, one finds that

$$T(K = 1) \approx \frac{104{,}000}{21.6}$$

$$= 4800°\text{K}.$$

* ΔE_σ and $\Delta S^0{}_\sigma$ do not vary at all with T if the products and reactants have identical heat capacities.

† This, of course, is a very long extrapolation.

Problems for Chapter 13

1. The universe of a chemical reaction that occurs at constant volume is composed of ——— (a) thermodynamically significant parts. The part labeled σ is called the ——— (b). The part labeled ——— (c) is called the ——— (d) of the ——— (e). For the universe of a constant volume reaction ΔS_{total} is equal to ——— (f). The entropy change in σ can be written as the difference between the entropy of the ——— (g) and the ——— (h); symbolically ——— (i). The entropy change in θ is always equal to ——— (j); the basis of this relation goes back to the definition of the ——— (k). By the First Law ——— (l). Therefore, the entropy change in θ can also be written as ——— (m). This means that ΔS_{total} can be expressed entirely in terms of quantities that refer directly to ——— (n); this expression for ΔS_{total} is ——— (o). At low temperatures the sign of ΔS_{total} is usually determined by the sign of ——— (p); this sign is positive for ——— (q) reactions and negative for ——— (r). At high temperatures the ——— (s) term is less important and the sign of ΔS_{total} is usually determined by the sign of ——— (t); this sign is positive if the entropy of the ——— (u) is ——— (v) than the ——— (w) of the ——— (x). Forms of matter stable at low temperatures tend to be ——— (y) forms. Forms of matter stable at high temperatures tend to be ——— (z) forms. One way to increase the entropy of a chemical species is to ——— (a′) its ——— (b′).
2. If the partial pressure of molecular hydrogen in a flask is 1 atm., what is the equilibrium partial pressure of atomic hydrogen at (a) 25°C? (b) 4800°K? (c) 10,000°K?
3. Does the value of the equilibrium constant for the reaction $H_2 = 2H$ depend upon the partial pressures of H_2 and H? Upon the total pressure? The volume? The temperature?
4. The reaction $H_2 = 2H$ is an ——— (a) reaction. In such reactions the sign of ΔE_θ is ——— (b). Hence (volume constant), the sign of ΔE_σ is ——— (c). This means that as the temperature increases the logarithm of the equilibrium constant for such reactions becomes less ——— (d), or more ——— (e). That is to say, as T increases, K for an ——— (f) reaction ——— (g). Does this conclusion agree with the principle of ——— (h)? ——— (i)?
5. Over modest temperature intervals the equilibrium constants of many reactions can be accurately represented by a two-parameter equation of the form

$$K = Ae^{-B/T}.$$

What is the thermodynamic significance of the parameters A and B?

What are their values in the neighborhood of 25°C for the reaction $H_2 = 2H$?

6. When the logarithm to the base ten of an equilibrium constant is plotted against the reciprocal of the absolute temperature, a nearly straight line is obtained. What is the slope of this line and what is its intercept on the $1/T = 0$ axis?
7. Once the equilibrium constant of a reaction is known at two temperatures, a calculation can be made of both $(\Delta E_\sigma)_V$ and $(\Delta S^0{}_\sigma)_V$. Show how this might be done.

14

Chemical Equilibrium (II)

$$(\Delta S_\sigma)_{T,P_\sigma} - \frac{(\Delta E_\sigma)_{T,P_\sigma} + P(\Delta V_\sigma)_{T,P_\sigma}}{T} = 0$$

For a chemical system and its surroundings to be at equilibrium with respect to an event for which they are the universe, the event must not change their total entropy; otherwise the contemplated event, or its reverse, could occur spontaneously.

Exchange of thermal energy between two parts of a universe, for example, or expansion of one part of a universe at the expense of the volume of another part, or occurrence of a chemical reaction within one part of a universe with a simultaneous change in the thermal energy of another part —were any such event able to increase the total entropy of the universe, then, given time, the event should occur spontaneously. If, on the other hand, the event were to lead to a decrease in the total entropy, the reverse event would increase the total entropy and, given time, it should occur spontaneously.

A plot of the total entropy of the universe against some appropriate parameter that measures the degree of advancement of an event—the thermal energy or volume exchanged between two parts of the universe, for example, or the number of moles of a particular chemical consumed in a chemical reaction—must have the general shape shown in Fig. 14.1.

As the degree of advancement parameter increases, the total entropy rises to a maximum and then decreases. An illustrative example is the transfer of thermal energy between two objects. So long as the donor of thermal energy is warmer than the acceptor, the total entropy increases as thermal energy flows from donor to acceptor; but when the donor becomes cooler than the acceptor (not likely, of course), the total entropy decreases. Similarly, expansion of one part of the universe at the expense of the volume of some other part of the universe increases the total entropy of the universe so long as the pressure of the part expanding exceeds the pressure of the

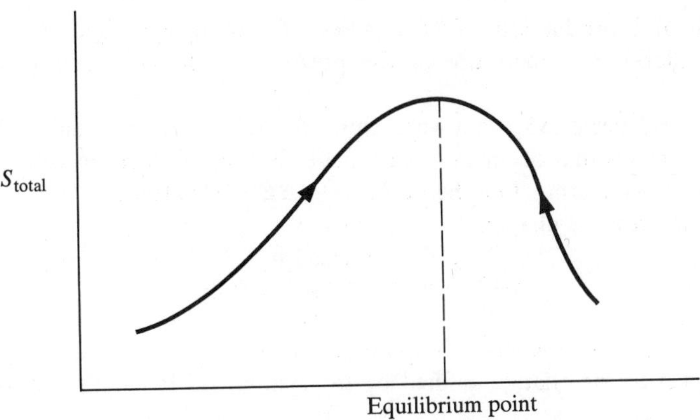

Fig. 14.1 Graphical representation of the Second Law: The entropy of the universe tends to increase.

part contracting; but when the former pressure becomes less than the latter, the total entropy decreases.

To the left of the maximum (Fig. 14.1), the total entropy increases as the event proceeds in the forward direction; in this region the event can proceed forward but not backward. To the right of the maximum the total entropy decreases as the event proceeds forward; in this region the event can proceed backward but not forward. Thus, regardless of where the system starts, given time, it will eventually proceed to that point where the total entropy is a maximum. This point is called the point of equilibrium. At this point the total entropy neither increases nor decreases as the variable of advancement changes. At the equilibrium point the rate of change of the total entropy with respect to the variable of advancement is zero. For a system at equilibrium with respect to a chemical reaction, this fact may be expressed as follows:

$$\Delta S_\sigma + \Delta S_\theta = 0.$$

At first glance the variable of advancement seems to be missing from this expression. It is in the statement implicitly, however, and shows up explicitly in the units. For the reaction $H_2 = 2H$, for example, the units used in Chapter 13 for ΔS_σ and ΔS_θ were cal per deg *per mole of molecular hydrogen consumed.*

ΔS_σ, it should be noted, is equal to the change that would occur in the entropy of a *very large* equilibrium mixture of molecular and atomic hydrogen when one mole of molecular hydrogen dissociates. In a very large system (an ocean, so to speak) consumption of one mole of molecular hydrogen and production of two moles of atomic hydrogen would not sensibly alter the temperature or the pressure or the composition of the system.*

The second term, ΔS_θ, is always equal to $\Delta E_\theta / T$. Expression of ΔE_θ in terms of changes that occur in σ is particularly simple when the volume of σ is considered constant. For then $\Delta E_\theta = -\Delta E_\sigma$. This leads to the equation used in the previous chapter.

$$(\Delta S_\sigma)_{T,V_\sigma} - \frac{(\Delta E_\sigma)_{T,V_\sigma}}{T} = 0. \qquad (\dagger)$$

This equation is very general. It applies to all systems that are at equilibrium with respect to the change defined by the symbol Δ. It has one drawback, however.

This drawback does not concern mixtures of ideal gases; nor is it of serious concern for reactions in condensed systems at ordinary pressures. What it mostly concerns are reactions at high pressures—the synthesis of diamonds, for example, and many other geochemical processes. The drawback is this.

$(\Delta E_\sigma)_{V_\sigma}$—we could equally well speak about $(\Delta S_\sigma)_{V_\sigma}$—represents the difference between the energies of the products of a reaction and the corresponding reactants. The energy of the reactants (the same remarks apply to the energy of the products) is the sum of the changes in the energy of σ that occur as the reactants are added to σ without any net change in the temperature of σ or the volume of σ. It is the last condition that causes difficulty. As each individual component is added to σ at constant volume, the pressure in σ will generally increase. This will generally affect the molar energies of all the substances in σ. How much these energies are affected will depend upon how much the pressure increases; this in turn, will depend upon the compressibilities of the components of σ. This is an awkward situation. Effective energies and entropies for general use in the above equation cannot easily be tabulated. For in each specific application of this equation, the required contribution of the relevant chemicals to E_σ and S_σ will depend to some extent upon what other chemicals are in the system. A more useful equation would be one that permitted direct use to be made of the ordinary molar energies and entropies of chemical substances. Such an equation is not difficult to obtain.

* Alternatively, ΔS_σ may be regarded as the value of the ratio $\delta S_\sigma / \delta n_{H_2}$, where δS_σ is the small change produced in the entropy of a finite system σ when the number of moles of molecular hydrogen in the system decreases by the small amount δn_{H_2}. If σ is a whole ocean of H_2 and H, δn_{H_2} may be taken equal to 1 mole.

† The subscript T indicates that the chemical system σ is surrounded by a thermostat with which it is in thermal equilibrium. The subscript V_σ indicates that the volume of σ is held constant.

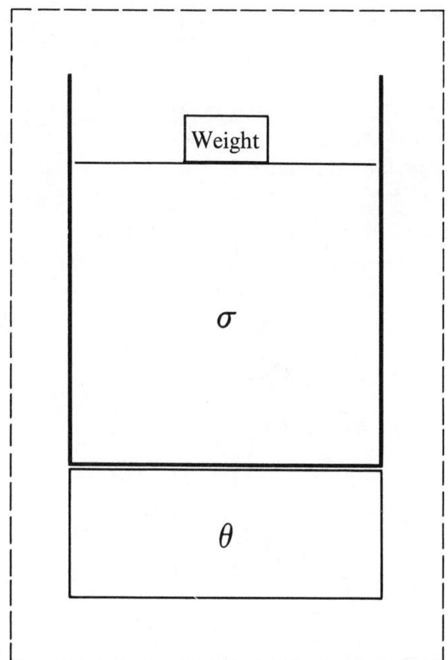

Fig. 14.2 The universe of a reaction that occurs at constant pressure without the production or consumption of useful work.

In Fig. 14.2 is shown schematically a thermostated chemical system confined to a chamber by a frictionless piston that has a weight resting upon it. The thermostat θ maintains constant the temperature of σ. The piston and weight Wt. maintain constant the pressure of σ. If m is the mass of the weight (plus piston), if the acceleration imparted to a freely falling object by the gravitational attraction between the object and the earth is g (at sea level 32 ft/sec^2 = 980 cm/sec^2), and if the cross-sectional area of the piston is A, this pressure—the force on the system per unit area—is mg/A.

Suppose, next, that this little universe—the chemical system, the thermostat, and the weight—is at equilibrium with respect to a chemical reaction. For this reaction, then,

$$\Delta S_{\text{total}} = 0.$$

Now, by the additivity of S,

$$S_{\text{total}} = S_\sigma + S_\theta + S_{\text{Wt.}}\,.$$

Therefore

$$\Delta S_{\text{total}} = \Delta S_\sigma + \Delta S_\theta + \Delta S_{\text{Wt.}}\,.$$

But $\Delta S_{\text{Wt.}} = 0$. Thus,

$$\Delta S_\sigma + \Delta S_\theta = 0.$$

As usual, $$\Delta S_\theta = \Delta E_\theta / T. \qquad (*)$$

Hence, as usual,

$$\Delta S_\sigma + \Delta E_\theta / T = 0.$$

To express ΔE_θ in terms of changes that occur within σ, we turn to the First Law. (In the preceding chapters it was generally assumed that $\Delta E_\theta = -\Delta E_\sigma$. This is true if $\Delta E_{\text{Wt.}} = 0$, that is, if $\Delta V_\sigma = 0$. If, as in this chapter, it is assumed that $\Delta P_\sigma = 0$, it will generally be true that $\Delta V_\sigma \neq 0$.)

By the First Law,

$$\Delta E_{\text{total}} = 0.$$

By the additivity of E

$$E_{\text{total}} = E_\sigma + E_\theta + E_{\text{Wt.}}\,.$$

Therefore, $$\Delta E_\sigma + \Delta E_\theta + \Delta E_{\text{Wt.}} = 0.$$

Hence, $$\Delta E_\theta = -(\Delta E_\sigma + \Delta E_{\text{Wt.}}).$$

To express $\Delta E_{\text{Wt.}}$ in terms of changes that occur within σ, we note that

$$\Delta E_{\text{Wt.}} = mg \cdot \Delta h,$$

where Δh is the change in the weight's height. Now

$$P = mg/A;$$

that is, $$mg = P \cdot A.$$

Therefore, $$\Delta E_{\text{Wt.}} = P \cdot A \cdot \Delta h.$$

But (see Fig. 14.2)

$$A \cdot \Delta h = \Delta V_\sigma\,.$$

Thus, $$\Delta E_{\text{Wt.}} = P \cdot \Delta V_\sigma\,.$$

ΔE_θ may now be expressed entirely in terms of changes that occur in σ:

$$\Delta E_\theta = -(\Delta E_\sigma + P\Delta V_\sigma).$$

At equilibrium, therefore,

$$\Delta S_\sigma - \frac{\Delta E_\sigma + P\Delta V_\sigma}{T} = 0.$$

More precisely, at equilibrium

$$(\Delta S_\sigma)_{T,P_\sigma} - \frac{(\Delta E_\sigma + P_\sigma \Delta V_\sigma)_{T,P_\sigma}}{T} = 0.$$

* Throughout Parts I, II, and III of this book the assumption is made that the volume of the thermal surroundings is held constant. This assumption is convenient, but not necessary.

An advantage this constant-temperature, constant-pressure formulation of chemical equilibrium enjoys over the previously derived constant-temperature, constant-volume formulation,

$$(\Delta S_\sigma)_{T,V_\sigma} - \frac{(\Delta E_\sigma)_{T,V_\sigma}}{T} = 0,$$

is that once the temperature and the pressure and the concentrations of the chemical species in the system have been specified, the individual molar entropies and molar energies that comprise ΔS_σ and ΔE_σ are determined. For the constant-temperature, constant-volume formulation the corresponding statement is not always valid. Properly used, however, the two formulations lead to the same conclusions. To illustrate, for the dissociation of hydrogen at 25°C,

$$\begin{aligned}
(\Delta S_\sigma)_{T,P} &= S_{\text{products}} - S_{\text{reactants}} \\
&= 2S_{\text{H}} - S_{\text{H}_2} \\
&= 23.6 - R\ln(P_{\text{H}}{}^2/P_{\text{H}_2}) \\
(\Delta E_\sigma)_{T,P} &= 104 \text{ kcal (per mole H}_2\text{ consumed)} \qquad (*) \\
(\Delta V_\sigma)_{T,P} &= V_{\text{products}} - V_{\text{reactants}} \\
&= 2V_{\text{H}} - V_{\text{H}_2} \\
&= \frac{2RT}{P} - \frac{RT}{P} \\
&= RT/P \\
P(\Delta V_\sigma)_{T,P} &= RT \text{ (per mole H}_2\text{ consumed).}
\end{aligned}$$

Thus, $\Delta S_{\text{total}} = 0$ at 25°C for the reaction $H_2 = 2H$ whenever P_{H} and P_{H_2} are such that

$$23.6 - R\ln(P_{\text{H}}{}^2/P_{\text{H}_2}) - \frac{104{,}000 + R\cdot 298}{298} = 0;$$

that is, whenever, as found in Chapter 13,

$$R\ln(P_{\text{H}}{}^2/P_{\text{H}_2}) = -327.$$

More generally, for the reaction of a moles of A and b moles of B to give c moles of C and d moles of D according to the equation

$$aA + bB = cC + dD,$$

$$\begin{aligned}
(\Delta S_\sigma)_{T,P} &= S_{\text{products}} - S_{\text{reactants}} \\
&= (cS_C + dS_D) - (aS_A + bS_B).
\end{aligned}$$

* If the products and reactants behave as ideal gases, E_{products} and $E_{\text{reactants}}$, and hence, ΔE_σ, are independent of the pressure (or volume). In the present case, therefore, $(\Delta E_\sigma)_{T,P} = (\Delta E_\sigma)_{T,V}$. Generally, however, $(\Delta E_\sigma)_{T,P} \neq (\Delta E_\sigma)_{T,V}$.

Similar expressions hold for $(\Delta E_\sigma)_{T,P}$ and $(\Delta V_\sigma)_{T,P}$. If A, B, C, and D behave as ideal gases, their entropies S_A, S_B, S_C, and S_D at partial pressures P_A, P_B, P_C, and P_D may be expressed, respectively, as

$$S_A = S_A{}^0 - R \ln P_A \qquad (*)$$

$$\vdots$$

$$S_D = S_D{}^0 - R \ln P_D,$$

where $S_A{}^0$, $S_B{}^0$, $S_C{}^0$, and $S_D{}^0$ are the molar entropies of A, B, C, and D, respectively, when P_A, P_B, P_C, and P_D are (respectively) 1 atm. Using these expressions for S_A, S_B, S_C, S_D, one finds that

$$\begin{aligned}(\Delta S_\sigma)_{T,P} &= (cS_C{}^0 + dS_D{}^0) - (aS_A{}^0 + bS_B{}^0) - R \ln (P_C{}^c P_D{}^d / P_A{}^a P_B{}^b) \\ &= \Delta S^0 - R \ln (P_C{}^c P_D{}^d / P_A{}^a P_B{}^b).\end{aligned}$$

At equilibrium P_A, P_B, P_C, and P_D must be such that

$$\Delta S^0 - R \ln \frac{P_C{}^c P_D{}^d}{P_A{}^a P_B{}^b} - \frac{(\Delta E_\sigma)_{T,P} + P(\Delta V_\sigma)_{T,P}}{T} = 0.$$

That is to say, at equilibrium

$$RT \ln \frac{P_C{}^c P_D{}^d}{P_A{}^a P_B{}^b} = -[(\Delta E_\sigma)_{T,P} + P(\Delta V_\sigma)_{T,P}] + T\Delta S^0 .$$

$(\Delta E_\sigma)_{T,P}$, $P(\Delta V_\sigma)_{T,P}$, and ΔS^0 are independent of P_A, P_B, P_C, P_D. (By definition, ΔS^0 is simply the algebraic difference between the standard entropies of the products and the reactants. By assumption, A, B, C, and D are ideal gases; hence ΔE_σ and $P\Delta V_\sigma$ depend upon the absolute temperature only.) At a given temperature, therefore, the quantity

$$\frac{P_C{}^c P_D{}^d}{P_A{}^a P_B{}^b}$$

must at equilibrium always work out to be the same number. This number is called the equilibrium constant K for the reaction. K can be calculated

* It was found in Chapter 12 that for any ideal gas (say A), $S_A = S_A{}^0 - R \ln P_A$. P_A in this equation was the total gas pressure, i.e. the external pressure. The assumption that the same equation holds when P_A represents the partial pressure of gas A in a gaseous mixture—the total pressure, by Dalton's Law, being here $P_A + P_B + P_C + P_D$—is an assumption that cannot be logically proven within the framework of classical thermodynamics.

from two kinds of data: from the values of ΔE and ΔS^0 or from the values at any equilibrium point of P_A, P_B, P_C, P_D.

$$K = e^{-\frac{\Delta E_\sigma + P\Delta V_\sigma - T\Delta S^0}{RT}}$$

$$\overset{\Delta S_{\text{total}}=0}{=} \frac{P_C{}^c P_D{}^d}{P_A{}^a P_B{}^b}.$$

The calculation of specific equilibrium constants from tabulated thermodynamic data will be illustrated in Part IV.

Concluding remarks. In the preceding chapter it was found that

$$\Delta S_{\text{total}} = (\Delta S_\sigma)_{T,V_\sigma} - \frac{(\Delta E_\sigma)_{T,V_\sigma}}{T}.$$

In this chapter, it was found that

$$\Delta S_{\text{total}} = (\Delta S_\sigma)_{T,P_\sigma} - \frac{(\Delta E_\sigma)_{T,P_\sigma} + P_\sigma(\Delta V_\sigma)_{T,P_\sigma}}{T}.$$

These two expressions for ΔS_{total} are equivalent if the chemical system σ is in hydrostatic equilibrium with its surroundings; for then small changes in the volume of σ leave unchanged the entropy of the universe (if this were not so, the system by expanding or contracting could produce an increase in S_{total}, contrary to the assumed condition of hydrostatic equilibrium).

To show that the assumption of hydrostatic equilibrium between σ and its surroundings implies that

$$(\Delta S_{\text{total}})_{T,V_\sigma} = (\Delta S_{\text{total}})_{T,P_\sigma},$$

it may be supposed, arbitrarily, that the reaction in σ occurs at constant volume. Depending upon whether or not the system is at equilibrium, this may or may not produce a change in S_{total}. In any event, the final state of σ can be compressed or expanded to make it identical to what would have been the final state if the reaction had occurred at constant pressure. But, if the system is in hydrostatic equilibrium with its surroundings, this last step leaves S_{total} unchanged. Thus, as stated,

$$(\Delta S_{\text{total}})_{T,V_\sigma} = (\Delta S_{\text{total}})_{T,P_\sigma}.$$

Multiplying both sides of this equation by T, one finds that

$$(\Delta E_\sigma)_{T,V_\sigma} - T(\Delta S_\sigma)_{T,V_\sigma} = (\Delta E_\sigma)_{T,P_\sigma} + P_\sigma(\Delta V_\sigma)_{T,P_\sigma} - T(\Delta S_\sigma)_{T,P_\sigma}.$$

This may be written

$$\Delta(E_\sigma - TS_\sigma)_{T,V_\sigma} = \Delta(E_\sigma + P_\sigma V_\sigma - TS_\sigma)_{T,P_\sigma}\,.$$

Introducing the symbols

$$A \equiv E - TS$$
$$G \equiv E + PV - TS$$

called, respectively, the Helmholtz and Gibbs free energies, one may write, for a system in hydrostatic equilibrium with its surroundings,

$$\Delta(A_\sigma)_{T,V_\sigma} = \Delta(G_\sigma)_{T,P_\sigma}\,.$$

Either expression, $\Delta(A_\sigma)_{T,V_\sigma}$ or $\Delta(G_\sigma)_{T,P_\sigma}$, gives the value of $T\Delta S_{\text{total}}$ for the change symbolized by the symbol Δ. At equilibrium these expressions vanish.

Sometimes the Gibbs free energy is written in the form

$$G = H - TS,$$

where the symbol H, the enthalpy, has the definition

$$H \equiv E + PV.$$

For chemical reactions that occur at constant pressure without the production of useful mechanical work, the change in enthalpy of the reaction mixture is equal, after a sign change, to the heat of the reaction.

The phrase "heat of reaction," as commonly used, refers to the thermal energy absorbed by the thermal surroundings of a chemical system during the course of a chemical reaction within the chemical system. The symbol often used for "heat of reaction" is Q. Thus, in this terminology,

$$Q = \Delta E_\theta\,. \qquad (*)$$

Defined in this way Q is positive for exothermic reactions and negative for endothermic reactions. For reactions at constant volume, $\Delta E_\theta = -\Delta E_\sigma$. Thus, when V_σ is constant,

$$Q_{V_\sigma} = -(\Delta E_\sigma)_{V_\sigma}\,.$$

For reactions at constant pressure, $\Delta E_\theta = -(\Delta E_\sigma + P_\sigma \Delta V_\sigma)$. Thus, when P_σ is constant,

$$\begin{aligned} Q_{P_\sigma} &= -[\Delta E_\sigma + P_\sigma \Delta V_\sigma]_{P_\sigma} \\ &= -[\Delta(E_\sigma + P_\sigma V_\sigma)]_{P_\sigma} \\ &= -[\Delta(H_\sigma)]_{P_\sigma}\,. \end{aligned}$$

* In another context the symbol Q (or q) is used to represent the thermal energy absorbed *by* a chemical system *from* its thermal surroundings.

Problems for Chapter 14

1. Under what conditions are the following equations valid?
   (a) $\Delta S - (\Delta E/T) = 0$
   (b) $\Delta S - [(\Delta E + P\Delta V)/T] = 0$
2. Explain briefly how the $P\Delta V$ term gets into equation 1(b).
3. In many practical applications of thermodynamics, what would the object labeled "Weight" in Fig. 14.2 represent? What would be the corresponding value of mg/A?
4. Consider the reaction $2H = H_2$. Suppose this reaction were to occur at constant temperature in a vessel that is open to the atmosphere. The volume of the product is less than that of the reactant. Thus, as the reaction progresses the air mantle about the earth will sink closer to the center of the earth. In other words, during the reaction $2H = H_2$ the potential energy of the atmosphere decreases. What happens to this energy?
5. When is the heat of a reaction equal to $-\Delta E_\sigma$? When is it not? When is it greater (more positive) than $-\Delta E_\sigma$? When is it less than $-\Delta E_\sigma$?
6. What is the difference between the enthalpy and energy of
   (a) substances in general?
   (b) an ideal gas at (i) 1 atm.? (ii) 10 atm.? (iii) 300°K?
   (c) liquid water at (i) 0 atm.? (ii) 1 atm.?
   Are these differences large or small compared to the heats of ordinary chemical reactions?
7. At 25°C the heat of the reaction $2H = H_2$ at constant pressure is 104.178 kcal per mole of H_2 produced. (a) What is the value at 25°C of $2H_{\mathrm{H}} - H_{\mathrm{H}_2}$? (b) If the enthalpy of molecular hydrogen at 25°C (and 1 atm.) is set equal to zero, what is H_{H}? (c) What is the value at 25°C of $P\Delta V_\sigma$? (d) Of $2E_{\mathrm{H}} - E_{\mathrm{H}_2}$? (e) What is the heat of the reaction at 25°C at constant volume?
8. What are the units on the equilibrium constant for the reaction $aA(\text{gas}) + bB(\text{gas}) = cC(\text{gas}) + dD(\text{gas})$? What is the value of $P\Delta V_\sigma$? (Assume that A, B, C, and D behave as ideal gases.)
9. Under what conditions is the reaction in problem 8 thermodynamically possible?
10. For a reversible reaction, what is the relation between the heat of reaction and the entropy change in the chemical system? What relation holds for an irreversible reaction?

15

Addendum to Part I

$$\frac{d(\ln x)}{dx} = \frac{1}{x}$$

Central to the derivation of the equilibrium constant expressions given in the two preceding chapters was the equation $\mathrm{S} = \mathrm{S}^0 - R \ln P$, which follows directly from the definition of S^0, the equation of state of an ideal gas, and the equation

$$\Delta S_\sigma = nR \ln (V_2/V_1).$$

The latter equation was obtained in Chapter 12 from an analysis of the entropy and energy changes in the universe during a reversible, isothermal expansion of an ideal gas (Fig. 12.1). If the expansion is reversible ($\Delta S_{\text{total}} = 0$) and isothermal ($T$ constant), and if the gas is ideal ($\Delta E_\sigma = 0$; $\Delta E_\theta = -\Delta E_{\text{Wt.}}$),

$$\Delta S_\sigma = -\Delta S_\theta = -\frac{\Delta E_\theta}{T} = \frac{\Delta E_{\text{Wt.}}}{T}.$$

The relation $\Delta S_\sigma = nR \ln (V_2/V_1)$ will therefore be valid if in a reversible, isothermal expansion of an ideal gas

$$\Delta E_{\text{Wt.}} = nRT \ln (V_2/V_1).$$

The purpose of this addendum is to verify that this is so. Use will be made of the fact (which will thereafter be derived) that for small changes in a quantity x, here V_σ (later T_σ), the fractional change in x is equal to the change in the natural logarithm of x:

$$\frac{\delta x}{x} = \delta(\ln x).$$

The addendum concludes with several additional applications of this useful mathematical relation.

The change in potential energy of a weight whose altitude changes by the small amount δh is $mg \cdot \delta h$.* If the change occurs slowly, any change in the kinetic energy of the weight is negligible; under these conditions

$$\delta E_{\text{Wt.}} = mg \cdot \delta h.$$

If the weight is resting through a frictionless piston on a gas, and if the cross-sectional area of the piston is A, the pressure the weight exerts on the gas is mg/A. In a reversible expansion, this external pressure must almost equal the gas pressure $P(P = mg/A)$. Thus, in a reversible expansion

$$\delta E_{\text{Wt.}} = P \cdot A \cdot \delta h.$$

But, as before (see Chapter 14), $A \cdot \delta h$ is just the volume change of the gas δV. Hence

$$\delta E_{\text{Wt.}} = P\delta V.$$

Now, as a gas expands isothermally, P decreases.† At any point in the expansion of the gas, if it is an ideal gas,

$$P = nRT/V.$$

Therefore, in terms of the volume of the gas and the change in its volume (and, also, in terms of n, R, and T, which do not change during the expansion),

$$\delta E_{\text{Wt.}} = (nRT)\frac{\delta V}{V}.$$

Since the fractional change in the volume, $\delta V/V$, is equal to the change in the natural logarithm of the volume $[\delta V/V = \delta(\ln V)]$,

$$\delta E_{\text{Wt.}} = (nRT)\delta(\ln V).$$

This equation states that the change in the energy of the weight is proportional to the change in the logarithm of the volume of the expanding gas. The sum of all the infinitesimal changes in the energy of the weight during an expansion from V_1 to V_2 is just the total change in the energy of the weight, $E_2 - E_1$. Similarly, the sum of all the infinitesimal changes in the logarithm of V is just the total change in the logarithm of V, $\ln V_2 - \ln V_1$. Thus

$$E_2 - E_1 = (nRT)(\ln V_2 - \ln V_1),$$

or

$$\Delta E_{\text{Wt.}} = (nRT)\ln(V_2/V_1).$$

* The symbol δ, rather than Δ, is used here to signify "a small (or infinitesimal) change" since in thermodynamics Δ often signifies a finite change, or a finite rate of change (p. 20).

† To permit the gas to expand, the mass of the weight on the piston must be continually diminished.

It remains to be shown that $\delta V/V = \delta(\ln V)$; or, what is the same thing, that

$$\frac{\delta(\ln V)}{\delta V} = \frac{1}{V}.$$

If V changes from V to $(V + \delta V)$, the change in the logarithm of V (to any base) may be written as follows:

$$\begin{aligned}\delta(\log V) &= \log(V + \delta V) - \log V \\ &= \log\left(\frac{V + \delta V}{V}\right) \\ &= \log\left(1 + \frac{\delta V}{V}\right).\end{aligned}$$

Therefore,

$$\begin{aligned}\frac{\delta(\log V)}{\delta V} &= \frac{1}{\delta V}\log\left(1 + \frac{\delta V}{V}\right) \\ &= \log\left(1 + \frac{\delta V}{V}\right)^{1/\delta V} \\ &= \frac{1}{V}\log\left(1 + \frac{\delta V}{V}\right)^{V/\delta V} \\ &= \frac{1}{V}\log(1 + x)^{1/x},\end{aligned}$$

where $x = \delta V/V$.

As x approaches zero, $(1 + x)^{1/x}$ approaches the number 2.718..., called e. This can be seen as follows. By the binomial theorem

$$(a + b)^n = a^n + \frac{na^{n-1}b}{1!} + \frac{n(n-1)a^{n-2}b^2}{2!} + \cdots.$$

Hence,

$$(1 + x)^{1/x} = 1^{1/x} + \frac{\frac{1}{x}\,1^{(1/x)-1}\,x}{1!} + \frac{\frac{1}{x}\left(\frac{1}{x} - 1\right)1^{(1/x)-2}\,x^2}{2!} + \cdots.$$

As x approaches zero, the right-hand side of this expression approaches the value of the sum

$$1 + \frac{1}{1!} + \frac{1}{2!} + \frac{1}{3!} + \frac{1}{4!} + \frac{1}{5!} + \frac{1}{6!} + \cdots.$$

The sum of the first seven terms is 2.718.

If the expression $d(\log V)/dV$ is used to denote the limiting value of $\delta(\log V)/\delta V$ as δV approaches zero, it follows that

$$\frac{d(\log V)}{dV} = \frac{1}{V}\log e.$$

Since $\ln e \equiv \log_e e = 1$, $d(\ln V)/dV = 1/V$.

The expression

$$\lim_{x\to 0}(1 + x)^{1/x} = e$$

can also be written, through the substitution $x = 1/y$, as

$$\lim_{y\to\infty}\left(1 + \frac{1}{y}\right)^y = e.$$

With the aid of the latter expression, it can be shown that the contribution per mole that an ideal gas makes to the entropy of a system when the gas is added at constant pressure exceeds by R the contribution the gas makes when added at constant volume. This fact was mentioned in Chapter 13 in the discussion of the equilibrium $H_2 = 2H$.

Let the gas be added at constant temperature and constant volume to a system that already contains n moles of ideal gas. The pressure in the system will increase from nRT/V to $(n + 1)RT/V$. To make the final pressure the same as the initial pressure, the system must be expanded isothermally from V to a new volume V' which is such that

$$(n + 1)\frac{RT}{V'} = n\frac{RT}{V}.$$

Thus,
$$V' = V\left(1 + \frac{1}{n}\right).$$

This isothermal expansion will increase the entropy of the system by the amount

$$\begin{aligned}(n + 1)R\ln(V'/V) &= (n + 1)R\ln\left(1 + \frac{1}{n}\right)\\ &= R\ln\left(1 + \frac{1}{n}\right)^{(n+1)}\\ &= R\ln\left(1 + \frac{1}{n}\right)^{n} + R\ln\left(1 + \frac{1}{n}\right).\end{aligned}$$

As n becomes very large the first term approaches the value $R\ln e = R$ and the second term approaches the value $R\ln 1 = 0$. Thus the amount by which the entropy increases approaches the value R. This represents the difference between the contribution the gas makes at constant pressure and at constant volume to the entropy of the system.*

Summation of both sides of the expression

$$\frac{\delta x}{x} = \delta(\ln x)$$

from $x = x_1$ to $x = x_2$ yields this result:

$$\lim_{\delta x \to 0} \sum_{x=x_1}^{x=x_2} \frac{\delta x}{x} = \ln x_2 - \ln x_1$$

$$= \ln (x_2/x_1).$$

With this expression it is possible to calculate how a substance's entropy changes as its temperature changes; and from this, how an ideal gas's temperature changes as the gas expands reversibly and adiabatically. The latter result can be expressed in several equivalent forms if use is made of the fact that the molar heat capacity of an ideal gas at constant pressure exceeds by R its molar heat capacity at constant volume. The addendum concludes by considering each of these items in turn.

1. *Change in entropy with temperature.* When a small quantity of thermal energy δE is added to a substance at constant volume, the entropy of the substance changes by the amount

$$\delta S = \frac{\delta E}{T}. \qquad (V \text{ constant})$$

Simultaneously, the temperature of the substance changes. This temperature change δT is related to δE and the substance's heat capacity at constant volume by the equation

$$C_V = \frac{\delta E}{\delta T}. \qquad (V \text{ constant})$$

Alternatively,

$$\delta E = C_V \delta T.$$

* Mathematically,

$$\left(\frac{\partial S}{\partial n}\right)_{T,P} = \left(\frac{\partial S}{\partial n}\right)_{T,V} + \left(\frac{\partial S}{\partial V}\right)_{T,n} \left(\frac{\partial V}{\partial n}\right)_{T,P}.$$

For an ideal gas,

$$\left(\frac{\partial S}{\partial V}\right)_{T,n} = \frac{P}{T} \quad \text{and} \quad \left(\frac{\partial V}{\partial n}\right)_{T,P} = \frac{RT}{P}.$$

Substitution from the last equation into the first one above yields

$$\delta S = C_V \frac{\delta T}{T}. \qquad (V \text{ constant})$$

Summing both sides from $T = T_1$ to $T = T_2$, V constant, and assuming that over this temperature interval C_V is constant, one finds that

$$\begin{aligned} S(T_2,V) - S(T_1,V) &= \lim_{\delta T \to 0} \sum_{T=T_1}^{T=T_2} C_V \frac{\delta T}{T} \\ &= C_V \lim_{\delta T \to 0} \sum_{T=T_1}^{T=T_2} \frac{\delta T}{T} \\ &= C_V \ln (T_2/T_1). \end{aligned}$$

If, on the other hand, thermal energy is absorbed by a system σ from its thermal surroundings θ and if P_σ, not V_σ, is constant, and if σ and θ are at essentially the same temperature so that the process is reversible ($\delta S_{\text{total}} = 0$),

$$\begin{aligned} \delta S_\sigma &= -\delta S_\theta \\ &= -\frac{\delta E_\theta}{T} = \frac{\delta E_\sigma + P_\sigma \delta V_\sigma}{T} \qquad (P_\sigma \text{ constant}) \\ &= \frac{\delta H_\sigma}{T}. \qquad (*) \end{aligned}$$

The change in the enthalpy of σ is related to its change in temperature and its heat capacity at constant pressure by the equation

$$C_P = \frac{\delta H}{\delta T}. \qquad (P \text{ constant})$$

Alternatively, $$\delta H = C_P \delta T.$$

Thus, at constant pressure

$$\delta S = C_P \frac{\delta T}{T}.$$

If C_P is constant over the interval T_1 to T_2,

$$S(T_2,P) - S(T_1,P) = C_P \ln (T_2/T_1).$$

* For the absolute temperature of a system σ one may write the following mathematical expressions:

$$T_\sigma = \left(\frac{\partial E_\sigma}{\partial S_\sigma}\right)_{V_\sigma} = \left(\frac{\partial H_\sigma}{\partial S_\sigma}\right)_{P_\sigma}.$$

For liquid water in its standard state (25°C and 1 atm.), C_P (or C_V)* is about 1 cal/deg-gram, or 18 cal/deg-mole. Thus, in going from 0 to 25°C the entropy of liquid water changes by about

$$\begin{aligned}(18\text{ cal/deg-mole}) \ln (298/273) &= (18\text{ cal/deg-mole}) \ln (1.092)\\ &\approx (18\text{ cal/deg-mole})(0.092)\\ &= 1.66\text{ cal/deg-mole.}\end{aligned}$$

2. *Reversible, adiabatic expansion of an ideal gas.* An adiabatic change is a change in a chemical system that is thermally insulated from its surroundings. In any adiabatic change, therefore, $\Delta E_\theta = 0$; therefore, too, $\Delta S_\theta = 0$. Thus, in an adiabatic change $\Delta S_{\text{total}} = \Delta S_\sigma$. Now, if the change is reversible, $\Delta S_{\text{total}} = 0$. Hence, $\Delta S_\sigma = 0$. In other words, in any reversible, adiabatic change,

$$S_\sigma = \text{constant.}$$

Reversible, adiabatic changes are sometimes called isentropic changes.†

In the isentropic expansion of an ideal gas against a restraining piston (Fig. 9.2, step *BC*), the gas must cool; otherwise its entropy would not remain constant. To calculate the final temperature of the gas, given its initial condition and its final volume or its final pressure, one may think of passing from the initial state P_1, V_1, T_1 to the final state P_2, V_2, T_2 along either one of two paths (Fig. 15.1).

Along one path it is supposed that the gas first expands isothermally from its initial volume V_1 to its final volume V_2 (point *A*, Fig. 15.1); during this step its entropy increases by the amount $nR \ln (V_2/V_1)$. Then the gas is cooled at constant volume from its initial temperature T_1 to its final temperature T_2; during this step its entropy changes by the amount $C_V \ln (T_2/T_1)$. If the over-all change is to be isentropic, T_2 must be such that

$$nR \ln (V_2/V_1) + C_V \ln (T_2/T_1) = 0.$$

As V_2 is greater than V_1, the first term is positive; therefore, the second term must be negative; that is, T_2 must be less than T_1.

Along the other path one supposes that the gas first expands isothermally from its initial pressure P_1 to its final pressure P_2 ($P_2 < P_1$) (point *B*, Fig 15.1); during this step its entropy increases by the amount $-nR \ln (P_2/P_1)$ Following this the gas is cooled at constant pressure from its initial to it

* At ordinary pressures C_P and C_V are nearly the same for condensed phases whic (compared to gases) have relatively small coefficients of thermal expansion. For ideal gase (see below, item 3) $C_P = C_V + nR$.

† From the point of view of the universe, all reversible changes are isentropic. From th point of view of the thermal surroundings, all adiabatic changes are isentropic. From th point of view of the chemical system, all reversible, adiabatic changes are isentropic.

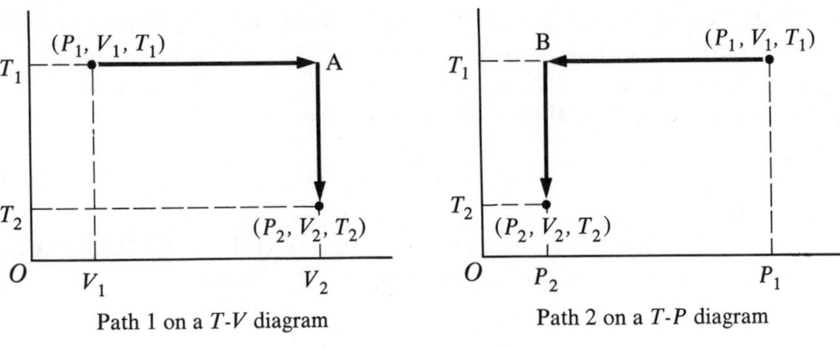

Path 1 on a *T-V* diagram

Path 2 on a *T-P* diagram

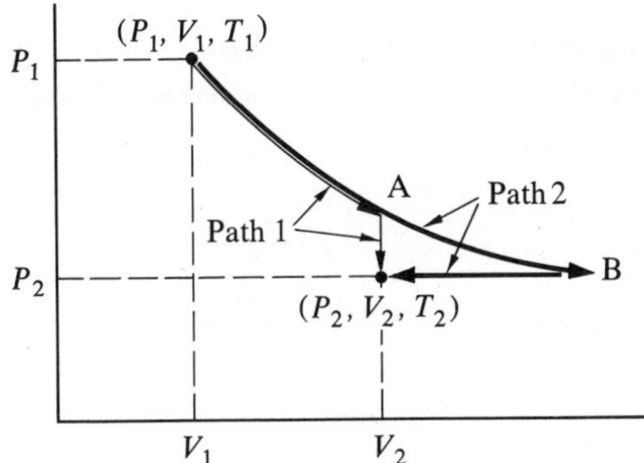

Paths 1 and 2 on a *P-V* diagram

Fig. 15.1

final temperature; during this step its entropy changes by the amount $C_P \ln (T_2/T_1)$. For this path T_2 must be such that

$$-nR \ln (P_2/P_1) + C_P \ln (T_2/T_1) = 0.$$

Since for an ideal gas

$$\frac{P_2}{P_1} = \left(\frac{T_2}{T_1}\right)\left(\frac{V_1}{V_2}\right), \qquad (n \text{ constant})$$

one may write for the second path that

$$nR \ln (V_2/V_1) + (C_P - nR) \ln (T_2/T_1) = 0.$$

The calculations for the two paths should yield identical results. This will be true if $C_V = C_P - nR$.

3. $C_P - C_V$ *for an ideal gas.* This paragraph is a recapitulation of the previous discussion as it bears directly upon the relation between C_P and

C_V for an ideal gas. The heat capacity of a substance σ is defined as the energy *absorbed by σ from* its thermal surroundings θ divided by σ's change in temperature: $C = -\delta E_\theta/\delta T_\sigma$, where $\delta T_\sigma = T_2 - T_1$. If the volume of σ is maintained constant during warm-up, $\delta E_\theta = -\delta E_\sigma$. Thus,

$$C_V = \left(\frac{\delta E_\sigma}{\delta T_\sigma}\right)_{V_\sigma}.$$

On the other hand, if P_σ is held constant, $\delta E_\theta = -(\delta E_\sigma + P_\sigma \delta V_\sigma)$. Thus, the constant-pressure heat capacity is given by the expression

$$C_P = \left(\frac{\delta E_\sigma}{\delta T_\sigma}\right)_{P_\sigma} + P_\sigma \left(\frac{\delta V_\sigma}{\delta T_\sigma}\right)_{P_\sigma}. \quad (*)$$

Consider in turn each term on the right-hand side of this equation. The change in internal energy of an ideal gas depends solely upon its change in temperature; i.e. for an ideal gas

$$\left(\frac{\delta E_\sigma}{\delta T_\sigma}\right)_{P_\sigma} = \left(\frac{\delta E_\sigma}{\delta T_\sigma}\right)_{V_\sigma} = C_V.$$

This takes care of the first term. For the second term one has, by definition,

$$\delta V_\sigma = V_{\text{at } T_2} - V_{\text{at } T_1},$$

where for an ideal gas

$$V_{\text{at } T_2} = \frac{nRT_2}{P_2} \quad \text{and} \quad V_{\text{at } T_1} = \frac{nRT_1}{P_1}.$$

Therefore, if, as here, $P_2 = P_1 = P_\sigma$,

$$\begin{aligned} \delta V_\sigma &= \frac{nR}{P_\sigma}(T_2 - T_1) \\ &= \frac{nR}{P_\sigma} \cdot \delta T_\sigma. \end{aligned} \qquad (P_\sigma \text{ constant})$$

Thus, for an ideal gas

$$\left(\frac{\delta V_\sigma}{\delta T_\sigma}\right)_{P_\sigma} = \frac{nR}{P_\sigma}, \quad \text{or} \quad P_\sigma \left(\frac{\delta V_\sigma}{\delta T_\sigma}\right)_{P_\sigma} = nR.$$

Hence, for an ideal gas

$$C_P = C_V + nR.$$

The heat capacity of helium, for example, is at constant volume (3/2)R and at constant pressure (5/2)R.

* The right-hand side of this equation could be written $(\delta H_\sigma/\delta T_\sigma)_{P_\sigma}$.

II

Free Energy and Phase Stability

16

Introduction and Review

If the absolute temperature of an object and its change in thermal energy are known, its change in entropy can be calculated by the formula

$$\Delta S = \frac{\Delta E}{T}.$$

When ice melts, for example, thermal energy equal by the First Law to the difference between the internal energy of liquid and solid water is absorbed by the melting water from its thermal surroundings. By the formula given above, the change in the entropy of the thermal surroundings is

$$\frac{-(E^{\text{liq}}_{\text{H}_2\text{O}} - E^{\text{solid}}_{\text{H}_2\text{O}})}{T}. \qquad (*)$$

This expression is strictly correct if the energy changes that occur in the universe when ice melts are confined strictly to the melting water and its thermal surroundings. In fact, because the volume of liquid water is slightly less than the volume of the same mass of solid water, the melting of ice in an open beaker allows the earth's atmosphere to sink slightly closer to the center of the earth. During melting, therefore, the potential energy of the atmosphere decreases slightly. How much this decrease in potential energy amounts to depends upon the magnitude of the earth's attraction for the atmospheric mass above each square centimeter of the earth's surface—i.e. upon the atmospheric pressure—and upon the volume difference between liquid and solid water. With an allowance for this normally relatively small change in potential energy (see Chapter 14), the expression for the entropy change in the thermal surroundings when ice melts becomes

$$\frac{-[(E^{\text{liq}}_{\text{H}_2\text{O}} - E^{\text{solid}}_{\text{H}_2\text{O}}) + P(V^{\text{liq}}_{\text{H}_2\text{O}} - V^{\text{solid}}_{\text{H}_2\text{O}})]}{T}.$$

* Since $E^{\text{liq}}_{\text{H}_2\text{O}} > E^{\text{solid}}_{\text{H}_2\text{O}}$, the numerator of this ratio is a negative number, in agreement with the statement made earlier that when ice melts the entropy of its thermal surroundings decreases.

In the melting water itself there occurs the positive entropy change

$$(S_{H_2O}^{liq} - S_{H_2O}^{solid}).$$

This entropy change must equal or exceed in absolute value the entropy change in the thermal surroundings if ice is to melt—otherwise the total entropy of the universe (the water and its surroundings) would decrease. This means that if melting is to occur spontaneously, if the change from ice to pure water is to avoid diminishing the entropy of the universe, the absolute temperature must be such that

$$(S_{H_2O}^{liq} - S_{H_2O}^{solid}) - \frac{[(E_{H_2O}^{liq} - E_{H_2O}^{solid}) + P(V_{H_2O}^{liq} - V_{H_2O}^{solid})]}{T} \geq 0.$$

Multiplying through by T, a positive number, and rearranging, one finds that for melting to occur the free energy of the solid must equal or exceed the free energy of the pure liquid.

$$E_{H_2O}^{solid} - TS_{H_2O}^{solid} + PV_{H_2O}^{solid} \geq E_{H_2O}^{liq} - TS_{H_2O}^{liq} + PV_{H_2O}^{liq},$$

or

$$G_{H_2O}^{solid} \geq G_{H_2O}^{liq}.$$

At 0°C and 1 atm. (as well as at a particular series of lower temperatures and higher pressures),

$$G_{H_2O}^{solid} = G_{H_2O}^{liq}.$$

More generally (Chapter 14), for reactions that can occur spontaneously ($\Delta S_{total} > 0$), the sum of the molar free energies of the reactants, $G_{reactants}$, is greater than the corresponding sum for the products.

$$G_{reactants} > G_{products}.$$

For reactions that are at equilibrium ($\Delta S_{total} = 0$),

$$G_{reactants} = G_{products}.$$

17

Thermodynamic Implications of Common Chemical Terminology

$$G_{\text{reactants}} \geq G_{\text{products}}$$

Many common chemical phrases have interesting thermodynamic implications. To say, for example, that

"Ice melts at 0°C"

is to say, in thermodynamic terminology, that

At 0°C (and 1 atmosphere)

$$G^{\text{solid}}_{H_2O} = G^{\text{liq}}_{H_2O}\,.$$

Similarly, to say that

"Water boils at 100°C"

is to say that

At 100°C (and 1 atmosphere)

$$G^{\text{liq}}_{H_2O} = G^{\text{gas}}_{H_2O}\,.$$

Many other examples might be cited. The statement

"The vapor pressure of water at 20°C is 17.36 mm Hg"

implies that

At 20°C and 17.36 mm Hg

$$G^{\text{gas}}_{H_2O} = G^{\text{liq}}_{H_2O}\,.$$

The statement

"A 1 molar aqueous glucose solution freezes at −1.86°C"

implies that

When $T = -1.86°\text{C}$, $N_{H_2O} = 55.5/56.5$, and $P = 1$ atm.

$$G^{\text{liq}}_{H_2O} = G^{\text{solid}}_{H_2O}\,.$$

The statement

"The triple point of CO_2 occurs at $-56.6°C$ and 3885 mm Hg"

implies that

At $-56.6°C$ and 3885 mm Hg

$$G_{CO_2}^{\text{solid}} = G_{CO_2}^{\text{liq}} = G_{CO_2}^{\text{gas}}.$$

And the statement

"The solubility of I_2 in water at 50°C is 0.078 grams/liter"

implies that pure iodine can coexist indefinitely with liquid water at 50°C if the water has dissolved in it 0.078 grams of I_2 per liter; that is to say,

At 50°C and a concentration of I_2 in water of 0.078 grams/liter

$$G_{I_2}^{\text{solid}} = G_{I_2}^{\text{aqueous}}.$$

The previous equations are of the form

$$G_{\text{some chemical}}^{\text{some phase}} = G_{\text{same chemical}}^{\text{some other phase}}.$$

More briefly

$$G_A{}^{\alpha} = G_A{}^{\beta}.$$

This equation expresses the necessary and sufficient condition for a system to be at equilibrium with respect to the distribution of chemical A between phases α and β. Expressed in words, a system is at equilibrium with respect to the change

$$A(\text{phase } \alpha) = A(\text{phase } \beta)$$

when (and only when) the contribution A makes to the free energy of phase α equals the contribution A makes to the free energy of phase β—or, more briefly, when (and only when) A's "free energy" in α equals its "free energy" in β.

Following Willard Gibbs, the phrase "the free energy of A" is often written "the chemical potential of A," symbol μ_A. In this phraseology the equation $G_A{}^{\alpha} = G_A{}^{\beta}$ becomes

$$\mu_A{}^{\alpha} = \mu_A{}^{\beta}. \qquad (*)$$

* The free energy, or chemical potential, of chemical A in phase α is defined as the rate at which the total free energy of phase α changes with respect to the number of moles of chemical A in phase α when there is no change in the temperature and pressure of phase α and no concomitant addition to phase α of other chemicals B, C, ... Expressed mathematically,

$$G_A{}^{\alpha} \equiv \mu_A{}^{\alpha} \equiv \left(\frac{\partial G^{\alpha}}{\partial n_A{}^{\alpha}}\right)_{T^{\alpha},\, P^{\alpha},\, n_B{}^{\alpha},\, n_C{}^{\alpha},\, \ldots}$$

In summary, for a system to be at equilibrium with respect to the distribution of chemical A among the system's several parts, the free energy, or chemical potential—or, speaking figuratively, the "escaping tendency"—of chemical A must be uniform throughout the system. This statement is analogous to the statement that for a system to be at equilibrium with respect to the distribution of thermal energy among its parts, the escaping tendency of thermal energy—i.e. the temperature—must be uniform throughout the system. Similarly, for a system to be at equilibrium with respect to the distribution of volume among its parts, the escaping tendency of volume—the pressure—must be uniform throughout the system.*

Note that it is not generally correct to write down an equation like

$$G_{I_2}^{solid} = G_{H_2O}^{liq}. \qquad \text{(Wrong)}$$

This equation corresponds to the alchemical transformation

$$I_2(solid) = H_2O(liq). \qquad \text{(Impossible!)}$$

On the other hand, it is legitimate to write down an equality between or among free energies or linear combinations of free energies if the chemical subscripts (together with the numerical coefficients of the free energy terms) correspond to a balanced chemical equation. The equation

$$G_{CaCO_3}^{solid} = G_{CaO}^{solid} + G_{CO_2}^{gas},$$

for example, expresses the condition that is and must be satisfied by a system that is at equilibrium with respect to the thermal decomposition of calcium carbonate according to the equation

$$CaCO_3(solid) = CaO(solid) + CO_2(gas).$$

At 1115.4°K the previous free energy equation is satisfied when P_{CO_2}, the partial pressure of CO_2 over a mixture of solid $CaCO_3$ and CaO, is 0.4513 atm.; at 1210.1°K the equilibrium partial pressure of CO_2 over the two solids is 1.770 atm.

Similarly, the equation

$$2G_{H_2O}^{aq} = G_{H_3O^+}^{aq} + G_{OH^-}^{aq}$$

is satisfied by systems that are at equilibrium with respect to the proton transfer reaction

$$2H_2O(aq) = H_3O^+(aq) + OH^-(aq). \qquad (\dagger)$$

* The conditions for chemical, thermal, and mechanical equilibrium were first stated in this form by Willard Gibbs (*Collected Works*, New Haven: Yale University Press, 1948, Vol. I, p. 65).

† Recent experimental data suggest this equation might better be written

$$8H_2O(aq) = H_9O_4^+(aq) + H_7O_4^-(aq).$$

The latter free energy equation is satisfied at 25°C and 1 atm. when pH = pOH = 7 or, more generally, whenever pH + pOH = 14.*

The free energy equation

$$G_A^\alpha = G_A^\beta,$$

which corresponds to the statement that a system is at equilibrium with respect to the phase change

$$A(\text{phase } \alpha) = A(\text{phase } \beta),$$

may be written out in full

$$E_A^\alpha + PV_A^\alpha - TS_A^\alpha = E_A^\beta + PV_A^\beta - TS_A^\beta$$

and solved for P.

$$P = \frac{\Delta S}{\Delta V} T - \frac{\Delta E}{\Delta V}.$$

From this equation and standard tabulations of thermodynamic data useful estimates can be made of solid-solid and solid-liquid coexistence curves in condensed systems, whose thermodynamic parameters V, E, and S generally vary only slowly with temperature and pressure.

For example, calcium carbonate exists in two common forms, calcite and aragonite. The free energies and enthalpies (both with respect to the elements in their standard states), Third Law entropies, and molar volumes of these two forms at 25°C and 1 atm. are listed in Table 17.1. From these values

TABLE 17.1

Substance	$G°$ kcal/mole	$H°$ kcal/mole	$S°$ cal/deg-mole	$V°$ cc/mole
Calcite	−269.78	−288.45	22.2	36.90
Aragonite	−269.53	−288.49	21.2	33.93

can be obtained the values of ΔG, ΔH, ΔS, and ΔV at 25°C and 1 atm. for the calcite-aragonite transition.

It may be noted, however, that the previous equation calls for ΔE, not ΔH. At low pressures the two terms are nearly the same. For by definition $H = E + PV$. Therefore,

$$\Delta E = \Delta H - \Delta(PV).$$

If the pressure is the same on both phases ($\Delta P = 0$),

$$\Delta E = \Delta H - P\Delta V.$$

* $pX \equiv -\log_{10}(X)$; (X) = concentration of X in moles/liter.

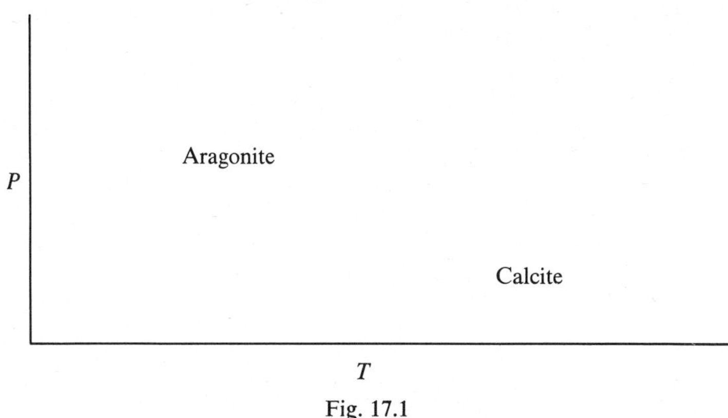

Fig. 17.1

Hence, for the phase change

$$CaCO_3(\text{calcite}) = CaCO_3(\text{aragonite})$$

and a pressure of 1 atmosphere,

$$\Delta E = [(-288{,}490) - (-288{,}450)]\,\frac{\text{cal}}{\text{mole}} - (1\text{ atm.})(33.93 - 36.90)\,\frac{\text{cc}}{\text{mole}}$$

$$= -40\,\frac{\text{cal}}{\text{mole}} + 2.97\,\frac{\text{cc-atm.}}{\text{mole}}$$

$$= -40\,\frac{\text{cal}}{\text{mole}} + 0.072\,\frac{\text{cal}}{\text{mole}} \qquad (*)$$

$$\approx -40\,\frac{\text{cal}}{\text{mole}} = \Delta H.$$

According to the table the free energy of calcite is less than the free energy of aragonite at 25°C and 1 atm.; under these conditions, therefore, calcite is more stable than aragonite—not because its energy is lower (it isn't), but because its entropy is larger. At low temperatures, however, where the TS term is less important than the energy term in the expression $E + PV - TS$, the lower energy form aragonite has the lower free energy and is stable with respect to calcite; also, at high pressures, where the PV term becomes important, the lower volume form aragonite has again the smaller free energy.

The region in which aragonite is more stable than calcite is thus the region of low temperatures and/or high pressures, i.e. the left-hand and upper portions of a conventional phase diagram (Fig. 17.1).

* $R = 1.987\,\frac{\text{cal}}{\text{deg-mole}} = 82.06\,\frac{\text{cc-atm.}}{\text{deg-mole}}$. Therefore, 1 cal = 41.3 cc-atm.

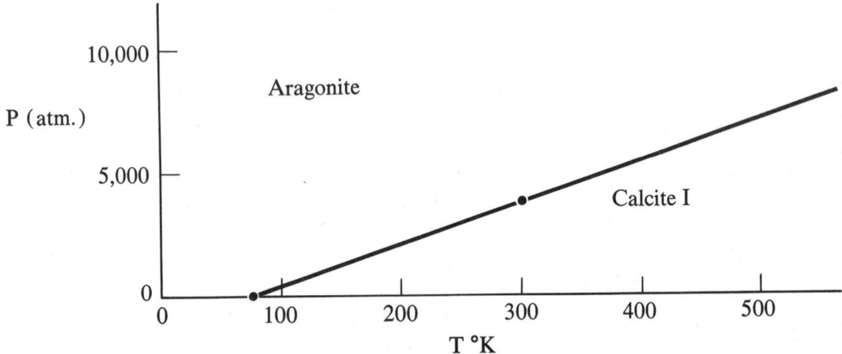

Fig. 17.2 Experimental phase diagram for $CaCO_3$.

These conclusions may be rendered more precise by substituting into the solid-solid coexistence equation

$$P = \frac{\Delta S}{\Delta V} T - \frac{\Delta E}{\Delta V}$$

the values (all per mole of aragonite formed at 25°C and 1 atm.) $\Delta V = -2.97$ cc, $\Delta E = -39.93$ cal, and $\Delta S = -1.0$ cal/deg.

One estimates in this way that the free energies of calcite and aragonite are equal when

$$P \approx 13.9\, T - 556. \qquad \begin{array}{l} P \text{ in atmospheres} \\ T \text{ in degrees Kelvin} \end{array} \qquad (*)$$

This relation will yield good estimates of equilibrium temperatures and pressures if the parameters ΔV, ΔE, and ΔS vary only slowly with temperature and pressure. For a pressure of zero atmospheres the estimated equilibrium temperature is 40°K. The estimated equilibrium pressure for a temperature of 298°K is 3590 atm. Both estimates could be improved at this point by substituting in the solid-solid coexistence equation in place of the values for ΔV, ΔE, and ΔS at 25°C and 1 atm. their actual values at $P = 0$ atm., $T = 40$°K and at $P = 3590$ atm., $T = 298$°K, respectively. This substitution would require a knowledge of the heat capacities and compressibilities of calcite and aragonite over the temperature and pressure intervals 40 to 298°K and 0 to 3590 atm. The experimentally determined values

* In mathematical notation,

$$P = \frac{\Delta S(T,P)}{\Delta V(T,P)} T - \frac{\Delta E(T,P)}{\Delta V(T,P)}$$
$$\approx \frac{\Delta S(298°\text{K, 1 atm.})}{\Delta V(298°\text{K, 1 atm.})} T - \frac{\Delta E(298°\text{K, 1 atm.})}{\Delta V(298°\text{K, 1 atm.})}.$$

(Fig. 17.2) for the equilibrium temperature at zero pressure and the equilibrium pressure at 298°K are 79°K and 3820 atm.

Problems for Chapter 17

1. What are the thermodynamic implications of the following statements?
   (a) Titanium melts at 1800°C.
   (b) Monoclinic sulfur melts under its own vapor pressure of 0.025 mm Hg at 120°C.
   (c) Gold has a higher melting point than silver.
   (d) Rhombic sulfur is the stable form of sulfur at room temperature and atmospheric pressure.
   (e) Hydrogen and oxygen form an explosive mixture.
   (f) Hydrochloric acid will dissolve zinc but not copper.
   (g) The melting point of ice under a pressure of 1120 atm. is −10°C.
   (h) At 22,400 atm. water freezes to a solid called Ice VII at +81.6°C.
   (i) The osmotic pressure of a solution at 25°C is 7.6 atm.
   (j) At atmospheric pressure the lead-antimony system has a simple eutectic point at 246°C and a lead mole fraction of 0.80.
   (k) The air is supersaturated with water vapor.
   (l) The dew point of water in the air is 10°C.
2. Is the equation
$$P = \frac{\Delta S}{\Delta V} T - \frac{\Delta E}{\Delta V}$$
   dimensionally homogeneous?
3. Using the equation in problem 2, estimate the freezing pressure of water at −10°C.
4. What calculations do the data in the text on the dissociation of calcium carbonate suggest?

18

Free Energy Functions and Phase Diagrams

$$G = E + PV - TS$$

Molecules adopt, broadly speaking, one of three arrangements. They may be ordered and touching, disordered and touching, or disordered and not touching.

The preferred arrangement at a given temperature and pressure is the arrangement with the lowest molar free energy. Generally this is the ordered - touching arrangement at low temperatures (and not too low pressures), the disordered - touching arrangement at intermediate temperatures (and not too low pressures), and the disordered - not touching arrangement at high temperatures (and not too high pressures). These facts are summarized diagrammatically in Figs. 18.1 through 18.7.

Let temperature be represented by the horizontal distance to the right of a vertical line. (Later pressure will be represented by the vertical distance above the horizontal line.) Position a indicates a low temperature, position b a high temperature.

In the neighborhood of a, T is small. Here $G \approx E$, unless (see below) the pressure is very large, or (for gases) very small.* Since the energy of a solid

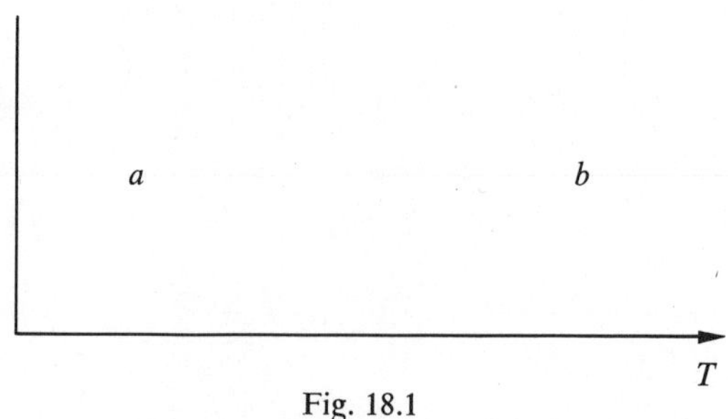

Fig. 18.1

* In this chapter and the chapter following the symbols G, E, S, and V, and the phrases "free energy," "energy," "entropy," and "volume" stand for molar quantities.

phase is less than the energies of the corresponding liquid and gas phases at the same temperature (and pressure), at a the free energy of the solid is less than the free energies of the latter phases.

In symbols,

$$G^{\text{solid}} < G^{\text{liq}} \text{ and } G^{\text{gas}} \qquad \text{at } a.$$

On the other hand, in the neighborhood of b, T is large. Here $G \approx -TS$. Since the entropy of a gas is greater than the entropy of the corresponding solid and liquid at the same temperature (and pressure),

$$G^{\text{gas}} < G^{\text{solid}} \text{ and } G^{\text{liq}} \qquad \text{at } b.$$

This is shown diagrammatically in Fig. 18.2.

This diagram summarizes two well-known facts about substances: at low temperatures the solid form is stable with respect to the corresponding liquid and gas; at high temperatures the gas is stable with respect to the corresponding solid and liquid. If the low-temperature region and the high-temperature region are designated by the name of the phase most stable in that region, the diagram takes the simple form shown in Fig. 18.3.

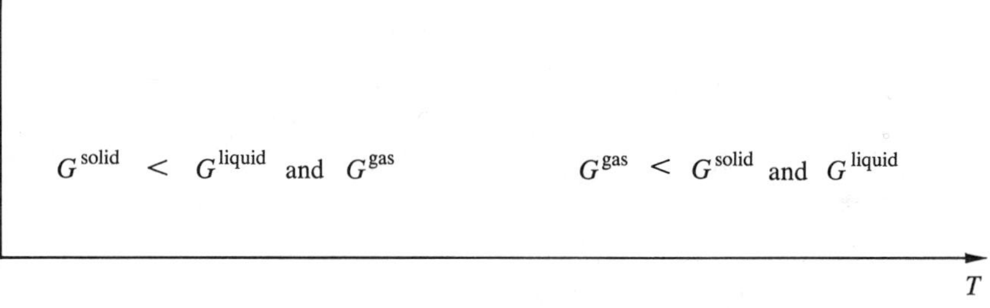

Fig. 18.2

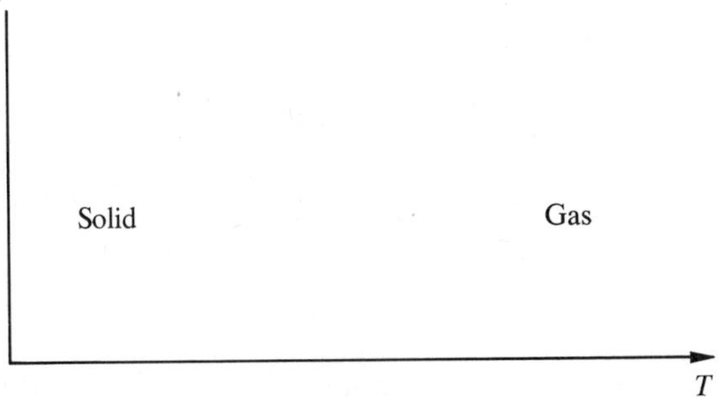

Fig. 18.3

One remark should be added concerning Fig. 18.3. At intermediate temperatures—and not too low pressures—the liquid phase with its intermediate energy ($E^{\text{solid}} < E^{\text{liq}} < E^{\text{gas}}$) and intermediate entropy ($S^{\text{solid}} < S^{\text{liq}} < S^{\text{gas}}$) has a free energy less than that of either of the other two phases ($G^{\text{liq}} < G^{\text{solid}}$ and G^{gas}); i.e. the phase most stable at intermediate temperatures—and not too low pressures—is the liquid (Fig. 18.4).

Temperature is not the only variable that affects the relative stabilities of phases, however. Pressure also has an effect. At low pressures, S^{gas} becomes very large, making $G^{\text{gas}} < G^{\text{solid}}$ and G^{liq}, even at low temperatures. At higher pressures, however, and at temperatures not so high as to make S^{gas} overridingly important, the relatively large value of E^{gas} makes $G^{\text{gas}} > G^{\text{solid}}$ and G^{liq}. These facts are summarized graphically in Fig. 18.5 in two different notations. Pressure is represented by the vertical distance above the horizontal line. The top inequality in the left diagram is valid only at low to moderate temperatures; at high temperatures (see Fig. 18.2) $G^{\text{gas}} < G^{\text{solid}}$ and G^{liq}.

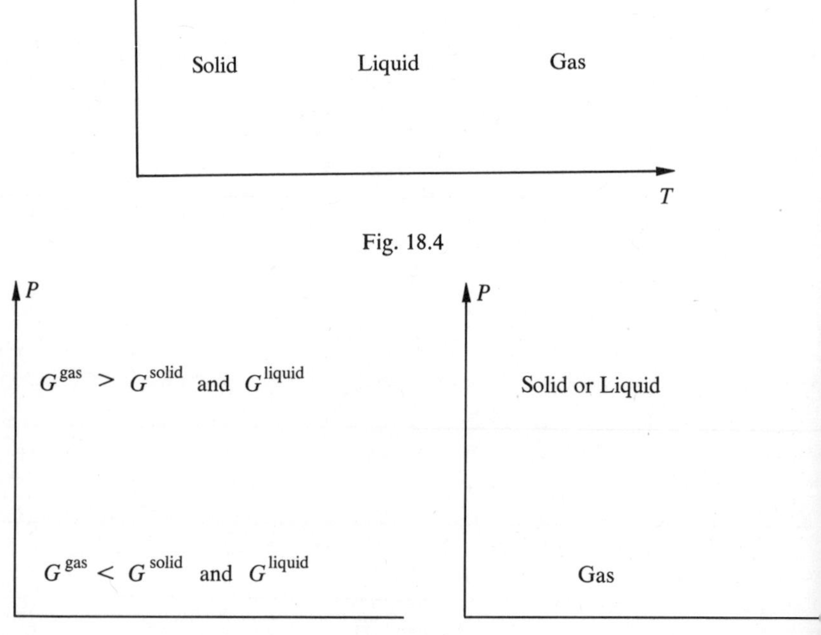

Fig. 18.4

Fig. 18.5

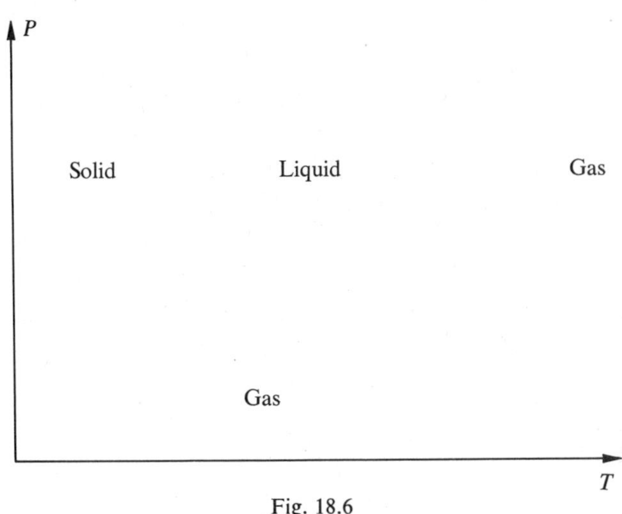

Fig. 18.6

Thus, underlying the region of stability of the solid and liquid phases is a low-pressure region where (owing to the large value of S^{gas}) the gas phase is at all temperatures above 0°K more stable than either of the condensed phases. Combined with Fig. 18.4, this fact yields the picture of phase stability shown in Fig. 18.6.

In summary, at high temperatures and, also, at low pressures, the term TS^{gas} is relatively large; under these conditions the gas is more stable than the corresponding condensed phases. At higher pressures, however, and at not too high temperatures, the term TS^{gas} is less important than E^{gas}, which, being relatively large, renders the gas less stable than the condensed phases.

Separating the regions of phase stability from each other are lines along which the molar free energies of adjacent phases are equal (Fig. 18.7). These lines—they may be called coexistence curves—have slopes equal to the entropy change ΔS that occurs in the substance on melting, evaporating, or subliming, divided by the corresponding volume change ΔV.*

For most substances the slopes of these lines are positive (ΔS and ΔV both positive or both negative, depending upon which way the phase change is taken). Water is an exception. The line separating the solid and liquid regions of water has a negative slope, owing to the fact that for the phase change $H_2O(\text{solid}) = H_2O(\text{liq})$ $\Delta S_{H_2O} \equiv S^{\text{liq}}_{H_2O} - S^{\text{solid}}_{H_2O}$ is positive, whereas $\Delta V_{H_2O} \equiv V^{\text{liq}}_{H_2O} - V^{\text{solid}}_{H_2O}$ is negative.

* Mathematically, $dP/dT = \Delta S/\Delta V$. The derivation of this equation is given in a later chapter.

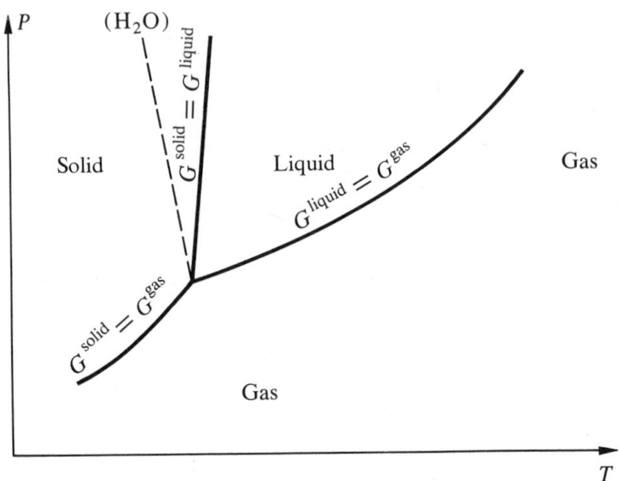

Fig. 18.7 Phase diagram for a one-component system.

Where the three lines $G^{\text{solid}}_{H_2O} = G^{\text{liq}}_{H_2O}$, $G^{\text{liq}}_{H_2O} = G^{\text{gas}}_{H_2O}$, and $G^{\text{gas}}_{H_2O} = G^{\text{solid}}_{H_2O}$ meet, $G^{\text{solid}}_{H_2O} = G^{\text{liq}}_{H_2O} = G^{\text{gas}}_{H_2O}$. This is called the "triple point."

Problems for Chapter 18

Molecules adopt, broadly speaking, one of three phases: the ———(a) phase, the ———(b) phase, or the ———(c) phase. The most stable phase at a given temperature and pressure is the phase with the ———(d) molar free energy. The free energy is equal to ———(e). When the temperature is low the most important term in this expression is generally ———(f). At low temperatures, therefore, the most stable phase is generally the phase with the lowest ———(g) ——— (h). This is the ———(i) phase. On a conventional *P-T* diagram the solid phase occupies, therefore, the region at the ———(j) [———(k) small, ———(l) important]. The *TS* term becomes important when its first factor is large, i.e. when the ———(m) is ———(n), especially if the second factor, the ——— ———(o) is ———(p); it becomes important, also, when the second factor is ———(q); for gases this occurs when the ———(r) is ———(s). On a conventional phase diagram the gas phase occupies, therefore, the region at the ———(t) (high temperatures) and ———(u) (low pressures). Between the regions of stability of the solid and gas phase is a region of temperatures and pressures where the ———(v) phase is more ———(w) than the other phases. On a *P-T* phase diagram the lines that separate from each other the regions of phase stability represent temperatures and pressures at which adjacent phases can ———(x). The slopes of these lines are equal to ———(y). From the densities of two coexisting phases and the slope of the corresponding coexistence curve can be determined the value of the ———(z) of ———(a′) and, hence, since at equilibrium ———(b′), the value of the ———(c′) of ———(d′). The curve that separates the region where the gas phase is the most stable phase from the region where the liquid phase is the most stable phase is commonly called the substance's ——— ———(e′) curve. From liquid densities, the equation of state ———(f′), and ——— ———(g′) data, can be calculated values for ———(h′) of ———(i′).

19

Free Energy Curves and Phase Transitions

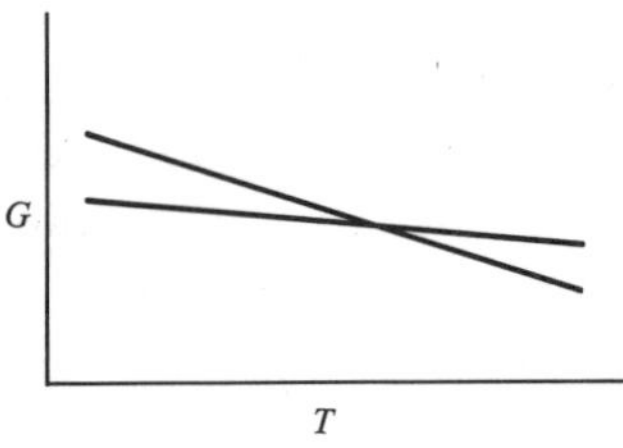

A substance's stability in a solid, a liquid, or a gaseous phase can generally be increased by decreasing its concentration. A decrease in concentration increases the contribution the substance makes per mole to the entropy of the phase in question and, consequently, decreases the substance's molar (or "partial molar") free energy in that phase. Addition of sugar, salt, alcohol, or some other soluble substance to liquid water, for example, lowers the concentration of the water in the liquid phase, increases its molar (or "partial molar") entropy in that phase, decreases its free energy, and increases the stability of the water in the liquid phase with respect to the formation of other phases, e.g. water vapor or ice.

Graphs of molar free energies against temperature express perhaps better than words and equations the relation between thermodynamic terminology and common chemical experience.

At absolute zero the free energy of a substance and its enthalpy are equal. From there their values diverge. With increasing temperature the enthalpy increases; the free energy decreases.

The enthalpy increases at a rate equal to the instantaneous value of the substance's heat capacity. The free energy decreases at a rate equal to the instantaneous value of the substance's entropy (on this, more later). Since a substance's entropy increases with increasing temperature, the free energy curve's downward trend increases as the temperature increases. A plot against temperature of the free energy of any substance, say ice, must have the general shape shown in Fig. 19.1.

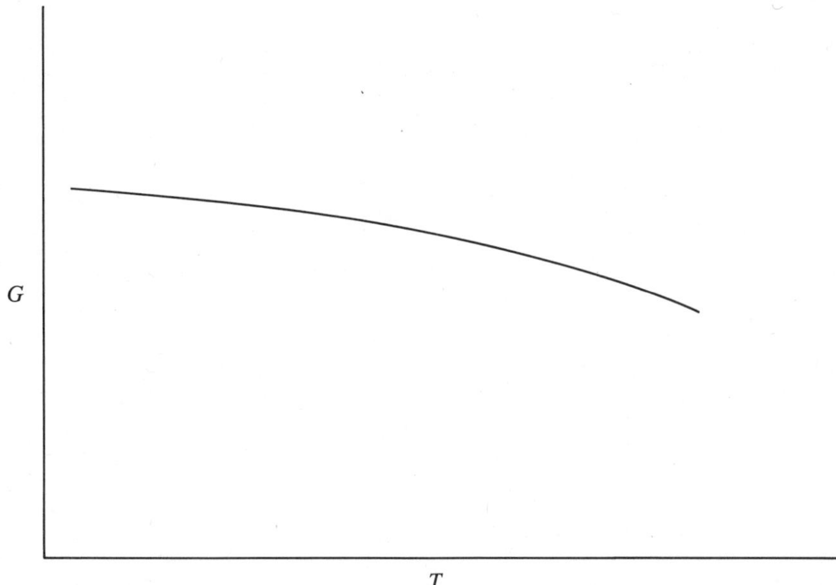

Fig. 19.1 Plot of the molar free energy of a substance against temperature (schematic).

By comparison with the free energy curve for ice, the free energy curve for liquid water starts higher ($H^{\text{liq}} > H^{\text{solid}}$) but falls faster ($S^{\text{liq}} > S^{\text{solid}}$). Eventually the curve for liquid water crosses the curve for ice, Fig. 19.2. At all temperatures below the temperature of the crossover point $G^{\text{solid}} < G^{\text{liq}}$; in this region, as indicated, the solid phase is more stable than the liquid phase. At all temperatures above the crossover temperature $G^{\text{liq}} < G^{\text{solid}}$; in this region, therefore, the liquid phase is more stable than the solid. Right at the crossover temperature $G^{\text{liq}} = G^{\text{solid}}$.

(For readability in the above discussion the subscript H_2O has been omitted from the symbols H, S, and G. These symbols are to be read as they would be read if this subscript were present—as the contribution water makes *per mole* to the enthalpy, entropy, and free energy of the superscripted phase. They are not to be read, as otherwise they would, as the *total* enthalpy, entropy, and free energy of the phase. A phase's total free energy has little significance in the present context. Adding water to pure liquid water increases, of course, the total free energy of the liquid phase; it does not, however, alter the molar free energy of liquid water—if the temperature remains unchanged—nor does it alter the stability of liquid water.)

For gaseous water the free energy curve starts still higher ($H^{\text{gas}} > H^{\text{liq}}$) and falls still faster ($S^{\text{gas}} > S^{\text{liq}}$), see Fig. 19.3. At very low pressures (S^{gas}

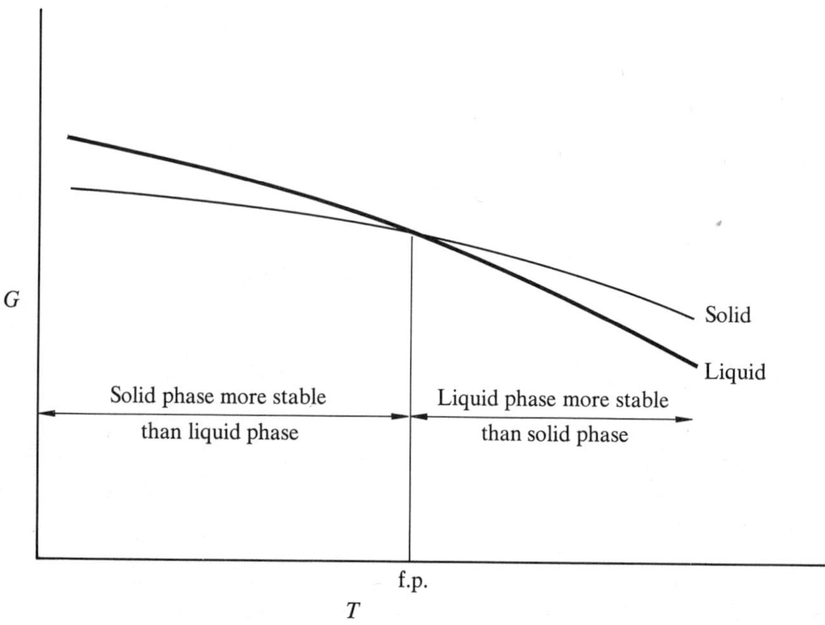

Fig. 19.2 Molar free energy curves for a substance's solid and liquid phases. "f.p." marks the substance's freezing point.

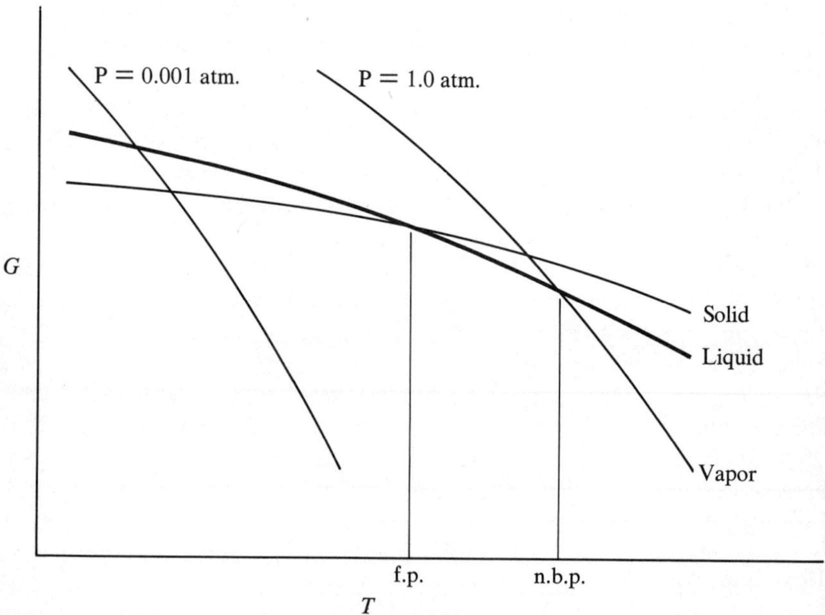

Fig. 19.3 Molar free energy curves for a substance's solid, liquid, and vapor phases, the latter at two pressures. "n.b.p." marks the substance's normal boiling point.

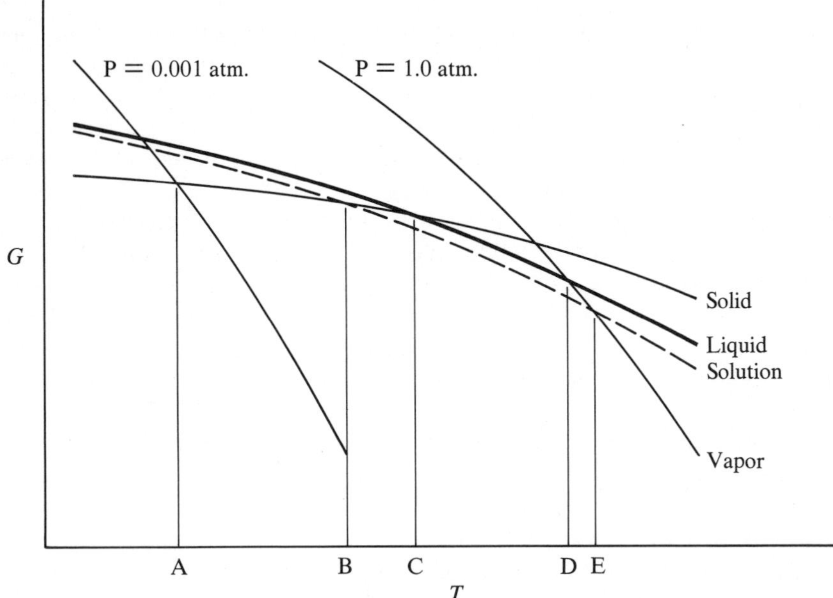

Fig. 19.4 Addition to Fig. 19.3 of the molar free energy curve for the substance in solution.

very large), the free energy curve for the gas crosses the other two free energy curves in a region where $G^{\text{solid}} < G^{\text{liq}}$; in other words, when the partial pressure of water vapor is very low, water on warm-up passes directly from the solid state to the gaseous state without melting: On cold days ice sublimes if the air is very dry. At higher pressures (S^{gas} not so large), the free energy curve for the gas falls less rapidly with increasing temperature; at sufficiently high pressures it crosses the other two free energy curves *above* the solid-liquid crossover temperature. Under these conditions the liquid phase has a finite range of stability.

Adding to water a solute—sugar, salt, alcohol, or some other water-soluble substance—increases the contribution water makes to the entropy of the liquid phase (a configurational entropy effect, see Chapter 25) and, as shown by the dashed line in Fig. 19.4, lowers the free energy of water in the liquid phase at all temperatures. If the added solute is insoluble in the solid phase (most substances are insoluble in ice*), the free energy curve for the solid phase will not be affected. The effect, therefore, of adding a solid-insoluble solute to liquid water is to extend at both ends the region in which G^{liq} is less than G^{solid} and G^{gas}. The freezing point is lowered, from C to B (Fig. 19.4), and the boiling point is raised, from D to E.

* Freezing water is a good way to purify it.

How much the liquid-solid and liquid-gas transition temperatures are altered depends upon how close to parallel the corresponding free energy curves are. The closer to parallel are the two free energy curves whose intersection defines the transition temperature under consideration, the greater the effect upon that transition temperature of a movement downward of the liquid phase's free energy curve.

Since the entropy of water in the solid phase does not differ greatly from the entropy of water in the liquid phase, the free energy curves for these two phases do not differ greatly in slope; broadly speaking the two curves are approximately parallel. Consequently a downward movement of the liquid phase free energy curve causes the intersection between itself and the free energy curve for the solid phase to move relatively rapidly to lower temperatures.

By contrast, at temperatures not too close to the critical temperature the entropy of water in the gas phase differs greatly from the entropy of water in the liquid phase; generally speaking, therefore, the slopes of the free energy curves for these two phases are very different (Fig. 19.4). A movement downward of the liquid phase free energy curve causes in this case the intersection between itself and the other free energy curve to move relatively slowly to higher temperatures.

These conclusions will be expressed more quantitatively in Part IV.

Problems for Chapter 19

1. Identify the numerical values of the abscissas A, B, C, D, E in Fig. 19.4. Assume the figure is drawn for water and that the dashed curve refers to a one molal sucrose solution.
2. On warm-up rhombic sulfur changes reversibly to monoclinic sulfur at 95.5°C. At 120°C monoclinic sulfur melts.

   (a) Show graphically approximately how the molar free energies of rhombic and monoclinic sulfur stand with respect to each other in the neighborhood of the rhombic-monoclinic transition temperature.
   (b) To the results in (a) add a curve for the molar free energy of liquid sulfur.
   (c) Which has the higher melting point, rhombic or monoclinic sulfur?
   (d) Which has the larger entropy?
   (e) Which of the two solids do you suppose is the more soluble in diethyl ether at 25°C?
   (f) Does the melting point of rhombic sulfur lie above or below the rhombic-monoclinic transition temperature?

3. Show that the melting points of two crystalline forms of a substance must both lie above or both lie below the solid-solid transition temperature.
4. Benzophenone, when heated, sometimes behaves in a curious fashion: it melts, recrystallizes, and later melts again. How do you account for this behavior?
5. Mole for mole, which depresses the freezing point of a solvent the most: a solute that is insoluble in the solvent's solid phase or a solute that is soluble in the solvent's solid phase? Or is there no difference?

III

An Introduction to Statistical Thermodynamics

General thermodynamics proceeds from the fact that, as far as we can tell from our experience up to now, all natural processes are irreversible. Hence according to the principles of phenomenology, the general thermodynamics of the second law is formulated in such a way that the unconditional irreversibility of all natural processes is asserted as a so-called axiom. . . . [However] general thermodynamics (without prejudice to its unshakable importance) also requires the cultivation of mechanical models representing it, in order to deepen our knowledge of nature—not in spite of, but rather precisely because these models do not always cover the same ground as general thermodynamics, but instead offer a glimpse of a new viewpoint.*

* Ludwig Boltzmann, *Lectures on Gas Theory*, trans. by Stephen G. Brush, Berkeley: University of California Press, 1964.

Introduction

Physical science is the study of inanimate objects, called particles. It may be divided into specialties according to the number of particles considered, the size of the particles, and their velocity.

	VERY SMALL	MODERATE	VERY LARGE
Particle Number	NM		SM Chemistry
Particle Size	QM Chemistry	NM	Classical Astronomy
Particle Velocity	Statics	NM	Relativity

NM = Newtonian Mechanics
QM = Quantum Mechanics
SM = Statistical Mechanics

Newtonian mechanics is the study of small numbers of particles of moderate size moving with moderate velocities. Quantum mechanics is the study of very small particles; statistical mechanics of very many particles; astronomy of very large particles; and relativity of very rapidly moving particles.

Chemistry is the study of very many very small particles. On the theoretical side chemistry is therefore a blend of statistical mechanics and quantum mechanics. Statistical mechanics asks, how are particles distributed among a system's allowed energy levels? Quantum mechanics asks, what are these allowed energy levels? The first question is discussed in Chapters 21, 22, and 23; the second in Chapter 24. There is a certain logic, however, in beginning a study of theoretical chemistry with a study of allowed energy levels and proceeding thence to their use in statistical mechanics. The reader who desires to pursue the subject in this order—quantum mechanics first, statistical mechanics second—may do so by reading the chapters of Part III in the order 24, 20, 21, 22, 23.

20

Boltzmann's Relation

$$S = k \ln \Omega.$$

Violent attacks upon the mathematical kinetic-molecular theory of heat of James Clerk Maxwell and Ludwig Boltzmann were made as late as 1900 by a school of scientists that included the famous Ostwald who, rejecting the atomic hypothesis for "science without suppositions," argued that since mechanical processes are reversible and heat conduction is not, thermal phenomena cannot possibly be explained in terms of hidden, internal mechanical variables.

The resolution of this "reversibility paradox" was given most clearly by Boltzmann. Mechanical processes, Boltzmann pointed out, are irreversible if the number of particles is sufficiently large. Shake 1000 black marbles with 1000 white marbles and one obtains a random mixture. Further shaking does not reverse this process. Elaborating upon this observation, Boltzmann formulated an equation that today provides a bridge between the concepts of classical thermodynamics and molecular statistics.

Entropy, it has been said, is related to microscopic disorder. The quantitative relation between entropy and disorder, known as Boltzmann's relation, is

$$S = k \ln \Omega.$$

This expression has been carved above Boltzmann's name on his tombstone in the Zentralfriehof in Vienna. k (Boltzmann's constant) is equal to the gas constant R divided by Avogadro's number N_0.

$$k = \frac{R}{N_0} = \frac{1.987 \text{ cal/deg-mole}}{6.023 \times 10^{23} \text{ mole}^{-1}} = 3.30 \times 10^{-24} \text{ cal/deg}$$

$$= 1.38 \times 10^{-16} \frac{\text{gm cm}^2}{\text{sec}^2 \text{ deg}}. \qquad (*)$$

* 1 cal = 4.18×10^7 (gm cm²/sec²). A derivation of the relation $k = R/N_0$ is given in Chapter 36.

Ω is the number of microscopically different configurations the system can adopt without appearing macroscopically different.*

The widespread use of Boltzmann's relation stems mainly from the simplicity of the relation and the good agreement between theoretical predictions based upon it and experimental observations. This agreement is not altogether surprising. A list of the prominent properties of the quantity $k \ln \Omega$ reads exactly like a list of the properties of S.

Like S, $k \ln \Omega$ is a measure of microscopic disorder. The greater our ignorance regarding the precise microscopic state of a macroscopic system, the greater Ω and the greater S.

Like S, $k \ln \Omega$ is an extensive property. This fact follows from a general property of Ω and a familiar property of the logarithmic operator. Suppose that a system is composed of two parts A and B, each part of which independently of the other part has accessible to it Ω_A and Ω_B microstates, respectively. This means that when part A is in any one of its Ω_A microstates part B by supposition can exist in any one of its Ω_B microstates, and vice versa. The total number of microstates accessible to the two parts considered jointly as a single macroscopic system is, therefore, $\Omega_A\Omega_B$. Hence

$$\begin{aligned} k \ln \Omega_{\text{total}} &= k \ln \Omega_A\Omega_B \\ &= k \ln \Omega_A + k \ln \Omega_B. \end{aligned}$$

In other words, because Ω is a multiplicative property ($\Omega_{\text{total}} = \Omega_A\Omega_B$), the logarithm of Ω (like S) is an additive property.

Like S_{total}, $k \ln \Omega_{\text{total}}$ always increases in a spontaneous process. This property of $k \ln \Omega$ follows from the fact that the logarithm of Ω increases as Ω itself increases,† plus the fact that in a spontaneous process the change is always from a less disordered, relatively small-Ω state of the universe to a more disordered, large-Ω state.

Like S, $k \ln \Omega$ vanishes for perfect crystals at 0°K. Removal of all thermal energy from a perfect crystal reduces the crystal to a single, well-defined micromolecular configuration. Under these conditions $\Omega = 1$ and $k \ln \Omega = 0$.

Like S, $k \ln \Omega$ is never negative. A system (to be a system) must possess at least *one* micromolecular configuration.

Like S, $k \ln \Omega$ increases as E increases. This behavior is illustrated with specific examples in the following chapter. In the present and three following chapters the symbol E is used to denote the thermal energy of a body, i.e. its energy with respect to the energy of the same mass (at the same volume) at 0°K.

* In Richard Feynman's words, "We measure 'disorder' by the number of ways that the insides can be arranged, so that from the outside it looks the same."

† $d(\ln \Omega)/d\Omega = 1/\Omega > 0$.

Like S, k ln Ω *increases at a decreasing rate as E increases.* This behavior corresponds to the previously noted fact that the absolute temperature

$$\frac{\delta E}{\delta S}$$

normally increases as E increases. The reason for this is that the greater E the smaller δS for given δE, for the change produced in the logarithm of Ω (by a small change in E) corresponds to the *fractional* change in Ω

$$\delta\,(\ln \Omega) = \frac{\delta\Omega}{\Omega},$$

and, not surprisingly, the larger E—and, hence, the larger Ω in the denominator of the fraction $\delta\Omega/\Omega$—the smaller is the fractional change in Ω for a given change in E; i.e. when E is large, δS [$= k\delta(\ln \Omega)$] is small and T ($= \delta E/\delta S$) is large.

Like S, k ln Ω *tends to be small for hard substances and large for soft substances.* Also, like S, the value of $k \ln \Omega$ for a substance increases when the substance melts and when it evaporates. More generally, like S, $k \ln \Omega$ tends to increase whenever there is an increase in the characteristic volume involved in the microscopic motions of the individual particles of matter. The theoretical reasons for this behavior are discussed with illustrative examples in Chapter 21.

Like S, k ln Ω *has the dimensions of energy per degree, usually expressed cal/°K.* The dimensions of S have been commented upon in Chapters 5 and 6. The dimensions of $k \ln \Omega$ come from the dimensions of k. The introduction of Boltzmann's constant k into Boltzmann's relation $S = k \ln \Omega$ is discussed most easily from a mathematical point of view (Part V).

Like S, k ln Ω *usually has a value in the range* 0–100 *cal/deg-mole.* A mole of particles each particle of which can adopt either one of two geometrical orientations independently of the orientation of its neighbors has 2^{N_0} microscopically distinct geometrical configurations. In this case

$$\begin{aligned} k \ln \Omega &= k \ln 2^{N_0} = kN_0 \ln 2 \\ &= R \ln 2 \quad = R(2.303) \log_{10} 2 \\ &= 1.38 \text{ cal/deg-mole.} \end{aligned}$$

This number represents the "configurational entropy" of the sample. To obtain the sample's total entropy, there should be added to this number the sample's "thermal entropy," i.e. entropy arising from the random thermal motions of the sample's individual particles. The contribution thermal motion makes to Ω is the subject of the next chapter.

Problems for Chapter 20

1. What is the value of Ω for a substance whose entropy is 10 cal/deg?
2. Suppose an isolated system—something which may properly be called a "universe"—changes spontaneously from a macroscopic state that has Ω_1 micromolecular configurations and an entropy S_1 to a macroscopic state that has Ω_2 micromolecular configurations and an entropy S_2. Derive for this case a formula that expresses the value of the ratio Ω_2/Ω_1 in terms of the entropies S_1 and S_2. What is the value of Ω_2/Ω_1 when $\Delta S_{\text{total}} \equiv S_2 - S_1 = 10$ cal/deg? 1 cal/deg? 10^{-6} cal/deg? 0 cal/deg?
3. What is the chance of converting 300 microcalories of thermal energy at 27°C entirely to work?
4. What is the chance that a one-hundred-gram weight resting on a table whose temperature is 27°C might spontaneously spring one millimeter into the air at the expense of the thermal energy of the table?
5. The entropy of an ideal gas can be calculated from theoretical equations if the temperature of the gas and its pressure are known and if, in addition, the moments of inertia and vibrational frequencies of the individual molecules are known; usually the latter are deduced from spectroscopic data. For carbon monoxide gas at 298.15°K and 1 atmosphere the calculated or "spectroscopic" entropy is 47.3 cal/deg-mole.

   The entropy of a substance can also be determined by cooling it to a temperature near absolute zero, where by the Third Law its entropy should be vanishingly small, and then warming it up and applying at every stage of warm-up the equation $\delta S = (1/T) \times$ (thermal energy absorbed) [see Chapter 6]. For carbon monoxide gas, again at 298.15°K and 1 atmosphere, this calorimetric or "Third Law" entropy comes out to be 46.2 cal/deg-mole.

   Why do you suppose these two values are different?
6. For an ensemble of very weakly interacting point particles, an ideal gas, for example,

$$\Omega = f(T,M,n) \cdot V^{nN_0}.$$

   In this expression $f(T,M,n)$ indicates a function that depends only on the temperature, the molecular weight, and the number of moles of gas; V is the volume of the gas and N_0 is Avogadro's number. By what factor does Ω change when the volume is doubled? By what amount does the entropy change in an isothermal expansion from V_1 to V_2?
7. Show that if thermal energy can flow spontaneously between two objects in the direction $A \rightarrow B$, the fractional decrease in Ω_A must be less than the fractional increase in Ω_B.
8. The logarithmic operator was introduced into Boltzmann's relation to produce an additive function from the multiplicative function Ω. One

may ask, "Is the logarithm the only operator that will do this?" Suppose, for example, there exists another operator O, which, like the logarithm, is such that

$$O(xy) = O(x) + O(y).$$

Show by setting $y = 1, x, x^2, x^3, \ldots$ and $x^{-1}, x^{-2}, x^{-3}, \ldots$ that, like the logarithm, $O(1) = 0$ and $O(x^n) = nO(x)$, for n any positive or negative integer. Show, further, that this last result can be generalized to cover cases where the exponent is any rational number, by making the substitution $y = x^{n/m}$, where n and m are integers. Finally, to discover how rapidly the function $O(x)$ changes in the neighborhood of the point $x = 1$, where $O(x)$ and $\log(x)$ are equal, consider the substitution $y = 1 + \delta x/x$. What are your conclusions?

Planck's Description of the Introduction into Science of the Constants *k* and *h*

"Generalization naturally starts from the simplest, the most transparent particular case."* Classical thermodynamics, we have seen, began from observations on two familiar processes—heat conduction and heat production.

Next to heat, light is one of the simplest—and most transparent—of all phenomena. Just as in classical mechanics and classical thermodynamics, in the study of light physical science achieved early quantitative successes. Then, unexpectedly, at the turn of the present century, a baffling problem arose.

Objects emit electromagnetic energy. That was not unexpected. What was unexpected was the wave length of the radiant energy. Warm objects radiate mainly in the infrared; hot objects ("red hot," "white hot") in the visible. Why? Max Planck has described in a charming manner how he solved this perplexing problem. His solution to the "black-body radiation problem" marked the birth of quantum mechanics.

So [following initial failures] I had no other alternative than to tackle the problem once again—this time from the opposite side, namely, from the side of thermodynamics, my own home territory where I felt myself to be on safer ground. In fact, my previous studies of the Second Law of Thermodynamics came to stand me in good stead now, for at the very outset I hit upon the idea of correlating not the temperature but the entropy of the [radiating atomic] oscillator with its energy. It was an odd jest of fate that a circumstance which on former occasions I had found unpleasant, namely, the lack of interest of my colleagues in the direction taken by

* George Polya, *Mathematics and Plausible Reasoning. Volume II. Patterns of Plausible Inference*, Princeton: Princeton University Press, 1954.

my investigations, now turned out to be an outright boon. While a host of outstanding physicists worked on the problem of spectral energy distribution, both from the experimental and theoretical aspect, every one of them directed his efforts solely toward exhibiting the dependence of the intensity of radiation on the temperature. On the other hand, I suspected that the fundamental connection lies in the dependence of entropy upon energy. As the significance of the concept of entropy had not yet come to be fully appreciated, nobody paid any attention to the method adopted by me, and I could work out my calculations completely at my leisure, with absolute thoroughness, without fear of interference or competition.

. . . Since the entropy S is an additive magnitude but the probability Ω is a multiplicative one, *I simply postulated that* $S = k \cdot \ln \Omega$, *where* k *is a universal constant; and I investigated whether the formula for* Ω, *which is obtained when* S *is replaced by its value corresponding to the above radiation law* [which had been previously discovered empirically], *could be interpreted as a measure of probability* [italics mine].

As a result [in December 1900], I found that this was actually possible, and that in this connection k represents the so-called absolute gas constant, referred not to gram-molecules or moles, but to the real molecules. It is, understandably, often called Boltzmann's constant. However, this calls for the comment that Boltzmann never introduced this constant, nor, to the best of my knowledge, did he ever think of investigating its numerical value. . . .

Now as for the magnitude of Ω, I found that in order to interpret it as a probability, it was necessary to introduce a universal constant, which I called h. Since it had the dimensions of action (energy $\times$ time), I gave it the name, *elementary quantum of action*. Thus the nature of entropy as a measure of probability, in the sense indicated by Boltzmann, was established in the domain of radiation, too. . . .

While the significance of the quantum of action for the interrelation between entropy and probability was thus conclusively established, the part played by this new constant in the uniformly regular occurrence of physical processes still remained an open question. I therefore tried immediately to weld the elementary quantum of action h somehow into the framework of the classical theory. But in the face of all such attempts, this constant showed itself to be obdurate. . . .

My futile attempts to fit the elementary quantum of action somehow into the classical theory continued for a number of years, and cost me a great deal of effort. Many of my colleagues saw in this something bordering on a tragedy. But I feel differently about it. For the thorough enlightenment I thus received was all the more valuable. I now know for a fact that the elementary quantum of action played a far more significant part in physics than I had originally been inclined to suspect, and this recognition made me see clearly the need for the introduction of totally new methods of analysis and reasoning in the treatment of atomic problems. The development of such methods—in which, however, I could no longer take an active part—was advanced mainly by the efforts of Niels Bohr and Erwin Schrödinger.

Ludwig Boltzmann (1844–1906)

In his early twenties Boltzmann, a skillful and imaginative laboratory assistant to Josef Stefan (Chapter 40), but suffering from growing near-sightedness, gave up a promising career in experimental research to devote himself entirely to theoretical work, where his contributions are considered today among the finest in physics.

In his own lifetime, however, Boltzmann's work was not in style. He was strongly criticized from many directions. Zermelo and Loschmidt attacked the mathematical and philosophical foundations of his work. Ostwald, Duhem, and Mach criticized the underlying physical model of atoms and molecules. Acutely aware of the growing hostility to his work, Boltzmann wrote in 1898, "I am conscious of being only an individual struggling weakly against the stream of time." Eight years later, suffering from severe headaches and greatly depressed, he committed suicide. Ironically, within a few years, following particularly Perrin's work on Brownian motion, the scientific world made a complete turnabout. By 1909 even Ostwald had been converted to Boltzmann's point of view.

21

Statistical Mechanics of Very Small Systems

$$\Omega = \frac{(n + N - 1)!}{(N - 1)!\, n!}$$

What study is the worthiest of all? The study of theories applicable to the widest range of experience, it has been said. Many thermodynamicists, no doubt, are of this persuasion. Laws may exist whose scope exceeds the scope of the Second Law; however their number is small.

Yet the existence of several laws of the physical universe demonstrates that none of them is all embracing. This fact suggests another type of study. This would be a study of phenomena to which many theories apply. Biology and geology are examples. So are chemistry and medicine and engineering. Another example, one close to the central theme of this volume, is the study of the number of microstates possessed by macroscopic systems. This study brings to a focus three disciplines: classical thermodynamics, statistical mechanics, and quantum mechanics. It is to this study that we now turn.

The calculation of Ω is one of the signal achievements of theoretical chemistry. Two ideas paved the way for this calculation. One was the atomic hypothesis, which "quantized" matter. The other was the quantum hypothesis, which quantized energy. The union of these two ideas yields many interesting insights into the nature of Ω. Many of these insights can be attained rather simply by considering the statistical distribution of energy in very small systems.

We begin by considering an idealized solid in which each atom is able to execute simple harmonic oscillations about its mean position in the solid almost independently of the motions of its neighbors. Like all constrained motion, harmonic motion is quantized. The quantization of energy is in this case particularly simple. If the energy of the lowest state is taken as zero, and the energy of the next state up—the first excited state—as ε, then 2ε is the energy of the second excited state, 3ε is the energy of the third excited state, and so forth.

The energy-level spacing ε is proportional to the mechanical frequency of the oscillators:

$$\varepsilon = h\nu.$$

The higher the mechanical frequency ν—i.e. in general, the harder the solid —the larger ε, the wider the spacing between adjacent energy levels, the more difficult is thermal excitation of the oscillations, the smaller Ω for the ensemble of oscillators, and the smaller the entropy. The constant of the proportionality between ε and ν is called Planck's constant. It enters into all of the equations of quantum physics; it has the value

$$6.62 \times 10^{-27} \text{ erg sec}^{-1} \text{ (1 erg} = 1 \text{ gm cm}^2 \text{ sec}^{-2}).$$

To summarize, the energies that the individual oscillators can have are $0, \varepsilon, 2\varepsilon, 3\varepsilon, \ldots$ (or $0, h\nu, 2h\nu, 3h\nu, \ldots$).* No values between these are allowed. This fact greatly simplifies the computation of Ω.

Suppose our solid contains only three atoms. For convenience the three atoms may be visualized as strung out on a line.

• • •

Each atom is distinguished from its neighbor by the fact that it occupies a unique position in space. At absolute zero each atom is in its vibrational ground level and the little ensemble is as ordered as it can be: it has no thermal energy, one microstate, and no entropy.

$$S = k \ln \Omega = k \ln 1 = 0.$$

Suppose that the system absorbs from its thermal surroundings a quantity of energy ε. This unit of energy—ε shall be referred to as one unit of energy —might be absorbed by the solid in three ways, depending upon whether it is the first, second, or third atom that is set to oscillating.

↕• • •   • ↕• •   • • ↕•

These three microstates may be designated, respectively, (100), (010), and (001).

* For completeness, it should be remarked that, like atomic electrons, an harmonic oscillator in its lowest quantum-mechanically allowed state is not completely at rest. For if it were at rest, its uncertainty in position would be zero and its uncertainty in momentum (by Heisenberg's principle) would be infinite. Like atomic electrons (and all other constrained systems), an harmonic oscillator has a zero point energy. If the energy of the hypothetical state of absolute rest is taken as zero, this zero point energy, the energy of the lowest quantum-mechanically allowed state, the so-called ground state, is $(\frac{1}{2})h\nu$, and the energies of the first, second, and third excited states are, respectively, $[1 + (1/2)]h\nu$, $[2 + (1/2)]h\nu$, and $[3 + (1/2)]h\nu$. Each level is shifted upward by the amount $(1/2)h\nu$. This has no effect upon a substance's entropy and thermal energy, which, we shall see, depend solely upon the temperature and the *spacing* of the allowed levels.

Two units of energy absorbed would be sufficient to promote one oscillator to its second excited state, or two oscillators to their first excited states. Each distribution of energy might be realized in three ways: in the first case (see below), any one of the three oscillators might be the one with the two units of energy; in the second case, any one of the three oscillators might be the one left in the ground state.

Another way to represent the same information is shown below.

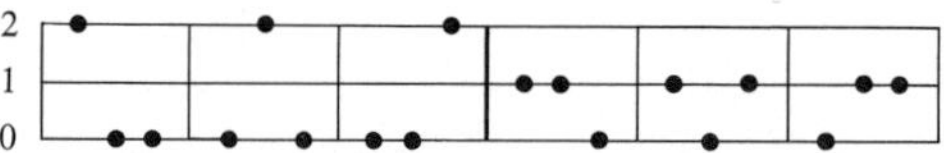

The horizontal lines in this diagram represent energy levels. A dot on the level labeled 0 represents an atom in its ground state; one on the level labeled 2, an atom in its second excited state. This diagram really contains more information than is necessary, however. Usually the question of interest is not, which atom is excited to, say, the second excited state? but, how many atoms are in the second excited state? The following short form of the previous diagram contains sufficient information to answer this question.

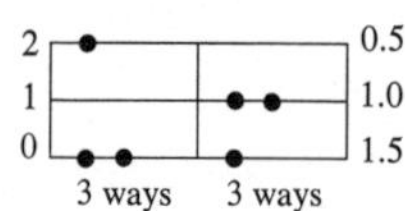

The short diagram shows that, from the standpoint of the population of the energy levels, three harmonic oscillators can share two units of energy in essentially two different ways. Let n_0, n_1, and n_2 represent the number of oscillators in the ground, first, and second excited states, respectively.

For the distribution shown on the left	For the distribution shown on the right
$n_2 = 1$	$n_2 = 0$
$n_1 = 0$	$n_2 = 2$
$n_0 = 2$	$n_0 = 1$

No other distributions exist that have different numbers for n_0, n_1, and n_2 but the same values for

$$N = n_0 + n_1 + n_2 = 3$$

and

$$E = 2\varepsilon.$$

Each distribution, as indicated, can arise in three ways. The weighted averages of n_0, n_1, and n_2 over these two distributions are shown in the column to the right of the diagram. The second excited state is half the time populated by one particle and half the time empty; this produces an average occupancy number of 0.5. The first excited state is half the time doubly occupied and half the time empty; its average occupancy number is, therefore, 1.0. The ground state is half the time doubly occupied and half the time singly occupied; its average occupancy number is 1.5.

In computing these averages, it has been assumed that each one of the six microstates of the system is as likely to occur as any one of the other five. This is the fundamental assumption of statistical mechanics. *The individual microstates of a system are assumed to have equal a priori probabilities.*

The significance of the numbers calculated for the case $E = 2\varepsilon$ becomes clearer when these numbers are compared with the corresponding numbers for the cases $E = 0$ and $E = \varepsilon$. The short diagrams for the latter two cases are:

0 —●●●— 3.0
1

1 ● 1.0
0 ● ● 2.0
3

It is instructive to add one more example, $E = 3\varepsilon$. This example presents a total of ten microstates distributed unequally over three different distributions.

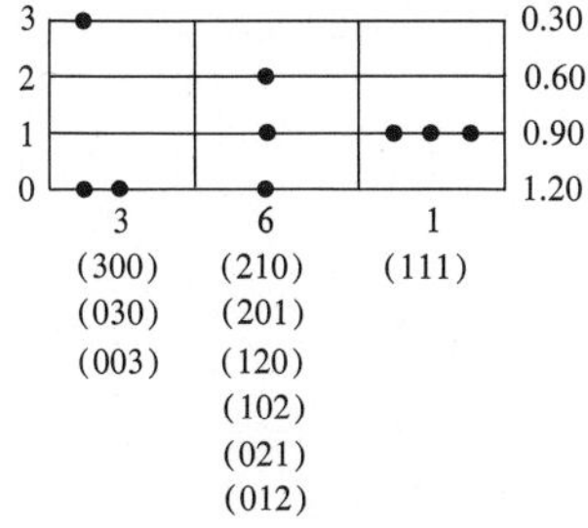

The average occupancy number of the ground level is calculated as follows:

$$\frac{3 \times 2 + 6 \times 1 + 1 \times 0}{10} = 1.20.$$

The values for the other levels are calculated in a similar fashion.

Summarized here are the values of Ω for three particles sharing 0 to 3 units of thermal energy:

E (*in units of* ε)	Ω	*Ratio of successive* Ω's
0	1	
		3.0
1	3	
		2.0
2	6	
		1.67
3	10	

Three things stand out in this summary.

I. Ω increases as E increases. This result is not unexpected. As Ω increases, $\ln \Omega$ increases, and as $\ln \Omega$ increases, the entropy ($k \ln \Omega$) increases. In short, E and S increase (or decrease) together. If δE is positive, δS is positive; if δE is negative, δS is negative. The ratio $\delta E/\delta S$ is always positive.* This ratio is the absolute temperature. To find that Ω increases as E increases is equivalent, therefore, to asserting that the absolute temperature of the system is positive.

II. The ratio of successive Ω's decreases as E increases. Now, the logarithm of this ratio is the difference between the logarithms of successive Ω's; and k times this difference is the change in S per unit change in E.

$$k \ln \left(\frac{\Omega_{E'}}{\Omega_E}\right) = k \ln \Omega_{E'} - k \ln \Omega_E = \delta S$$

From these relations and the numbers in column 3, one sees that the change in S per unit change in E decreases as E increases, i.e. the ratio $\delta E/\delta S$ increases as E increases. The system gets hotter as thermal energy is added to it.†

III. Ω can be expressed as the sum of a simple arithmetic series. Thus,

$$1 = 1$$
$$3 = 1 + 2$$
$$6 = 1 + 2 + 3$$
$$10 = 1 + 2 + 3 + 4$$

The last term of each sum is one larger than the number of units of thermal energy. Let the latter number be designated by n.

$$E = n\varepsilon$$
$$n = E/\varepsilon$$

* This conclusion is based upon a premise that may not be valid if the number of energ[y] levels available to the system is finite (problem 14).

† The statement that T increases as E increases is more generally valid than the statemen[t] that S increases as E increases. Usually both statements are valid.

It would appear that in general

$$\Omega_n = 1 + 2 + \cdots + (n + 1),$$

or

$$\Omega_n = (n + 1) + n + \cdots + 1.$$

Adding, one finds that for the three-particle system

$$2\Omega_n = (n + 2) + (n + 2) + \cdots + (n + 2)$$
$$= (n + 2)(n + 1)$$
$$\Omega_n = \frac{(n + 2)(n + 1)}{2}. \qquad (*)$$

We shall return to this result later in the chapter.

An examination of the average occupancy numbers n_0, n_1, n_2, n_3 suggests the following generalizations.

The ground level is always the most heavily populated level.

The first excited state is never less populated than the second excited state; the second excited state is never less populated than the third. Briefly, the population appears to be graded downward. Each level is on the average less heavily populated than the ones beneath it. The ratios n_1/n_0, n_2/n_1, n_3/n_2 are less than unity.

The number of particles in the ground level decreases as E increases.

The ratios n_1/n_0, n_2/n_1, n_3/n_2 increases as E increases.

An examination of the distribution of energy among four particles increases the force and scope of these generalizations. Short diagrams for $n = 4$ and $E = 0, 1, 2, 3$ are given below.

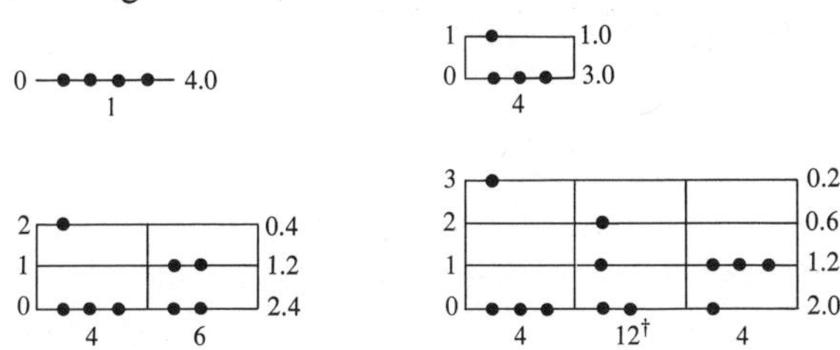

* That the difference between the difference of successive Ω's is constant implies that Ω can be expressed as a quadratic in n.

† There are four different ways to select a particle for the second excited state from four distinguishable particles; these four alternatives may be designated (2 . . .), (.2. .), (. .2.), (. . .2). Following this, there remain three ways to select a particle for the first excited state; from alternative (2. . .), for example, one obtains these three microstates: (2100), (2010), (2001). Thus, altogether there are $4 \times 3 = 12$ microstates for the distribution $n_0 = 2$, $n_1 = 1$, $n_2 = 1$, $n_3 = 0$. In general, the number of microstates included under the heading of a single distribution n_0, n_1, n_2, n_3 is

$$\frac{N!}{n_0!\, n_1!\, n_2!\, n_3!},$$

where $N = n_0 + n_1 + n_2 + n_3$. It is understood that $0! = 1$.

A graded population is again obtained and, as before, n_0 decreases and n_1/n_0 increases as E increases. Interestingly, for a given value of E, n_1/n_0 is smaller for the four-particle system than it is for the three-particle system (the values of n_1/n_0 for $E = 0, 1, 2, 3$ for $n = 4$ and $n = 3$ are, respectively, 0, 0.33, 0.50, 0.60 and 0, 0.50, 0.67, 0.75). This implies, as we shall see later, that for a given value of E the four-particle system is colder than the three-particle system.* One can see that this is so by looking at Ω. Summarized below are the values of Ω for four particles that share 0 to 3 units of thermal energy.

E	Ω	*Ratio of successive* Ω's
0	1	
		4.0
1	4	
		2.5
2	10	
		2.0
3	20	

For the four-particle system the ratios of successive Ω's are larger than the corresponding ratios for the three-particle system. The logarithms of these ratios are inversely proportional to the absolute temperature. The larger the ratios, the lower the temperature.

Another interesting fact about Ω emerges from a study of the previous table. The differences between successive Ω's for the four-particle system are the values of Ω for the three-particle system.† This implies that Ω for the four-particle system can be expressed as a cubic in n:

$$\Omega = an^3 + bn^2 + cn + d.$$

The coefficients can be evaluated by setting $n = 0$, 1, 2, and 3; this procedure produces four linear equations in a, b, c, d. The values of a, b, c, d that satisfy these equations are 1/6, 1, 11/6, 1, respectively. The cubic form associated with these coefficients has three roots: -1, -2, -3. It can be written:

$$\Omega = \frac{(n + 1)(n + 2)(n + 3)}{6}.$$

It is instructive to compare this expression with the expression for Ω when the number of particles in the system—let us call this number N—is 1, 2,

* For a macroscopic system, $n_1/n_0 = e^{-\varepsilon/kT}$ (Chapter 23). For given ε, a small value of T implies a small value of n_1/n_0, and vice versa.

† For the four-particle system the difference between the difference between the difference of successive Ω's is constant.

and 3. For $N = 1$, $\Omega = 1$. For $N = 2$, one finds that $\Omega = n + 1$. For $N = 3$,

$$\Omega = \frac{(n+1)(n+2)}{2}.$$

These results are tabulated below.

N	Ω
1	$\frac{1}{0!}$
2	$\frac{(n+1)}{1!}$
3	$\frac{(n+2)(n+1)}{2!}$
4	$\frac{(n+3)(n+2)(n+1)}{3!}$

It would appear that in general

$$\Omega_N = \frac{(n+N-1)\ldots(n+1)}{(N-1)!}$$

$$= \frac{(n+N-1)!}{(N-1)!\,n!}.$$

A formal proof of this formula will be developed in the following chapter. Since $n = E/\varepsilon$ and $\varepsilon = h\nu$, one may write

$$\Omega = \frac{\left(\frac{E}{h\nu} + N - 1\right)!}{(N-1)!\left(\frac{E}{h\nu}\right)!}.$$

This formula displays explicitly for an ensemble of oscillators the exact form of the functional dependence of the number of microstates (Ω) upon the ensemble's size (N), its energy (E), the frequency of its oscillators (ν), and Planck's constant (h).

This chapter concludes with an observation that will later be useful. When three units of energy are shared by three particles, or by four particles, here is one distribution of energy—the distribution that is graded downward most smoothly—that is several times more probable than the other distributions. The error made if the number of microstates for this most probable distribution is used to calculate the logarithm of Ω is, for the two cases just mentioned, 22 and 17 per cent, respectively. For seven particles the error drops to 13 per cent; for a hundred particles the error is forty-five hundredths of 1 per cent. For 6×10^{23} particles the error is infinitesimal. For systems this large, the logarithm of the number of microstates associated with the most probable distribution may for practical purposes be taken to be the logarithm of Ω.

Problems for Chapter 21

The following problems are concerned with the properties of ensembles of loosely coupled harmonic oscillators. Symbols used are:

E = thermal energy.
ε = spacing between adjacent energy levels.
n = number of units of thermal energy, E/ε.
N = total number of particles.
n_i = number of particles in the ith excited state ($i = 0, 1, 2, 3, \ldots$).
W = number of microstates associated with the most probable distribution of energy.
Ω = total number of microstates.

(The spirit behind these problems has been aptly described by Kenneth Grahame, in *The Wind in the Willows.* To Mole, who is about to begin a new life on the River, Water Rat says, "Believe me, my young friend, there is *nothing*—absolutely nothing half as much worth doing as simply messing about in boats.")

1. For an ensemble of seven loosely coupled harmonic oscillators sharing 4 units of energy, determine
   (a) The number of different distributions (i.e. the number of different sets of values for n_0, n_1, n_2, n_3, n_4).
   (b) The number of microstates associated with the most probable distribution.
   (c) Ω.
   (d) The value of $\dfrac{\log \Omega - \log W}{\log \Omega} \times 100$.
   (e) The value of n_1/n_0 (i) averaged over all distributions (ii) for the most probable distribution.
2. For N particles sharing 4 units of energy ($N > 4$), determine
   (a) The number of different distributions.
   (b) The values of W and Ω as N becomes very large.
   (c) The value of W/Ω as N becomes very large.

(d) The value of n_1/n_0 as N becomes very large.

3. Compare the value of n_1/n_0 for N particles sharing n units of energy with the value of Ω_n/Ω_{n+1} when (a) $N = 7$, $n = 4$ (b) $N =$ a very large number, $n = 4$. What generalization do these comparisons suggest?
4. Derive an expression for the (approximate) temperature of an ensemble of oscillators in terms of h, ν, k, and the value of the ratio Ω_n/Ω_{n+1}.
5. Derive an expression for the value of n_1/n_0 in terms of h, ν, k, and T.
6. Express W in terms of the occupation numbers $n_0, n_1, n_2, n_3, \ldots$ for the most probable distribution of energy.
7. In terms of $n_0, n_1, n_2, \ldots$, what is the value of W_n/W_{n+1}? Assume (for the moment) that the unit of energy added promotes an oscillator from its ground level to its first excited state.
8. Consider a solid that is in one of the

$$\frac{(n_0 + n_1 + n_2 + n_3 + n_4 + \cdots)!}{n_0!\, n_1!\, n_2!\, n_3!\, n_4!}$$

microstates associated with the most probable distribution of energy $n_0, n_1, n_2, n_3, n_4, \ldots$, and suppose that in this solid a particle A in its first excited state drops to its ground level as, simultaneously, some other particle B jumps from its first excited state to its second excited state.

A + B → A + B

Does this event alter the energy of the solid? How might the new distribution of energy be described? How many microstates are associated with this distribution? Is this distribution more or less probable than the previous one?

9. Show that when n_0, n_1, and n_2 are large,

$$\frac{n_1}{n_0} = \frac{n_2}{n_1}.$$

What generalization does this result suggest?

10. Using the result derived in problem 4, calculate in terms of ε and k the (approximate) temperature of an N-particle system containing 0 – 5 units of energy when (a) $N = 2$ (b) $N = 3$ (c) $N = 4$.
11. Using the data calculated in problem 10, estimate the heat capacities at high temperatures of an N-particle system for $N = 2, 3, 4$. "Normalize" these values by dividing each one by N; then multiply by Avogadro's number. This converts them all to a molar basis. Do you see any trend in these numbers?

12. A 1-oscillator system is in its second excited state. Is this system hot or cold? What is its temperature?
13. What does a plot of the entropy of a substance against its internal energy look like?
14. Consider a system of N independent particles, and suppose that each particle has accessible to it only two energy states: a ground level and one excited state ε energy units above the ground level. (These particles might be the odd electrons in non-interacting free radicals in a magnetic field, with which the electron spins can be aligned either parallel or anti-parallel.) What is the maximum energy the system can have? What is its entropy at that point? Plot the entropy of the system against its thermal energy.

A Note by Boltzmann on the Calculus of Probabilities

"From an urn, in which many black and an equal number of white but otherwise identical spheres are placed, let 20 purely random drawings be made. The case that only black balls are drawn is not a hair less probable than the case that on the first draw one gets a black sphere, on the second a white, on the third a black, etc. The fact that one is more likely to get 10 black spheres and 10 white spheres in 20 drawings than one is to get 20 black spheres is due to the fact that the former event can come about in many more ways than the latter. The relative probability of the former event as compared to the latter is the number 20!/10!10!, which indicates how many permutations one can make of the terms in the series of 10 white and 10 black spheres, treating the different white spheres as identical, and the different black spheres as identical. Each one of these permutations represents an event that has the same probability as the event of all black spheres."*

* Ludwig Boltzmann, *Lectures on Gas Theory*, trans. by Stephen G. Brush, Berkeley: University of California Press, 1964.

22

Einstein's Solid

$$E = \frac{3N_0 h\nu}{e^{h\nu/kT} - 1}$$

In the left-hand column below are listed the ten ways three localized oscillators can share 3 units of energy. On the right is another representation of the same information.

(0 0 3)	•	•	•	ϵ	ϵ	ϵ
(0 1 2)	•	•	ϵ	•	ϵ	ϵ
(0 2 1)	•	•	ϵ	ϵ	•	ϵ
(0 3 0)	•	•	ϵ	ϵ	ϵ	•
(1 0 2)	•	ϵ	•	•	ϵ	ϵ
(1 1 1)	•	ϵ	•	ϵ	•	ϵ
(1 2 0)	•	ϵ	•	ϵ	ϵ	•
(2 0 1)	•	ϵ	ϵ	•	•	ϵ
(2 1 0)	•	ϵ	ϵ	•	ϵ	•
(3 0 0)	•	ϵ	ϵ	ϵ	•	•

In the representation on the right dots represent oscillators. Dots followed immediately on the right by another dot (. .) represent oscillators in their ground states. Dots followed on the right by the symbol ε once (. ε), twice (. ε ε), and three times (. ε ε ε) represent oscillators in their first, second, and third excited states, respectively.* Thus, the horizontal array . . . ε ε ε represents the microstate: first oscillator in its ground state, second oscillator

* "Epsilon" the world over stands for a difference between two energy levels in a quantum mechanical system. The symbol of choice for "epsilon" in the British Isles, where this book was set, is ε. The usual symbol in the United States, where the figure above was drawn, is ϵ. Parenthetically, it may be remarked that an occasional exposure to the use of strange signs for familiar quantities may be very valuable. It may facilitate the transition from a possibly unconscious, superficial recognition of symbols to a deeper, mental recognition of concepts. Particularly useful to the modern reader is this regard are the works of J. Willard Gibbs.

in its ground state, third oscillator in its third excited state; similarly, the array . . ε . ε ε represents the microstate: first oscillator in its ground state, second oscillator in its first excited state, third oscillator in its second excited state; and so forth. Permutation of the ε's with the dots to the right of the dashed line yields every microstate listed at the left—no more, no less. Thus, for N oscillators sharing n units of energy,

$$\Omega = \frac{(N + n - 1)!}{(N - 1)!\,n!}.$$

All the thermodynamic properties of the N-oscillator system can be obtained from this equation. It will be called the ensemble's fundamental equation.

The ensemble's fundamental equation yields with Boltzmann's relation an expression for the ensemble's entropy in terms of the number of particles in the ensemble and the ensemble's thermal energy (in units of ε).

The thermal energy, however, is an awkward variable in terms of which to express the ensemble's entropy. To determine the ensemble's thermal energy the ensemble must be cooled to nearly absolute zero and the energy absorbed during warm-up measured. From the experimental point of view it would be convenient to have the ensemble's entropy expressed in terms of its temperature.

The ensemble's temperature can be introduced into its thermodynamic formulas by combining the fundamental equation and Boltzmann's relation with the thermodynamic definition of temperature. The procedure is to determine by the fundamental equation and Boltzmann's relation how the ensemble's entropy changes as its thermal energy changes. The ratio of these two changes, by definition, is the ensemble's absolute temperature.

When one unit of thermal energy is added to an ensemble of oscillators, the parameter n changes from n to $n + 1$ and the ensemble's entropy changes by the amount

$$\begin{aligned} S_{n+1} - S_n &= k \ln \frac{(N + n)!}{(N - 1)!(n + 1)!} - k \ln \frac{(N + n - 1)!}{(N - 1)!n!} \\ &= k \ln \frac{N + n}{n + 1} \\ &\approx k \ln \frac{N + n}{n}. \qquad (n \gg 1) \end{aligned}$$

Introducing the thermodynamic definition of temperature

$$T \equiv \frac{\delta E}{\delta S}$$

with
$$\begin{aligned}\delta E &= E_{n+1} - E_n \\ &= (n+1)\varepsilon - n\varepsilon \\ &= \varepsilon\end{aligned}$$

and
$$\begin{aligned}\delta S &= S_{n+1} - S_n \\ &= k \ln \frac{N+n}{n},\end{aligned}$$

one finds that

$$T = \frac{\varepsilon}{k \ln \dfrac{N+n}{n}}.$$

It may be seen that this relation is physically reasonable.

An ensemble with relatively little thermal energy per particle should be cold, unless the spacing between allowed energy levels is very large—in which case the absorption of thermal energy by the ensemble, even at high temperatures, is not great. Now, for $n \ll N$, $(N + n)/n$ is large, $\ln [(N + n)/n]$ is large, and consequently T, as expected, is small, unless ε is very large. On the other hand, for an ensemble with many units of thermal energy per particle ($n \gg N$), $(N + n)/n \approx 1$, $\ln [(N + n)/n]$ is small, and T, as expected, is large, unless ε is very small.

The equation expressing T in terms of the thermal-energy-parameter n may be solved for n.

$$\ln \frac{N+n}{n} = \frac{\varepsilon}{kT}$$

$$\frac{N+n}{n} = e^{\varepsilon/kT}$$

$$n = \frac{N}{e^{\varepsilon/kT} - 1}.$$

Multiplication by ε, the spacing between adjacent energy levels, $h\nu$, yields the following expression for the thermal energy of an ensemble of N loosely coupled harmonic oscillators:

$$\frac{N\varepsilon}{e^{\varepsilon/kT} - 1} = \frac{Nh\nu}{e^{h\nu/kT} - 1}.$$

When T is small compared to ε/k ($kT \ll \varepsilon$), ε/kT is large, the quantity $(e^{\varepsilon/kT} - 1)$ is large, and, as expected, the ensemble's thermal energy is small. On the other hand, when T is large compared to ε/k ($kT \gg \varepsilon$), ε/kT is small,

$e^{\varepsilon/kT} \approx 1$, the quantity $(e^{\varepsilon/kT} - 1)$ is small, and the ensemble's thermal energy is large.

In the previous equations the symbol N stands for the number of (nearly) independent, one-dimensional oscillators in the ensemble. By contrast, real particles are able to oscillate in three directions. From the point of view of the number of accessible microstates, N three-dimensional oscillators are equivalent to $3N$ one-dimensional oscillators. Hence, if N_0 stands for Avogadro's number,

$$E(\text{per mole of real particles}) = \frac{3N_0h\nu}{e^{h\nu/kT} - 1}.$$

This equation is due to Einstein.

At high temperatures, the exponent $h\nu/kT$ will be small. Now, for small values of x,

$$(1 + x)^{1/x} \approx e.$$

Hence, for small x,

$$e^x \approx 1 + x.$$

Therefore, when $kT \gg h\nu$,

$$E(\text{per mole of particles}) = \frac{3N_0h\nu}{1 + \dfrac{h\nu}{kT} - 1}$$

$$= 3RT. \qquad (R = N_0k)$$

From this result we can immediately obtain the heat capacity of an ensemble of loosely coupled harmonic oscillators at high temperatures.

$$C(\text{per mole of particles}) = \frac{E_{T_2} - E_{T_1}}{T_2 - T_1} = \frac{3RT_2 - 3RT_1}{T_2 - T_1}$$

$$= 3R \qquad (*)$$

$$= 6 \text{ cal/deg/mole of particles.}$$

This result may be compared with the law of Dulong and Petit, which states that for many metals at room temperature the specific heat (in cal/deg C per gram) times the atomic weight (in grams per mole) yields a number close to 6.

The entropy of an Einsteinian solid can be calculated, in two steps, as follows. Imagine, in the first step, that one oscillator in its ground vibrational

* This result is valid when $kT \gg h\nu$; at 0°K, the heat capacity is zero. Heat capacity behavior at low temperatures is well described for many, but not all, substances by a modification of Einstein's equation due to Debye.

state is added to the system. This increases the system's entropy because it increases the total number of oscillators among which the system's thermal energy can be shared.* The change in S can be calculated from the expression obtained for Ω at the beginning of this chapter.

$$(S_{N+1} - S_N)_{n \text{ constant}} = k \ln \frac{(N+n)!}{N!n!} - k \ln \frac{(N+n-1)!}{(N-1)!n!}$$

$$= k \ln \frac{N+n}{N} = k \ln \left(1 + \frac{n}{N}\right).$$

Now, referring to the expression obtained earlier for n, one finds that

$$1 + \frac{n}{N} = 1 + \frac{1}{e^{\varepsilon/kT} - 1} = \frac{e^{\varepsilon/kT}}{e^{\varepsilon/kT} - 1} = (1 - e^{-\varepsilon/kT})^{-1}.$$

Therefore,

$$(S_{N+1} - S_N)_{n \text{ constant}} = -k \ln (1 - e^{-\varepsilon/kT}).$$

Since the quantity $(1 - e^{-\varepsilon/kT})$ is always a number less than 1, $-k \ln (1 - e^{-\varepsilon/kT})$ is always a positive quantity.

Next, to maintain constant the system's temperature, imagine there is added to the system a quantity of thermal energy equal to the average energy per oscillator: E/N. By the definition of T, this energy increment increases the system's entropy by the amount

$$\frac{(E/N_0)}{T}.$$

For simplicity, it has been assumed that $N = N_0$. The sum of these two entropy increments, times $3N_0$, is the entropy contributed to the system by a mole of three-dimensional oscillators.

$$S(\text{per mole of particles}) = -3R \ln (1 - e^{-\varepsilon/kT}) + \frac{E(\text{per mole of particles})}{T}$$

For temperatures large compared to ε/k this equation assumes a simpler form.

When $T \gg \varepsilon/k$, ε/kT is small, $e^{-\varepsilon/kT} \approx 1 - \varepsilon/kT$, and consequently

$$S(\text{per mole of particles}) \approx -3R \ln (\varepsilon/kT) + 3R.$$

* It also diminishes the system's temperature slightly.

This equation expresses semiquantitatively a statement made in Chapter 6 regarding the relation between a system's entropy and its energy-level spacing. As the energy-level spacing ε decreases, ε/kT decreases, ln ε/kT becomes smaller (possibly more negative), and S increases.

The vibrational frequency of atoms in solids that begin to obey the law of Dulong and Petit around room temperature is about 10^{12} cycles per second. Using this value for ν, one finds that at 300°K,

$$\frac{\varepsilon}{kT} = \frac{h\nu}{kT} \approx \frac{\left(6.62 \times 10^{-27}\,\frac{\text{gm cm}^2}{\text{sec}}\right)(10^{12}\ \text{sec}^{-1})}{\left(1.38 \times 10^{-16}\,\frac{\text{gm cm}^2}{\text{sec}^2\ {}^\circ\text{K}}\right)(300^\circ\text{K})} = 0.16.$$

For such solids, therefore,

$$\begin{aligned} S &\approx -\,3R\,(2.303)\log 0.16 + 3R \\ &= 17\ \text{cal/deg-mole.} \end{aligned}$$

Problems for Chapter 22

1. The law of Dulong and Petit is not valid at absolute zero. Under what general conditions is this law probably not valid?
2. Is the law of Dulong and Petit valid at room temperature for a solid whose characteristic frequency is 3×10^{13} cycles per second?
3. Calculate the thermal energy and entropy of an Einsteinian solid when $h\nu/kT = 4$ and $T = 300^\circ$K.
4. Show that when $e^{h\nu/kT} \gg 1$, the molar energy and entropy of an Einsteinian solid can be written as follows:

$$E = 3RT\left(\frac{h\nu}{kT}\right)e^{-h\nu/kT}$$

$$S = 3R\left(1 + \frac{h\nu}{kT}\right)e^{-h\nu/kT}$$

   What values are obtained from these formulas when $T = 300^\circ$K and $h\nu/kT = 3, 4, 5$?
5. If distortion of a bond in a molecule produces a change in the molecule's dipole moment, the molecule may absorb electromagnetic radiation of a frequency equal to the natural frequency of vibration of the molecule. Molecules containing carbon-carbon single bonds often absorb radiation

whose frequency corresponds to about 1000 waves per centimeter.* Estimate from this fact the molar entropy of diamond at 300°K.

6. At approximately what temperature does diamond obey the law of Dulong and Petit?
7. Each degree of freedom associated with a molecule in the gas phase can theoretically contribute to the heat capacity of the gas. According to classical theory, which is valid at high temperatures (but not at low temperatures, where the heat capacity approaches zero), the contribution is $R/2$ for each degree of translational freedom, of which there are three (for molecular motion in the x, y, z directions); $R/2$ for each rotational degree of freedom, of which there are none for monatomic gases, two for diatomic gases and other linear molecules, and three for non-linear molecules; and R for each vibrational degree of freedom (this figure is twice the others, owing to the fact that vibrational motion involves terms in both kinetic energy and potential energy; translational and rotational motion involve only terms in kinetic energy). The number of vibrational degrees of freedom can be found as follows. N independent atoms have $3N$ degrees of freedom (3 translational degrees of freedom per atom). In an N-atomic molecule, three of these $3N$ degrees of freedom are assigned to motion of the molecule's center of mass and two (if the molecule is linear) or three (if it is not) to rotation of the molecule as a rigid body; motions involving these degrees of freedom do not alter bond angles and bond lengths. The remainder of the $3N$ degrees of freedom are assigned to the vibrational degrees of freedom of the molecule. How many vibrational degrees of freedom has (a) Ar? (b) N_2? (c) CO_2? (d) H_2O?
8. Derive a general expression for the number of vibrational degrees of freedom of a molecule.
9. The heat capacities at room temperature and at constant pressure of Ar, N_2, CO_2, and H_2O are, respectively, 4.97, 6.96, 8.87, and 8.02 cal/deg-mole. Do these seem to be reasonable values?
10. What is the heat capacity at constant volume of 1 mole of a monatomic gas? Of one-half a mole of a diatomic gas? (Suppose that all translational, rotational, and vibrational degrees of freedom are classically excited.) Of one-third a mole of a non-linear triatomic gas? Of one-fourth a mole of a non-linear tetratomic gas? Of one-fifth a mole of a non-linear pentatomic gas? Of one-twelfth a mole of a non-linear 12-atomic gas? Of $1/N^{\text{th}}$ a mole of a non-linear N-atomic gas? Summarize your results in the form of a table.

* This figure is called the wave number of the radiation. It is the reciprocal of the radiation's wave length, λ. 1000 cm^{-1} falls in about the middle of the infrared region of the spectrum. In a vacuum the frequency and wave length of electromagnetic radiation are related by the formula $\nu\lambda = 3 \times 10^{10}$ cm/sec.

11. Consider the last entry in your table for problem 10. What is the heat capacity of the system when $N = 100$? 1000? 10^6? 6.023×10^{23}? What does the last case represent?
12. When N is a large number,

$$\ln (N!) \approx N \ln (N) - N. \qquad (*)$$

Show, hence, that

$$S = k\left[N \ln \frac{N+n}{N} + n \ln \frac{N+n}{n}\right].$$

13. Show that the entropy formula derived in the previous problem is consistent with the entropy formula derived in the text.
14. Show that on a molar basis the heat capacity of an Einsteinian ensemble of N one-dimensional oscillators that share n units of thermal energy is given by the following expression:

$$\frac{R\left(\ln \dfrac{N+n+1}{n+2}\right)\left(\ln \dfrac{N+n}{n+1}\right)}{N\left(\ln \dfrac{N+n}{n+1} - \ln \dfrac{N+n+1}{n+2}\right)}.$$

What does this expression become (a) as $n \to 0$ and $N \to \infty$? (b) as $n \to \infty$?

* This formula is known as Stirling's approximation. It may be derived as follows.

$$\ln (N!) = \ln (N) + \ln (N-1) + \cdots + \ln (2) + \ln (1)$$

$$= \sum_{x=1}^{N} \ln x \approx \int_1^N \ln x \, dx$$

By parts

$$= x \ln x - x \Big|_1^N = N \ln N - N \qquad (N \gg 1).$$

23

Boltzmann's Factor and Partition Functions

$$n_i = \frac{N}{Q} e^{-\varepsilon_i/kT}$$

Several formulas derived in the previous chapter can be generalized quite simply.

For large systems, the entropy is equal to Boltzmann's constant times the logarithm of the number of microstates W associated with the most probable distribution $n_0, n_1, n_2, n_3 \ldots$.

$$\begin{aligned} S &= k \ln W \\ &= k \ln \frac{N!}{n_0!n_1!n_2!n_3! \ldots} . \qquad N \text{ large} \end{aligned}$$

In this equation N stands for the total number of particles in the system:

$$\begin{aligned} N &= n_0 + n_1 + n_2 + n_3 + \cdots \\ &= \sum_{i=0}^{\infty} n_i . \end{aligned}$$

The values of $n_0, n_1, n_2, n_3, \ldots$ can be found as follows.

Imagine there is added to the system a quantity of energy just sufficient to promote one particle from its first to its second excited state. The energy required is

$$\varepsilon_2 - \varepsilon_1 = \delta E_\sigma .$$

This addition of energy alters the occupancy numbers of the first and second excited states from n_1 and n_2 to $n_1 - 1$ and $n_2 + 1$, respectively; all other occupancy numbers remain the same. The change in the system's entropy is

$$k \ln \frac{N!}{n_0!(n_1 - 1)!(n_2 + 1)!n_3!\ldots} - k \ln \frac{N!}{n_0!n_1!n_2!n_3!\ldots}$$

$$= k \ln \frac{n_1}{n_2 + 1}$$

$$= k \ln \frac{n_1}{n_2} \qquad (n_2 \gg 1)$$

$$= \delta S_\sigma .$$

Therefore,
$$T \equiv \frac{\delta E_\sigma}{\delta S_\sigma}$$

$$= \frac{\varepsilon_2 - \varepsilon_1}{k \ln \dfrac{n_1}{n_2}};$$

or
$$\frac{n_2}{n_1} = e^{-\frac{(\varepsilon_2 - \varepsilon_1)}{kT}} .$$

The expression on the right is called Boltzmann's factor.

Generally, for any pair of levels i and j

$$\frac{n_i}{n_j} = e^{-\frac{(\varepsilon_i - \varepsilon_j)}{kT}} .$$

Setting $j = 0$ and $\varepsilon_0 = 0$, one obtains

$$n_i = n_0 \, e^{-\varepsilon_i/kT} .$$

The value of n_0 in this expression may be determined as follows. Since

$$\sum_{i=0}^{\infty} n_i = N,$$

one finds on substituting $n_0 \, e^{-\varepsilon_i/kT}$ for n_i that

$$\sum_{i=0}^{\infty} (n_0 \, e^{-\varepsilon_i/kT}) = n_0 \sum_{i=0}^{\infty} (e^{-\varepsilon_i/kT}) = N.$$

Hence,
$$n_0 = \frac{N}{\sum_{i=0}^{\infty} (e^{-\varepsilon_i/kT})} .$$

For convenience, let

$$Q \equiv \sum_{i=0}^{\infty} (e^{-\varepsilon_i/kT}).$$

Then

$$n_0 = \frac{N}{Q},$$

and consequently

$$n_i = \frac{N}{Q} e^{-\varepsilon_i/kT}.$$

Q is called the partition function. It determines—with Boltzmann's factor—how the particles are partitioned among the energy levels.

The system's energy is

$$\begin{aligned} E &= n_0\varepsilon_0 + n_1\varepsilon_1 + n_2\varepsilon_2 + n_3\varepsilon_3 + \cdots \\ &= \sum_{i=0}^{\infty} n_i\varepsilon_i \\ &= \sum_{i=0}^{\infty} \left(\frac{N}{Q} e^{-\varepsilon_i/kT}\right) \varepsilon_i \\ &= \frac{N}{Q} \sum_{i=0}^{\infty} \varepsilon_i\, e^{-\varepsilon_i/kT}. \qquad (*) \end{aligned}$$

The system's entropy, per particle, is equal to the entropy change produced by the addition to the system of a particle to the ground level,

$$\begin{aligned} k \ln \frac{(N+1)}{(n_0+1)!n_1!n_2!} - k \ln \frac{N!}{n_0!n_1!n_2!\ldots} &= k \ln \frac{N+1}{n_0+1} \\ &= k \ln \frac{N}{n_0} \qquad (n_0 \gg 1) \\ &= k \ln Q, \end{aligned}$$

plus the entropy change produced by the addition to the system of a quantity of thermal energy equal to the average thermal energy per particle; this entropy change per particle is E(per particle)$/T$. Thus,

* By calculus,

$$\frac{dQ}{dT} = \frac{1}{kT^2} \Sigma\, \varepsilon_i\, e^{-\varepsilon_i/kT}.$$

Therefore, one may also write

$$E = \frac{NkT^2}{Q}\frac{dQ}{dT} = NkT^2 \frac{d \ln Q}{dT}.$$

$$S(\text{per mole}) = R \ln Q + \frac{E(\text{per mole})}{T}. \qquad (*)$$

The numerical value of the partition function Q depends upon two factors: the temperature and the spacing of the allowed energy levels. Of considerable chemical interest, therefore, are the factors that determine the spacing of the allowed energy levels. These factors are considered in the following chapter.

Problems for Chapter 23

1. What value does the ratio n_i/n_j approach as T approaches zero? As T approaches infinity? Assume that $\varepsilon_i > \varepsilon_j$.
2. What is the expression for n_i when ε_0 is not set equal to zero? What is the corresponding expression for n_0?
3. What is the value of the first term in the series expansion of the partition function?
4. What value does the partition function approach as T approaches zero? As T approaches infinity?
5. What value does n_0 approach as T approaches zero? As T approaches infinity?
6. What value does n_i approach as T approaches zero? As T approaches infinity?
7. What effect has diminishing the spacing between energy levels on the partition function? On n_0?
8. What effect has doubling N (T constant) on the partition function? On n_0? On n_i, $i \neq 0$?
9. What are the entropy changes in σ and its thermal surroundings θ for the following process?

$$\begin{pmatrix}\text{Particle of } \sigma \text{ in}\\ 1^{\text{st}} \text{ excited state}\end{pmatrix} + \begin{pmatrix}\text{Energy}\\ \text{from } \theta\end{pmatrix} = \begin{pmatrix}\text{Particle of } \sigma \text{ in}\\ 2^{\text{nd}} \text{ excited state}\end{pmatrix}$$

   Under what conditions is $\Delta S_{\text{univ.}} \geq 0$?
10. Using Stirling's approximation, show that

$$S = -k\left[n_0 \ln \frac{n_0}{N} + n_1 \ln \frac{n_1}{N} + n_2 \ln \frac{n_2}{N} + \cdots\right].$$

11. Show that the entropy formula derived in the previous problem is consistent with the entropy formula derived in the text.

*The energy term in this expression stands for the *thermal* energy per mole. Note that the free energy, $E - TS$, is equal to $-RT \ln Q$.

12. Derive an expression for the partition function of an harmonic oscillator whose energy levels are 0, $h\nu$, $2h\nu$, $3h\nu$, ... What does this expression become when $T = 0$? When $kT \gg h\nu$?
13. Starting with the equation

$$E = \sum_{i=0}^{\infty} n_i \varepsilon_i \, ,$$

show that the thermal energy of an ensemble of loosely coupled one-dimensional harmonic oscillators is equal to $Nh\nu(e^{h\nu/kT} - 1)^{-1}$.
14. What percentage of the molecules of a diatomic gas whose vibrational frequency is 667.3 cm^{-1} are in their first excited state at 300°K? At 1000°K?
15. If the number of molecules in the first excited state of a diatomic gas whose vibrational frequency is 667.3 cm^{-1} is found, by spectroscopic measurements, to be $1/e$ or 36.8 per cent of the number in the ground level, what is the gas's temperature?
16. From the standpoint of the Boltzmann factor, a negative temperature defines what kind of population?

24

Interlude on Quantum Mechanics

$$\varepsilon \sim \frac{h^2}{mL^2}$$

In 1925 Louis Victor de Broglie suggested that moving particles might exhibit wave properties. He calculated that the wave length λ of a particle of mass m moving with a velocity v should be given by the expression

$$\lambda = \frac{h}{mv}. \qquad (*)$$

The de Broglie wave length of macroscopic objects is much too short to be detected experimentally. A 1 gram object moving with a velocity of 1 cm/sec, for example, has a de Broglie wave length of only

$$\frac{6.62 \times 10^{-27}\ \text{gm cm}^2\ \text{sec}^{-1}}{(1\ \text{gm})\ (1\ \text{cm sec}^{-1})} = 6.62 \times 10^{-27}\ \text{cm}.$$

On the other hand, an electron moving with the same velocity has a de Broglie wave length of

$$\frac{6.62 \times 10^{-27}\ \text{gm cm}^2\ \text{sec}^{-1}}{(9.11 \times 10^{-28}\ \text{gm})(1\ \text{cm sec}^{-1})} = 7.27\ \text{cm}.$$

Broglie's relation was soon verified experimentally for electrons; later for a variety of atoms such as hydrogen, helium, and neon; and more recently for neutrons. Practical applications of this relation are now very common, particularly in studies of the structure of matter with neutron beams and with beams of electrons. The resolving power of a microscope is limited by the wave length of the illuminating radiation; anything smaller than this wave length is not well resolved. Thus, with ordinary light details less than several thousand angstroms in size cannot be seen, no matter how great the

* The waves of which λ is the wave length are sometimes called "particle waves."

magnification. Illumination with a beam of high speed electrons, however, whose de Broglie wave length is less than that of visible light, can greatly improve the resolution of a microscope and has rendered the shapes of individual macromolecules clearly visible.

Broglie's relation is related closely to Heisenberg's uncertainty principle. In wave mechanics it is found that the uncertainty in our knowledge regarding the precise location of a particle is comparable to the particle's de Broglie wave length: $\Delta x \sim \lambda$. Heisenberg's principle states that the product of this uncertainty and the uncertainty in the particle's momentum, $\Delta(mv)$, is equal to or greater than a small constant times Planck's constant.

$$\Delta x \cdot \Delta(mv) \geq \frac{h}{4\pi}$$

This may be compared with de Broglie's relation,

$$\lambda \cdot mv = h.$$

Broglie's relation can be used to discover how the spacing of the allowed energy levels associated with constrained motion depends upon the mass and distance involved in the motion. Useful in this connection is a hypothesis first introduced by Bohr. Bohr suggested that the velocity of a particle involved in periodic motion and confined thereby to a path or orbit of length L must be such that its de Broglie wave length will fit nicely, once, twice, three times, ... into the orbit.

$$n\lambda = L \qquad\qquad n = 1, 2, 3, \ldots$$

The integer n is called the quantum number. Since $\lambda = h/mv$, the velocity must be such that

$$n\left(\frac{h}{mv}\right) = L.$$

Thus,

$$v = \frac{nh}{mL}.$$

The allowed values for the particle's kinetic energy, $1/2\, mv^2$, are then

$$\frac{1}{2} m \left(\frac{nh}{mL}\right)^2 = \frac{1}{2}\frac{n^2h^2}{mL^2}.$$

The corresponding potential energy is in general either constant—as it is in the translational motion of a gas molecule in a flask (small changes in potential energy due to changes in altitude are in general insignificant*), and

* One exception is a substance near its critical point.

as it is in the rotation of a gas molecule (if, as is often the case, centrifugal stretching is relatively unimportant)—or else it is related in a simple way to the kinetic energy—for an harmonic oscillator, for example, the potential energy is equal to the kinetic energy. The total energy is always the sum of the kinetic and potential energies.

To recapitulate,

$$
\begin{aligned}
E_{\text{total}} &= \text{K.E.} + \text{P.E.} && \text{always} \\
&\sim \text{K.E.} && \text{often} \\
&= \frac{mv^2}{2} = \frac{1}{2m}(mv)^2 && \text{classical physics} \\
&= \frac{1}{2m}\left(\frac{h}{\lambda}\right)^2 && \text{de Broglie} \\
&= \frac{1}{2m}\left(\frac{nh}{L}\right)^2 && \text{Bohr} \\
&= \frac{n^2h^2}{2mL^2}.
\end{aligned}
$$

The energy separation ε between the ground level (level $n = 1$) and the first excited state (level $n = 2$) is, then, approximately h^2/mL^2.

$$\varepsilon \sim \frac{h^2}{mL^2}$$

As stated earlier (Chapter 6), the larger the mass of the particle and the larger the distance involved in its periodic motion, the smaller is the spacing between adjacent energy levels.

The temperature at which thermal excitation of quantized motion begins to become important is the temperature at which the population of the first excited state begins to become significant. The population of the first excited state begins to become significant when the exponent in Boltzmann's relation,

$$\frac{\varepsilon_{n=2} - \varepsilon_{n=1}}{kT} = \frac{\varepsilon}{kT},$$

begins to reach the neighborhood of unity; this happens when the temperature is equal to approximately ε/k; i.e. when T is approximately

$$\frac{1}{k}\left(\frac{h^2}{mL^2}\right).$$

Values of this "characteristic temperature" T_c are tabulated in Table 24.1 for some typical values of m and L. At room temperature, molecular trans-

TABLE 24.1

Characteristic Temperatures for Thermal Excitation of Quantized Molecular Motion

Type of Motion	m (amu)	L (cm)	$T_c = h^2/(kmL^2)$ (°K)
Translation	50	10	10^{-17}
Rotation	10	10^{-7}	2
Vibration	10	3×10^{-9}	2,000
Electronic	$\frac{1}{1837}$	10^{-7}	40,000
Nuclear	1	10^{-12}	10^{11}

lations and rotations are easily excited to their full classical values: $E_{\text{trans.}} = (3/2)RT$; $E_{\text{rot.}} = R$ (or $3R/2)T$; molecular vibrations are, in general, only slightly excited; and thermal excitation of electronic motion and nuclear motion is generally insignificant.

Problems for Chapter 24

1. Calculate the speed of an electron whose de Broglie wave length is (a) 2000 Å (b) such as to fit once into Bohr's first orbit for the hydrogen atom; the radius of this orbit, the $1s$ orbit, is 0.529 Å. (c) What is the kinetic energy of a $1s$ electron in hydrogen?
2. If the uncertainty in the momentum of an object illuminated with photons is roughly equal to the momentum of the photons and if, simultaneously, the uncertainty in the object's position is roughly equal to the wave length of the photons, what, roughly, is the product of the two uncertainties? The momentum of a photon is $h\nu/c$.
3. What is the length of the "orbit" of a particle that is confined to vibrate back and forth in a one-dimensional box whose length is l? What are the allowed values of the kinetic energy of such a particle? What is the particle's zero point kinetic energy if m is equal to 1 amu and l is equal to (a) 10 cm? (b) 10^{-7} cm? (c) 10^{-12} cm?
4. Derive an expression for the allowed values of the kinetic energy of a particle confined to a circular orbit of radius r.
5. Show that the allowed values for the kinetic energy of a particle in a one-dimensional box can be written in the form K.E. $= (1/2)nh\nu$, where ν is the frequency of the periodic motion.

6. To excite a quantum-mechanical system from one energy level to a higher one with electromagnetic radiation, the wave length of the radiation must be short enough to make the energy of the photon, $h\nu$, at least as great as the spacing between the two energy levels. For the values of m and L listed in Table 24.1, calculate the values of λ that make

$$h\nu = h^2/mL^2 ,$$

and identify in each case the region of the spectrum in which the radiation falls.

DIGRESSION ON DIMENSIONAL ANALYSIS

The physical dimensions, or units, on quantities are frequently useful in checking a formula to see if it has been written down correctly. Is a particle's de Broglie wave length h/mv, mv/h, or hv/m, for example? A dimensional check shows that the first possibility is the only likely one. Wave length has the dimensions of length. The dimensions of h are (mass)(length)^2(time)$^{-1}$ (in the cgs system, for example, $h = 6.62 \times 10^{-27}$ gm cm^2 sec^{-1}); dimensions of m are mass and of v (length)(time)$^{-1}$. Of the three possibilities listed, only h/mv has the dimensions of length. The formula $\lambda = h/mv$ is said to be "dimensionally homogeneous."

7. Is the formula $\lambda = (1/2)h/mv$ dimensionally homogeneous?
8. Show that the equations

$$\text{Photon momentum} = h\nu/c$$
$$E = mc^2$$

are dimensionally homogeneous.

To carry through a dimensional check on an equation, it is necessary to know the dimensional formula of each quantity in the equation. The dimensions of mass, length, and time are usually taken as primary. In terms of these, the dimensions of other quantities are expressed. This was the procedure adopted above. Often useful in applications of this procedure to thermodynamics is a fourth primary quantity—temperature. The dimensions of these four primary quantities—mass, length, time, and temperature—are usually abbreviated M, L, T, and Θ, respectively. Using the symbol [ ] for "dimensions of," or "dimensional formula of," one may write

$$[m] = M$$
$$[l] = L$$
$$[t] = T$$
$$[T] = \Theta.$$

9. What are the dimensional formulas of ν, v, c, h, p (momentum), a (acceleration), A (area), V (volume)?

Dimensional formulas for force, charge, and other secondary quantities can be obtained from their defining equations, or from equations in which they appear, if the dimensional formulas of the other quantities in the equation are known. The dimensional formula for force can be obtained from Newton's law; this law states that force is equal to the acceleration it imparts to an object times the object's mass. Then, from Coulomb's law, which states that the force between two charges is equal to the product of the charges divided by the square of the distance between them, can be obtained the dimensional formula of electrical charge; and so forth.

10. What are the dimensional formulas for F(force), Q(charge), E(energy), W(work), P(pressure), S(entropy), R(the gas constant)?

The functional form of an equation can frequently be inferred from the dimensional formulas of the quantities that occur in the equation. For example, the wave length of light, dimensional formula L, depends upon its frequency, dimensional formula T^{-1}, and its velocity, dimensional formula LT^{-1}. The only combination of LT^{-1} and T^{-1} that yields L is LT^{-1}/T^{-1}. Therefore, $\lambda \propto c/\nu$. The constant of proportionality (here unity) is not obtained by this method of analysis.

11. The energy of a photon depends upon its frequency and Planck's constant. What is the functional form of the relation between these three quantities?
12. The allowed energy levels ε for a particle of mass m confined to a one-dimensional box of length l depend upon m, l, and (since this is a quantum phenomenon) Planck's constant h. Derive by the methods of dimensional analysis the functional form of the relation between ε and m, l, h.
13. The centrifugal force associated with an object of mass m moving in a circle of radius r with a velocity v depends upon m, v, and r. What is the functional form of this relation?

The procedure for inferring the functional form of a relation from dimensional formulas can be systematized. Consider the problem of ascertaining the manner in which a photon's momentum p depends upon its frequency, Planck's constant, and the velocity of light. Assume that

$$p \propto \nu^x h^y c^z \, .$$

The exponents x, y, z are to be selected to make the relation dimensiona homogeneous, i.e. to make

$$[p] = [\nu]^x[h]^y[c]^z$$
$$= (T^{-1})^x(ML^2T^{-1})^y(LT^{-1})^z .$$

In other words,

$$MLT^{-1} = M^yL^{2y+z}T^{-x-y-z} .$$

From the exponents of M, L, and T on the left and on the right are obtain the following conditions on x, y, z.

M:	$1 = y$	$\Rightarrow$	$y = 1$
L:	$1 = 2y + z$	$\Rightarrow$	$z = 1 - 2y = -1$
T:	$-1 = -x - y - z$	$\Rightarrow$	$x = 1 - y - z = 1$

Hence,

$$p \propto \nu h c^{-1} .$$

14. In what manner does (a) the radius and (b) the energy of a Bohr or for hydrogen depend upon the electronic charge e, the electronic ma m_e, and Planck's constant h?

THE BOHR ATOM

A circular electronic orbit about a massive nucleus is stable according classical mechanics if the electron's velocity is such as to make the centrifug force m_ev^2/r just balance the coulombic attraction between the nucleus ar the electron.

$$\frac{m_ev^2}{r} = \frac{(Ze)(e)}{r^2}$$

Z is the atomic number of the nucleus.

15. What happens to v as r increases?
16. What happens to v as Z increases (r constant)?

With respect to their potential energy at infinite separation the potenti energy of the nucleus and electron at separation r is $-Ze^2/r$.*

17. What happens to the potential energy as r approaches (a) infinity (zero?
18. Show that the kinetic energy of an electron in a classically stable orb is minus one-half its potential energy.

* $V(r) - V(\infty) = -\int_\infty^r F dr = \int_\infty^r \frac{Ze^2}{r^2} dr = -\frac{Ze^2}{r}\Big|_\infty^r = -\frac{Ze^2}{r}$.

19. Express the total energy of an electron in a classically stable orbit in terms of (a) its kinetic energy (b) its potential energy.
20. What is the potential energy of an electron in a 1*s* orbit of hydrogen? What is its total energy? The radius of the 1*s* orbit is 0.529 Å. The value to use for the charge of the electron when all other quantities are expressed in the cgs system of units is 4.803×10^{-10}.
21. If a circular orbit is to be stable in a classical sense, what values may r have?

According to Bohr, a circular electronic orbit is stable only if the electron's velocity is such as to make its de Broglie wave length fit nicely around the orbit; i.e. the orbit is "allowed" only if the velocity is such that

$$n\lambda = n\frac{h}{m_e v} = 2\pi r \qquad n = 1, 2, 3, \ldots.$$

22. Express, in terms of n, h, Z, m_e, and e, the values of v and r that satisfy both Bohr's condition and the classical condition for stable orbits. What are the corresponding values of E?
23. What happens to v, r, and E as Z increases, n constant? As n increases, Z constant?
24. What are the values of v and r when $n = Z = 1$?
25. When an electron in a hydrogen atom falls spontaneously from the level $n = 2$ to the level $n = 1$, its kinetic energy increases, its potential energy decreases by twice the amount the kinetic energy increases, and the total energy decreases. What happens to this energy?
26. What is the wave length of a photon whose energy is equal to the difference in energy between the ground and first excited states of a hydrogen atom?

Epilogue

FEYNMAN'S REMARKS ON QUANTUM BEHAVIOR

Coming to grips with a new idea—the theory of evolution, for example, or he quantum behavior of small particles—is not easy. Darwin's theory of volution confounded many of his contemporaries; some people are disurbed by it still. Recall, too (Chapter 1), the astonishment and incredulity vith which William Thomson greeted Joule's experiments, which today are amiliar to many schoolboys.

The problem of adapting to new ideas is a psychological problem. Merely aming the problem, however, does not solve the problem. Quantum theory strange. The reason for this is very simple. In Richard Feynman's words:

"Quantum mechanics" is the description of the behavior of matter in all i details and, in particular, of the happenings on an atomic scale. Things on a ve small scale behave like nothing that you have any direct experience about. They not behave like waves, they do not behave like particles, they do not behave li clouds, or billiard balls, or weights on springs, or like anything that you have ev seen. . . .

Because atomic behavior is so unlike ordinary experience, it is very difficult get used to and it appears peculiar and mysterious to everyone, both to the novi and to the experienced physicist. Even the experts do not understand it the w they would like to, and it is perfectly reasonable that they should not, because of direct, human experience and of human intuition applies to large objects. W know how large objects will act, but things on a small scale just do not act that wa So we have to learn about them in a sort of abstract or imaginative fashion and n by connection with our direct experience.

As Boltzmann has said, "How awkward is the human mind in divinin the nature of things, when forsaken by the analogy of what we see and tou directly."

AN INTERVIEW WITH PETER DEBYE

The origin of quantum mechanics was one of the topics of an interview 1964 by three Cornell scientists, Salpeter, Bauer, and Corson, with Pet Debye, Nobel prize winner in chemistry in 1936. Excerpts from this intervie are given below.*

SALPETER: You were in the middle, so to speak, of the early developments quantum mechanics. I am curious what your reaction was at that time.

DEBYE: By early developments you mean, of course, the proposals of Heisenbe and Schrödinger. Well, I think of "early" as the period 25 years before that wh the whole thing started with a kind of interpolation formula by Planck. Nobo wanted to accept it then because it did not appear logical. The only man wh appeared sensible was Einstein. He had the feeling that if there was anything Planck's idea then it should also appear in other parts of physics. Well, at that tim he talked about the photoeffect, specific heats, and so forth. Then I tried to form late the theory of specific heats in a more general way. Also about that time Plan introduced the zero point energy, perhaps around 1910 or '11, because he was n content with his original derivation of Wien's Radiation Law.

BAUER: This was a patchwork operation?

DEBYE: Yes. It was patchwork the whole time; many were trying to make t

* Dale R. Corson, *et al.*, "Peter J. W. Debye: An Interview," *Science*, 7 August 196 Vol. 145, pp. 554–9. Copyright 1964 by the American Association for the Advancement Science.

formulation a bit more general. Then de Broglie published his paper. At that time Schrödinger was my successor at the University of Zurich, and I was at the Technical University, which is a federal institute, and we had a colloquium together. We were talking about de Broglie's theory and agreed that we didn't understand it, and that we should really think about his formulations and what they mean. So I asked Schrödinger to give us a colloquium. And the preparation of that really got him started.

CORSON: What was the date?

DEBYE: Oh, I don't know—between 1924 and 1927. It was in the same year that he published his paper, because there were only a few months between his talk and his publication. Of course there was also Heisenberg's theory. Personally, I liked Schrödinger's formulation better, because I was more familiar with differential equations than with matrices. Very soon we saw that one followed from the other.

SALPETER: You say that when Schrödinger gave his talk it seemed to culminate directly in a theory?

DEBYE: It was all prepared really—by all the discussion which had been going on for years—and it was only a question of mathematical formulation of the ideas which were around. You should not forget that there was 25 years of discussion in back of these theories: and it was not merely personal discussions—it was a discussion that went on all over Europe. The atmosphere was full of questions to which every physicist was addressing himself. It was the main topic of conversation between physicists, even between those who were strictly experimentalists.

Debye's comments nicely illustrate the modern historical view of science as it has been described, for example, by E. G. Boring, with whose observations—slightly paraphrased—we conclude the account given in this book of the personalities of science and the evolution of scientific thought.

EPONYMITY*

The conventional view of the history of science is that science advances gradually by the hard work of many investigators but that its course involves sudden spurts when someone, who is eventually to become known as a "Great Man," has a revolutionary insight or makes a crucial discovery which changes the speed or direction of progress in scientific endeavor.

Even while we admit that some men make habitually greater contributions to knowledge than others and note that feeble-mindedness has done little to advance science, careful consideration must lead to the conclusion that eponymity—the Great-Man theory of history—is mostly a delusion. The course of science is gradual and continuous, as the occurrence of multiples in discovery and invention proves over and over again.

* Edwin G. Boring, "Cognitive Dissonance: Its Use in Science," *Science*, 14 August 1964, Vol. 145, pp. 680–85. Copyright 1964 by the American Association for the Advancement of Science.

Eponymity performs a useful service, however. It packages history for easy handling, just as science itself is said to exist to promote economy of thinking.

An unusually interesting account of the ideas and personalities involved in the development of quantum mechanics has been given by E. U. Condon in a talk "60 Years of Quantum Physics" published in *Physics Today*, October 1962, Vol. 15, pp. 37–47.

25

Configurational Entropy

$$(S_{cf})_i = -R \ln N_i$$

The discussion of entropy to this point in Part III has been concerned with entropy that arises from the random distribution of thermal energy among identical particles. This entropy is called thermal entropy. All substances above absolute zero have some thermal entropy. Mixtures of substances have, in addition to their thermal entropy, an interesting and chemically important entropy contribution that arises from the random distribution in space of non-identical particles.

That mixtures do have greater entropies than their components can be seen by considering, after the fashion of Count Rumford, an ordinary occupation of life.

When cream is added to coffee, diffusion—aided by convection—produces in time a homogeneous-looking mixture. Compared to its components, this mixture is very stable. Never does it revert back of its own accord to the initial state of pure cream and black coffee.

This familiar behavior illustrates a general rule: diffusion of two miscible liquids into each other is a spontaneous, irreversible, entropy-producing process.* The origin of this entropy of mixing is pictured below.

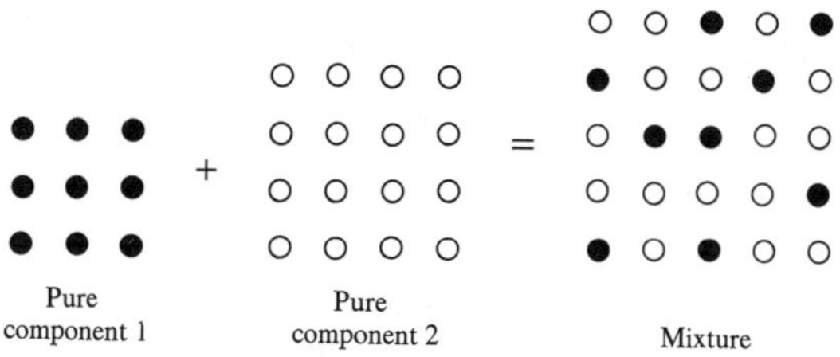

The mixture on the right has more microstates, and more entropy, than the pure components on the left. This is because an interchange with each other

* If the entropy did not increase during mixing, unmixing could occur spontaneously.

of two identical particles of the first component (●), or of the second component (○), in the pure reactants, or in the mixture, does not produce a new microstate. On the other hand, an interchange in the mixture of a particle of the first component (●) with one of the second (○) does produce a new microstate. If there are in the mixture n_1 particles of the first kind and n_2 of the second, the number of different microstates that can be produced in this fashion is

$$\frac{(n_1 + n_2)!}{n_1!\, n_2!}.$$

The natural logarithm of this number multiplied by R/N_0 is called the configurational entropy of the mixture (abbreviation: S_{cf}).

$$S_{cf} = \frac{R}{N_0} \ln \frac{(n_1 + n_2)!}{n_1!\, n_2!}$$

For a pure component n_1 or n_2 is zero and the configurational entropy vanishes.*

When a chemical reaction occurs in a mixture the number of particles of various types present in the mixture changes. Correspondingly, the configurational entropy of the mixture generally changes. Such entropy changes influence the magnitude of the total entropy change for the process. For this reason the rate at which the configurational entropy of a mixture changes as the number of particles of a particular kind in the mixture changes is an important number in the thermodynamic analysis of chemical reactions.

This number can be easily found, for, say, the first component of a two-component mixture, by calculating the configurational entropy of the mixture when it contains $n_1 + 1$ molecules of the first component and n_2 of the second—designated below $S_{cf}(n_1 + 1, n_2)$—and subtracting from this number the mixture's original configurational entropy, $S_{cf}(n_1, n_2)$. Remembering that $n! = n(n-1)(n-2)\ldots 1$, and that $\ln(a) - \ln(b) = \ln(a/b)$, one finds that

$$\begin{aligned} S_{cf}(n_1 + 1, n_2) - S_{cf}(n_1, n_2) &= \frac{R}{N_0} \ln \frac{n_1 + n_2 + 1}{n_1 + 1} \\ &= -\frac{R}{N_0} \ln \frac{n_1 + 1}{n_1 + n_2 + 1} \\ &= -\frac{R}{N_0} \ln \frac{n_1}{n_1 + n_2} \qquad (\text{if } n_1 \gg 1 \\ &= -\frac{R}{N_0} \ln N_1 . \end{aligned}$$

* $0! = 1$. $\ln 1 = 0$.

N_1 stands for the mole fraction of component 1 in the mixture. A similar expression holds for the second component. On a molar basis the contributions the first and second components make to the mixture's configurational entropy are $-R \ln N_1$ and $-R \ln N_2$. Calling these contributions $(S_{cf})_1$ and $(S_{cf})_2$, one has

$$(S_{cf})_1 = -R \ln N_1$$
$$(S_{cf})_2 = -R \ln N_2 \,.$$

Since the mole fractions N_1 and N_2 are never greater than one, the contribution a component makes to the configurational entropy of a mixture is always positive, or zero: zero when its mole fraction is one, i.e. when the mixture is not a mixture but a pure substance; and positive when its mole fraction is less than one. It is interesting to note that as the mole fraction of a component becomes very small, the component's contribution per mole to the configurational entropy of the mixture becomes very large. For this reason chemical reactions often do not go to completion. Instead states are reached in which the concentrations of reactants and products cease to change with time. These states are called states of chemical equilibrium. For such states there exists an interesting and very useful relation, to which we now turn in the early chapters of Part IV, between certain entropy changes that occur in the universe during a chemical reaction and the concentrations of the reactants and products of the chemical reaction.

IV

Applications

26

Equilibrium Constant Expressions (I)

$$K_{eq} = e^{-\Delta G^0/RT}$$

When a system is at equilibrium with respect to a chemical reaction, the changes produced by the reaction in the configurational entropy of the system, in the thermal entropy of the system, and in the thermal entropy of the surroundings of the system, add to zero:

$$\Delta S_{cf} + (\Delta S_{th})_\sigma + \Delta S_{surr} = 0.$$

For the sum of these three changes represents the change in the total entropy of the universe. And at equilibrium $\Delta S_{\text{total}} = 0$.

The three terms in the above sum may be considered separately.

1. ΔS_{surr}. The change in entropy of the thermal surroundings is equal to the change in the thermal energy of the surroundings divided by the absolute temperature.

$$\Delta S_{surr} = \frac{\Delta E_{surr}}{T}.$$

For reactions occurring at constant pressure,

$$\Delta E_{surr} = -(\Delta E_\sigma + P\Delta V_\sigma) \qquad \text{(see Chapter 14)}$$

$$= -\Delta H_\sigma. \qquad (P_\sigma \text{ constant})$$

Therefore, $$\Delta S_{surr} = -\frac{\Delta H_\sigma}{T}.$$

2. $(\Delta S_{th})_\sigma$. The change in the thermal entropy of the chemical system is equal to simply the difference between the thermal entropies of the products and the reactants.

3. ΔS_{cf}. The change in the configurational entropy of the chemical system is equal to the difference between the configurational entropies of the products and the reactants. For the reaction $HAc + H_2O = H_3O^+ + Ac^-$, for example,

$$\begin{aligned}\Delta S_{cf} &= [(S_{cf})_{H_3O^+} + (S_{cf})_{Ac^-}] - [(S_{cf})_{HAc} + (S_{cf})_{H_2O}] \\ &= [-R \ln N_{H_3O^+} - R \ln N_{Ac^-}] - [-R \ln N_{HAc} - R \ln N_{H_2O}] \\ &\qquad\qquad\qquad\qquad\qquad\qquad\qquad\qquad\qquad\qquad \text{(Chapter 25)} \\ &= -R \ln \frac{N_{H_3O^+} N_{Ac^-}}{N_{HAc} N_{H_2O}}.\end{aligned}$$

A system is at equilibrium with respect to the transfer of protons between acetate ions and water molecules, therefore, when

$$-R \ln \frac{N_{H_3O^+} N_{Ac^-}}{N_{HAc} N_{H_2O}} + (\Delta S_{th})_\sigma - \frac{\Delta H_\sigma}{T} = 0.$$

In dilute solutions the mole fractions of the solutes H_3O^+, Ac^-, and HAc are proportional to their molar concentrations (H_3O^+), (Ac^-), and (HAc). Absorbing R times the logarithms of the proportionality constants into $(\Delta S_{th})_\sigma$, and calling this new entropy term $\Delta S^0{}_\sigma$,* one finds that $\Delta S_{\text{total}} = 0$ for the reaction $HAc + H_2O = H_3O^+ + Ac^-$ when

$$-R \ln \frac{(H_3O^+)(Ac^-)}{(HAc)N_{H_2O}} + \Delta S^0{}_\sigma - \frac{\Delta H_\sigma}{T} = 0. \qquad (\dagger)$$

Setting

$$\left[\frac{(H_3O^+)(Ac^-)}{(HAc)N_{H_2O}}\right]_{\substack{\text{at equilibrium} \\ (\Delta S_{\text{total}}=0)}} = K_{eq}\,, \qquad (\ddagger)$$

and solving for K_{eq}, the "equilibrium constant" of the reaction, one finds that

$$K_{eq} = e^{\Delta S^0{}_\sigma/R} \cdot e^{-\Delta H_\sigma/RT}\,.$$

Introducing the so-called "standard free energy change"

$$\Delta G^0 = \Delta H_\sigma - T\Delta S^0{}_\sigma\,, \qquad (^{**})$$

* The numerical implications of this maneuver have been discussed in an article "Configurational Entropy and Choice of Standard States, Entropies of Formation of Complex Ions, and the Chelate Effect," H. A. Bent, *J. Phys. Chem.*, Vol. 60, p. 123, 1956.

† The $\Delta S^0{}_\sigma$ appearing in this equation is not precisely the quantity calculated from the S^0 values listed in Appendix I. The difference, however, is relatively small and, for our present purposes, unimportant. $\Delta S^0{}_\sigma$ is very nearly equal to the entropy change that would occur in the chemical system σ in the proton transfer reaction $HAc + H_2O = H_3O^+ + Ac^-$ if all solute species, HAc, H_3O^+, and Ac^-, were present at concentrations of 1 mole/liter. Such a solution (see later) would be far from an equilibrium state.

‡ In dilute solutions the mole fraction of the solvent, N_{H_2O}, is nearly 1.

** Strictly speaking $\Delta G^0 = \Delta H^0{}_\sigma - T\Delta S^0{}_\sigma$, where $\Delta H^0{}_\sigma$ is (very nearly) the difference between the enthalpies of the products and the reactants of a reaction when all solute concentrations are 1 mole/liter. In the range of dilute solutions ΔH_σ, unlike ΔS_σ, generally varies slowly, if at all, with concentration; i.e. in dilute solutions generally $\Delta H_\sigma \approx \Delta H^0{}_\sigma$. In so-called "ideal solutions" $\Delta H_\sigma = \Delta H^0{}_\sigma$.

one may write

$$K_{eq} = e^{-\Delta G^0/RT}$$
$$= 10^{-\Delta G^0/2.303RT}.$$

For the ionization of acetic acid in water at 25°C, $\Delta G^0 = 6{,}490$ cal per mole of acetic acid consumed. At this temperature, therefore, and for this reaction,

$$K_{eq} = 10^{-\frac{6{,}490 \text{ cal/mole}}{2.303(1.987 \text{ cal/deg-mole})(298 \text{ deg})}}$$
$$= 10^{-4.76}$$
$$= 1.8 \times 10^{-5}.$$

Problems for Chapter 26

For an exothermic reaction ΔE_{surr} is———(a). By the ———(b) Law, therefore, ΔH_σ is ———(c). Thus, for an exothermic reaction the exponent $-\Delta H_\sigma/RT$ in the expression $K_{eq} =$ ———(d) is ———(e). Consequently, as the temperature increases the algebraic value of the exponent $-\Delta H_\sigma/RT$ for an exothermic reaction ———(f) and, correspondingly, the value of the equilibrium constant ———(g), in agreement with the ———(h).

27

Equilibrium Constant Expressions (II)

$$\Delta H_\sigma = \frac{R \ln \dfrac{K_{eq}(T_2)}{K_{eq}(T_1)}}{\left(\dfrac{1}{T_1} - \dfrac{1}{T_2}\right)}$$

If the value of an ionization constant at one temperature is known, the value of the corresponding standard free energy change at that temperature can be calculated from the equation

$$\Delta G^0 = -RT \ln K_{eq} .$$

If the value of the ionization constant at two different temperatures is known, the standard free energy change at both temperatures can be calculated from the equations

$$\Delta G^0(T_1) = -RT_1 \ln K_{eq}(T_1)$$
$$\Delta G^0(T_2) = -RT_2 \ln K_{eq}(T_2).$$

And if the value of the standard free energy at two temperatures is known, the change in the enthalpy of the chemical system can be calculated. For at any temperature

$$\Delta G^0(T) = \Delta H_\sigma(T) - T \cdot \Delta S^0{}_\sigma(T).$$

And if $\Delta H_\sigma(T)$ and $\Delta S^0{}_\sigma(T)$ in this expression do not vary greatly with temperature over the interval T_1 to T_2, i.e. if

$$\Delta H_\sigma(T_1) \approx \Delta H_\sigma(T_2) \qquad \text{and} \qquad \Delta S^0{}_\sigma(T_1) \approx \Delta S^0{}_\sigma(T_2),$$

one may write that

$$\begin{aligned} \Delta G^0(T_1) &\equiv \Delta H_\sigma(T_1) - T_1 \cdot \Delta S^0{}_\sigma(T_1) \\ &\approx \Delta H_\sigma - T_1 \cdot \Delta S^0{}_\sigma \\ \Delta G^0(T_2) &\equiv \Delta H_\sigma(T_2) - T_2 \cdot \Delta S^0{}_\sigma(T_2) \\ &\approx \Delta H_\sigma - T_2 \cdot \Delta S^0{}_\sigma , \end{aligned}$$

where ΔH_σ and ΔS^0_σ in the above expressions for $\Delta G^0(T_1)$ and $\Delta G^0(T_2)$ represent the average values of $\Delta H_\sigma(T)$ and $\Delta S^0_\sigma(T)$ over the temperature interval T_1 to T_2. Solving for ΔH_σ, one finds that

$$\Delta H_\sigma = \frac{\left[\dfrac{\Delta G^0(T_1)}{T_1} - \dfrac{\Delta G^0(T_2)}{T_2}\right]}{\left(\dfrac{1}{T_1} - \dfrac{1}{T_2}\right)}$$

$$= \frac{-R \ln K_{eq}(T_1) + R \ln K_{eq}(T_2)}{\left(\dfrac{1}{T_1} - \dfrac{1}{T_2}\right)}$$

$$= \left[R \ln \frac{K_{eq}(T_2)}{K_{eq}(T_1)}\right] \Big/ \left(\frac{1}{T_1} - \frac{1}{T_2}\right).$$

For the self-ionization of water, for example, K_{eq} is 1.008×10^{-14} at 25°C and 2.087×10^{-14} at 35°C. Therefore, for the reaction $2H_2O = H_3O^+ + OH^-$,

$$\Delta H_\sigma = \left[R \ln \frac{2.087 \times 10^{-14}}{1.008 \times 10^{-14}}\right] \Big/ \left(\frac{1}{298.15°\text{K}} - \frac{1}{308.15°\text{K}}\right)$$

$$= +13{,}400 \text{ cal/mole.}$$

The reaction $2H_2O = H_3O^+ + OH^-$ absorbs from its thermal surroundings 13,400 calories per mole of H_3O^+ produced. The reverse reaction, the neutralization of H_3O^+ by OH^-, liberates to the thermal surroundings 13,400 calories per mole of H_3O^+ consumed.

Problems for Chapter 27

Suppose that T_2 is greater than T_1. Then $1/T_1$ is ———(a) than $1/T_2$ and the term $(1/T_1 - 1/T_2)$ is ———(b). If an ionization constant increases with temperature, $K_{eq}(T_2)$ will be ———(c) than $K_{eq}(T_1)$. Hence, in this case the term

$$R \ln \frac{K_{eq}(T_2)}{K_{eq}(T_1)}$$

is ———(d). Consequently, by the formula $\Delta H_\sigma =$ ———(e), ΔH_σ must be ———(f). One concludes from the equation for ΔH_σ, therefore, that a reaction whose ionization constant increases with temperature is an ———(g) reaction, in agreement with ———(h).

28

Equilibrium Constant Expressions (III)

Five Illustrative Examples

The equilibrium constant expressions obtained in Chapters 14 and 26 for gaseous and condensed phases, respectively, are of the same form. In each instance a free energy change is related to the logarithm of an equilibrium constant. In one instance the equilibrium constant was expressed in terms of partial pressures (Chapter 14), in the other, in terms of mole fractions or molar concentrations (Chapter 26). Introduction of a concentration term called the "activity," with certain associated conventions, permits the previous results to be summarized in a single expression.

The equilibrium constant for the reaction of a moles of a substance A with b moles of B to produce c moles of C and d moles of D, according to the equation

$$aA + bB = cC + dD,$$

may be written in the form

$$K_{eq} = \frac{a_C{}^c \cdot a_D{}^d}{a_A{}^a \cdot a_B{}^b}.$$

The symbols a_A, a_B, a_C, a_D in this expression for K_{eq} stand for the "activities" of substances A, B, C, D, respectively. Rules for expressing these activities are (for dilute solutions):*

$a_A = N_A$ if A is the solvent (for pure solvents, solid or liquid, $a_A = 1$);
$= C_A$ if A is a solute (concentrations to be expressed in moles/1000 gm solvent; or, what is nearly the same thing for dilute aqueous solutions, in moles/liter of solution);
$= P_A$ if A is a gas (pressures, or partial pressures, to be expressed in atmospheres).

* The specific forms of these rules are determined by the conventions adopted in tabulations of standard free energies, enthalpies, and entropies of chemical substances. If the reference state for an ideal gas, for example, is altered from $P^0 = 1$ atmosphere to $P^0 = 0.1$ atmosphere, the activity rule for ideal gases must be altered correspondingly, pressures, or partial pressures, now being expressed not in atmospheres but deciatmospheres.

Below are five examples of how these activity rules may be combined with the equation $K_{eq} = e^{-\Delta G^0/RT}$ and the data of Appendix I to obtain useful information about chemical systems. Calculations are made for the temperature 25°C. At this temperature

$$e^{-\frac{\Delta G^0}{RT}} = 10^{-\frac{\Delta G^0}{2.303RT}} = 10^{-\frac{\Delta G^0}{1364\ \text{cal/mole}}}.$$

It may be noted that when ΔG^0 is positive the equilibrium constant for the corresponding reaction is less than unity. For negative values of ΔG^0, the equilibrium constant is greater than unity.

1. For the self-ionization of liquid water according to the equation

$$2H_2O(\text{liq}) = H_3O^+(\text{aq}) + OH^-(\text{aq}),$$

$$\begin{aligned}\Delta G^0 &= G^0_{\text{products}} - G^0_{\text{reactants}}\\ &= G^0_{H_3O^+(\text{aq})} + G^0_{OH^-(\text{aq})} - 2G^0_{H_2O(\text{liq})}\\ &= (-56.6899) + (-37.595) - 2(-56.6899)\ \text{kcal/mole}\\ &= +19{,}095\ \text{cal/mole of } H_3O^+ \text{ produced,}\end{aligned}$$

and

$$\begin{aligned}K_{eq} &= \frac{a_{H_3O^+}a_{OH^-}}{a^2_{H_2O}}\\ &= \frac{(H_3O^+)(OH^-)}{N^2_{H_2O}}\\ &\approx (H_3O^+)(OH^-) \qquad N_{H_2O} \approx 1.\end{aligned}$$

Therefore, at equilibrium

$$(H_3O^+)(OH^-) \approx 10^{-\frac{19{,}095}{1364}} = 10^{-14}.$$

2. For the vaporization of liquid water according to the equation

$$H_2O(\text{liq}) = H_2O(\text{gas}),$$

$$\begin{aligned}\Delta G^0 &= G^0_{\text{products}} - G^0_{\text{reactants}}\\ &= G^0_{H_2O(\text{gas})} - G^0_{H_2O(\text{liq})}\\ &= (-54.6351) - (-56.6899)\ \text{kcal/mole}\\ &= +2054.8\ \text{cal/mole of } H_2O \text{ vaporized,}\end{aligned}$$

and

$$\begin{aligned}K_{eq} &= \frac{a_{H_2O(\text{gas})}}{a_{H_2O(\text{liq})}}\\ &= \frac{P_{H_2O(\text{gas})}}{N_{H_2O(\text{liq})}}.\end{aligned}$$

Therefore, at equilibrium

$$\frac{P_{H_2O(gas)}}{N_{H_2O(liq)}} = 10^{-\frac{2054.8}{1364}} = 10^{-1.506}$$

$$= 3.119 \times 10^{-2} \text{ atm.}$$

$$= 23.7 \text{ mm Hg;}$$

or, $$P_{H_2O(gas)} = (23.7 \text{ mm Hg})\, N_{H_2O(liq)} \cdot$$

When $N_{H_2O(liq)} = 1$, $P_{H_2O(gas)} = 23.7$ mm Hg.* Calling this value $P^0{}_{H_2O}$, one may write

$$P_{H_2O} = P^0{}_{H_2O} N_{H_2O}.$$

This is Raoult's Law.

3. For the dissociation of water vapor according to the equation

$$2H_2O(gas) = 2H_2(gas) + O_2(gas),$$

$$\begin{aligned}\Delta G^0 &= G^0{}_{\text{products}} - G^0{}_{\text{reactants}}\\ &= [2G^0{}_{H_2(gas)} + G^0{}_{O_2(gas)}] - [2G^0{}_{H_2O(gas)}]\\ &= [2(0.0000) + 1(0.0000)] - [2(-54.6351)] \text{ kcal/mole}\\ &= +109{,}270 \text{ cal/mole of } O_2 \text{ produced,}\end{aligned}$$

and

$$K_{eq} = \frac{a^2{}_{H_2(gas)} a_{O_2(gas)}}{a^2{}_{H_2O(gas)}} = \frac{P^2{}_{H_2} P_{O_2}}{P^2{}_{H_2O}}.$$

Therefore, at equilibrium

$$\frac{P^2{}_{H_2} P_{O_2}}{P^2{}_{H_2O}} = 10^{-\frac{109{,}270}{1364}} = 10^{-80.2}.$$

In a gas saturated with water vapor at 25°C, $P_{H_2O} = 10^{-1.506}$ atm. (example 2). If, also, $P_{H_2} = 2P_{O_2}$, then at equilibrium

$$\begin{aligned}4P^3{}_{O_2} &= 10^{-80.2} \times (10^{-1.506})^2 = 10^{-83.2}\\ P_{O_2} &= (10^{-83.8})^{1/3}\\ &= 10^{-27.9} \text{ atm.}\end{aligned}$$

There is little dissociation of water vapor at 25°C.

* This is the vapor pressure of pure water at 25°C.

4. For the solution of lead chloride in water to produce aquated lead ions and chloride ions according to the equation

$$PbCl_2(c) = Pb^{++}(aq) + 2Cl^-(aq)$$

$$\begin{aligned}\Delta G^0 &= G^0_{\text{products}} - G^0_{\text{reactants}}\\ &= [G^0_{Pb^{++}(aq)} + 2G^0_{Cl^-(aq)}] - [G^0_{PbCl_2(c)}]\\ &= [(-5.81) + 2(-31.35)] - [-75.04] \text{ kcal/mole}\\ &= +6{,}530 \text{ cal/mole of } PbCl_2 \text{ consumed,}\end{aligned}$$

and

$$K_{eq} = \frac{a_{Pb^{++}(aq)}a^2_{Cl^-(aq)}}{a_{PbCl_2(c)}} = \frac{(Pb^{++})(Cl^-)^2}{1}.$$

Therefore, at equilibrium

$$(Pb^{++})(Cl^-)^2 = 10^{-\frac{6530}{1364}} = 10^{-4.79} = 1.6 \times 10^{-5} \text{ (moles/liter)}^3.$$

If the chloride ion concentration in a solution that is saturated with respect to lead chloride is 0.2 molar,

$$(Pb^{++}) = \frac{1.6 \times 10^{-5} \text{ (moles/liter)}^3}{(0.2 \text{ moles/liter})^2} = 4 \times 10^{-4} \text{ moles/liter.*}$$

5. For the reaction of metallic zinc with a strong mineral acid to produce hydrogen gas and aquated zinc ions according to the equation

$$Zn(c) + 2H_3O^+(aq) = H_2(gas) + Zn^{++}(aq) + 2H_2O(aq), \qquad (\dagger)$$

$$\begin{aligned}\Delta G^0 &= G^0_{\text{products}} - G^0_{\text{reactants}}\\ &= [G^0_{H_2(gas)} + G^0_{Zn^{++}(aq)} + 2G^0_{H_2O(aq)}] - [G^0_{Zn(c)} + 2G^0_{H_3O^+(aq)}]\\ &= [(0.000) + (-35.184) + 2(-56.6899)] - [(0.000 + 2(-56.6899)] \text{ kcal/mole}\\ &= -35{,}184 \text{ cal/mole of Zn consumed,}\end{aligned}$$

and

$$K_{eq} = \frac{a_{H_2(gas)}a_{Zn^{++}(aq)}a^2_{H_2O(aq)}}{a_{Zn(c)}a^2_{H_3O^+(aq)}} = \frac{P_{H_2}(Zn^{++})N^2_{H_2O}}{1 \cdot (H_3O^+)^2}.$$

* Also present in a lead chloride-saturated solution 0.2 molar in Cl^- will be ionic species such as $PbCl^+$. The figure 4×10^{-4} moles/liter does not, therefore, represent the full solubility of lead chloride in such solutions.

† Complex ions may also be produced in which one (or more) of the co-ordination sites about the zinc ion is occupied by an anion of the acid.

Therefore, at equilibrium

$$\frac{P_{H_2}(Zn^{++})N^2_{H_2O}}{(H_3O^+)^2} = 10^{\frac{35,184}{1364}} = 10^{25.8} \text{ atm/(mole/liter)}.$$

From this one sees that the reaction as written proceeds essentially to completion, unless the pressure of the hydrogen is exceedingly high or, which is more likely, unless the concentration of the hydronium ion is very small. If $(H_3O^+) = 10^{-14}$ moles/liter, and if $P_{H_2} = 1$ atm. and $N_{H_2O} \approx 1$, $(Zn^{++}) \approx 10^{-2.2}$ moles/liter; however, in solutions this basic (when $(H_3O^+) = 10^{-14}$ moles/liter, $(OH^-) = 1$ mole/liter) the aquated zinc ion is not one of the more important zinc species.

Problems for Chapter 28

1. If the molar entropy of a substance may be written in the form

$$\mathsf{S} = \mathsf{S}^0 - R \ln a,$$

where, when the substance behaves as an ideal gas, the activity a is equal to the substance's partial pressure (Chapter 12) and when the substance exists in a condensed phase, solid or liquid, the activity a is equal to the substance's mole fraction—or, at low concentrations, to its molar (or molal) concentration (Chapter 25)—and where the term S^0 is independent of the concentration, but not the temperature, being essentially equal to the substance's thermal entropy; and if, in addition, the substance's enthalpy, like its thermal entropy, is independent of its concentration, show that the substance's molar free energy may be written in the form

$$\mathsf{G} = \mathsf{G}^0 + RT \ln a,$$

where G^0—the substance's "standard molar free energy" (values for which are tabulated in Appendix I)—is a function of the temperature, but not the concentration.

2. Derive the equilibrium constant expression

$$RT \ln \left[\frac{a_C{}^c \cdot a_D{}^d}{a_A{}^a \cdot a_B{}^b}\right]_{\text{At equil.}} = -\Delta G^0$$

for the reaction

$$aA + bB = cC + dD$$

from the relation (thermodynamically equivalent to the statement $\Delta S_{\text{total}} = 0$)

$$aG_A + bG_B = cG_C + dG_D$$

and the expansion (problem 1)

$$G_X = G^0{}_X + RT \ln a_X \qquad (X = A, B, C, D).$$

29

Electron-Transfer Reactions and Galvanic Cells (I)

$$n\mathcal{F}\mathcal{E}_{\max} = -\Delta G_\sigma$$

When hydrogen and oxygen react spontaneously to form liquid water at 25°C according to the equation

$$H_2(\text{gas, 1 atm.}) + \tfrac{1}{2}O_2(\text{gas, 1 atm.}) = H_2O(\text{liq}) + 68{,}300 \text{ cal},$$

the entropy of the reaction mixture decreases

$$\begin{aligned}\Delta S_\sigma &= S^{\text{liq}}_{H_2O} - (S^{\text{gas}}_{H_2} + \tfrac{1}{2}S^{\text{gas}}_{O_2})\\ &= [16.716 - (31.211 + \tfrac{1}{2} \times 49.003)]\,\frac{\text{cal}}{\text{deg}}\\ &= -39.0\,\frac{\text{cal}}{\text{deg}}\end{aligned}$$

the entropy of the thermal surroundings increases

$$\begin{aligned}\Delta S_\theta &= \frac{\Delta E_\theta}{T}\\ &= \frac{+68{,}300 \text{ cal}}{298°\text{K}}\\ &= +229\,\frac{\text{cal}}{\text{deg}}\end{aligned}$$

and the entropy of the universe increases.

$$\begin{aligned}\Delta S_{\text{total}} &= \Delta S_\sigma + \Delta S_\theta\\ &= [-39.0 + 229]\,\frac{\text{cal}}{\text{deg}}\\ &= +190\,\frac{\text{cal}}{\text{deg}}\,.\end{aligned}$$

By the First Law the 68,300 calories of energy liberated to the thermal surroundings in the reaction of hydrogen and oxygen is the difference between the enthalpies of the reactants and the products of the reaction.

$$\begin{aligned} 68{,}300 \text{ cal} &= \Delta E_\theta = -(\Delta E_\sigma + P\Delta V_\sigma) \\ &= -\Delta(E_\sigma + PV_\sigma) \qquad (P \text{ constant}) \\ &= -\Delta H_\sigma \\ &= H_{\text{reactants}} - H_{\text{products}} \,. \end{aligned}$$

Not all 68,300 calories need be delivered to the thermal surroundings, however. To compensate for the decrease in entropy of the reaction mixture, the entropy of the thermal surroundings need not increase by +190 cal/deg; an increase of +39 cal/deg would be sufficient. Since $\Delta E_\theta = T\Delta S_\theta$, the modest entropy increment +39 e.u. in S_θ requires at 298°K that of the 68,300 calories liberated by the reaction mixture, only

$$\begin{aligned} \Delta E_\theta &= (298°\text{K}) \left(+39.0 \frac{\text{cal}}{\text{deg}} \right) \\ &= +11{,}600 \text{ cal} \end{aligned}$$

need be delivered to the thermal surroundings to avoid a violation of the Second Law. The remainder of the energy,

$$(68{,}300 - 298 \times 39.0) \text{ cal} = 56{,}700 \text{ cal}$$

could be used in other ways. It could in principle be converted entirely into mechanical work—the lifting of a weight, for example. Call this work W; then, for each mole of water produced,

$$W_{\text{max}} = 56{,}700 \text{ cal.}$$

In tracing back the origin of the number 56,700, one discovers that the maximum work available in the reaction has a simple interpretation: it is equal to the decrease in free energy of the chemical system σ.

$$\begin{aligned} W_{\text{max}} &= 68{,}300 - 298(39) \\ &= -\Delta H_\sigma - T(-\Delta S_\sigma) \\ &= -\Delta G_\sigma \,. \qquad (*) \end{aligned}$$

To obtain this work, one can in principle use an electrochemical cell that has for reversible electrode reactions the reactions

$$H_2(\text{gas}) = 2H^+(\text{aq}) + 2e$$

$$\tfrac{1}{2}O_2(\text{gas}) + 2H^+(\text{aq}) + 2e = H_2O(\text{liq}).$$

* T and P constant.

The sum of these two reactions—the net change in the cell—is

$$H_2(\text{gas}) + (1/2)O_2(\text{gas}) = H_2O(\text{liq}).$$

For each mole of water produced in such a cell, two faradays of electricity pass through the external circuit.* If the voltage generated by the electrode reactions at the cell terminals is $\mathscr{E}$, this charge of 2 faradays could do an amount of work

$$W = 2\mathscr{F}\mathscr{E}. \qquad (\dagger)$$

Evidently for this cell

$$\begin{aligned}\mathscr{E}_{\max} &= \frac{W_{\max}}{2\mathscr{F}} \\ &= \frac{-\Delta G_\sigma}{2\mathscr{F}} \\ &= \frac{56{,}700 \text{ cal}}{2 \times 96{,}500 \text{ coul.}} \times \frac{4.184 \text{ joules}}{1 \text{ cal}} \times \frac{1 \text{ volt}}{1 \text{ joule/coul.}} \\ &= 1.23 \text{ volts}. \qquad (\ddagger)\end{aligned}$$

This is the maximum voltage that can be produced in a galvanic cell by the reaction $H_2 + (1/2)O_2 = H_2O$.

The number 1.23 has this significance also: It is the minimum voltage that will electrolyze water at 25°C. For to decompose water according to the equation

$$H_2O(\text{liq}) = H_2(\text{gas}) + (1/2)O_2(\text{gas})$$

requires 68,300 calories of energy. This is the amount the enthalpy of the products exceeds the enthalpy of the reactant. Since the reaction increases the entropy of the chemical system by +39 cal/deg for every mole of hydrogen produced, part of the 68,300 cal could be extracted from the thermal surroundings: to be exact, the thermal surroundings could supply without violating the Second Law up to 298 × 39 = 11,600 calories of energy for every mole of hydrogen produced. The remainder of the energy, 68,300 − 11,600 = 56,700 calories, must be obtained in some other fashion. If an electrical source is used, energy must be supplied at the minimum rate of 56,700 calories for the passage of every 2 faradays of electricity, i.e. at the rate of 1.23 joules per coulomb.

* 1 faraday, $\mathscr{F}$ = 1 mole of electrons = 96,500 coulombs of electricity.

† If W is expressed in calories and $\mathscr{E}$ in volts, $\mathscr{F}$ must include a factor for converting from joules to calories.

‡ This is the maximum cell voltage when P_{H_2} and P_{O_2} are 1 atm.

To recapitulate, work may or may not be produced during the spontaneou reaction $H_2 + (1/2)O_2 = H_2O$. It makes no difference to ΔH_σ and ΔS_σ their values are completely determined once the initial and final states of —the reaction mixture—have been specified. What is different in the tw cases is the manner in which the energy lost by the chemical system, and th atmosphere, is partitioned between the system's thermal surroundings an a weight. In the first case discussed above all 68,300 calories of ΔH_σ wer to the thermal surroundings; none went to the weight. In the second cas the thermal surroundings received only 11,600 calories; the weight receive the rest. In both cases, $\Delta H_\sigma = -68{,}300$ cal and $\Delta S_\sigma = -39$ cal/deg.

30

Electron-Transfer Reactions and Galvanic Cells (II)

Another Point of View

In principle, work can be obtained from any spontaneous reaction.* This work may be described in the general case by supposing that the change

$$\text{Reactants} = \text{Products} + \text{Work}$$

occurs in two simultaneous steps: a spontaneous, entropy-producing step

$$\text{Reactants} = \text{Products} + \text{No Work} \qquad \text{(Step 1)}$$

and a by-itself-impossible, entropy-diminishing step

$$\text{Complete conversion of thermal energy to work.} \qquad \text{(Step 2)}$$

Steps 1 and 2 alter the entropy of the universe in the following manner.

$$\begin{aligned} \Delta S_{\text{total}}\,(\text{Step 1}) &= \Delta S_\sigma + \Delta S_\theta \\ &= \Delta S_\sigma - \frac{\Delta H_\sigma}{T} \qquad \begin{pmatrix} W = 0 \\ P \text{ constant} \end{pmatrix} \\ &= -\frac{\Delta G_\sigma}{T}. \qquad (P, T \text{ constant}) \end{aligned}$$

$$\Delta S_{\text{total}}\,(\text{Step 2}) = -\frac{W}{T}.$$

W is the energy removed from a thermal reservoir at temperature T and converted completely to work. By itself this is impossible. Step 2 can occur only when coupled to an entropy-producing change, such as Step 1—and then only if W and T are such that

$$\Delta S_{\text{total}}\,(\text{Step 1}) + \Delta S_{\text{total}}\,(\text{Step 2}) = -\frac{\Delta G_\sigma}{T} - \frac{W}{T} \geq 0.$$

* The reader may have witnessed the explosion of hydrogen and oxygen in a vertical stoppered iron pipe. The rate of increase of the potential energy of the stopper is impressive. Less rapid, but in principle more efficient in extracting work from the reaction $H_2 + (1/2)O_2 = H_2O$, is the electrochemical cell previously described.

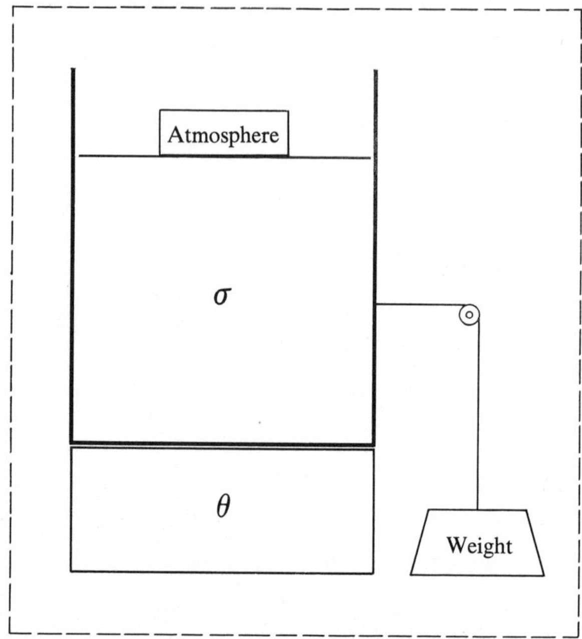

Fig. 30.1 The universe of a reaction that occurs at constant pressure with the production or consumption of useful work.

The over-all operation will be possible, therefore, only if

$$W \leq -\Delta G_\sigma \,.$$

It follows that

$$W_{\text{max}} = -\Delta G_\sigma \,.$$

Still another derivation of this relation may be offered.*

About any natural process it may be stated that

1. $\Delta E_{\text{total}} = 0$
2. $\Delta S_{\text{total}} \geq 0.$

Applied to a universe composed of a chemical system, its thermal surroundings, an atmosphere, and a weight, Fig. 30.1, these statements yield the following expressions, with the usual substitutions indicated.

$$1. \quad \Delta E_\sigma + \Delta E_\theta + \underset{\substack{\| \\ P\Delta V_\sigma}}{\Delta E_{\text{atm.}}} + \underset{\substack{\| \\ W}}{\Delta E_{\text{Wt.}}} = 0$$

* Learning has been described as comingling of reviews and new views.

2. $$\begin{array}{ccccccc} \Delta S_\sigma + & \Delta S_\theta & + & \Delta S_{\text{atm.}} & + & \Delta S_{\text{Wt.}} & \geq 0. \\ & \| & & \| & & \| & \\ & \dfrac{\Delta E_\theta}{T} & & 0 & & 0 & \end{array}$$

From the first expression it follows that

$$\Delta E_\theta = -(\Delta E_\sigma + P\Delta V_\sigma) - W$$
$$= -(\Delta H_\sigma + W). \qquad (P \text{ constant})$$

Substituting $-(\Delta H_\sigma + W)$ for ΔE_θ in the second expression, one finds that for any event for which Fig. 30.1 is a suitable abstraction

$$\Delta S_\sigma - \left(\frac{\Delta H_\sigma + W}{T}\right) \geq 0.$$

In other words,

$$W \leq T\Delta S_\sigma - \Delta H_\sigma = -\Delta G_\sigma . \qquad (T, P \text{ constant})$$

Therefore,

$$W_{\text{max}} = -\Delta G_\sigma . \qquad (*)$$

This expression has some interesting implications for galvanic cells.

* Said the Bellman, "What I tell you three times is true."

31

Electron-Transfer Reactions and Galvanic Cells (III)

The Nernst Equation

Consider a cell whose net cell reaction

$$aA + bB = cC + dD$$

corresponds to the passage through an external circuit of n faradays of electricity. The maximum voltage the cell can generate is given by the expression

$$n\mathscr{F}\mathscr{E}_{\text{max}} = W_{\text{max}} = -\Delta G_\sigma ;$$

i.e.

$$\mathscr{E}_{\text{max}} = -\frac{\Delta G_\sigma}{n\mathscr{F}},$$

where, as usual,

$$\begin{aligned}\Delta G_\sigma &\equiv G_{\text{products}} - G_{\text{reactants}}\\ &= (cG_C + dG_D) - (aG_A + bG_B).\end{aligned}$$

Recalling (problem 1, Chapter 28) that the molar free energies G_X ($X = A$, B, C, D) may be expressed in the form

$$G_X = G^0{}_X + RT\ln a_X,$$

where $G^0{}_X$ is chemical X's free energy in the system when X is present in the system at unit activity and a_X is X's actual activity in the system, one finds that the free energy change may be expressed in the form

$$\Delta G_\sigma = \Delta G^0{}_\sigma + RT\ln\frac{a_C{}^c a_D{}^d}{a_A{}^a a_B{}^b}$$

where

$$\Delta G^0{}_\sigma = (cG^0{}_C + dG^0{}_D) - (aG^0{}_A + bG^0{}_B). \quad (*)$$

* Standard free energies at 298°K are tabulated in Appendix I. Activities (Chapter 28) are equal to unity for pure solids and liquids, to solvent mole fractions and solute molalities for dilute solutions, and to gas partial pressures (in atmospheres) for ideal gas mixtures

Using this expression for ΔG_σ in the relation

$$\mathscr{E}_{\max} = -\Delta G_\sigma / n\mathscr{F},$$

one finds that

$$\mathscr{E}_{\max} = -\frac{\Delta G^0{}_\sigma}{n\mathscr{F}} - \frac{RT}{n\mathscr{F}} \ln \frac{a_C{}^c a_D{}^d}{a_A{}^a a_B{}^b}.$$

The leading term on the right is called the standard cell voltage, $\mathscr{E}^0$.

$$\mathscr{E}^0 \equiv -\frac{\Delta G^0{}_\sigma}{n\mathscr{F}}.$$

It is the value of $\mathscr{E}_{\max}$ when $a_A = a_B = a_C = a_D = 1$. When $T = 298°\text{K}$,

$$\frac{2.303RT}{n\mathscr{F}} = \frac{(2.303)\left(1.987 \dfrac{\text{cal}}{\text{deg-mole}}\right)(298°\text{K})}{n\left(96{,}500 \dfrac{\text{coulombs}}{\text{faraday}}\right)} \times \frac{4.184 \text{ joules}}{1 \text{ cal}} \times \frac{1 \text{ volt}}{1 \dfrac{\text{joule}}{\text{coulomb}}}$$

$$= \frac{0.0592}{n} \frac{\text{volts-faraday}}{\text{mole}}. \qquad (*)$$

Thus, for the reaction $aA + bB = cC + dD$, one may write

$$\mathscr{E}_{\max} = \mathscr{E}^0 - \frac{0.0592}{n} \log \frac{a_C{}^c a_D{}^d}{a_A{}^a a_B{}^b}. \qquad (T = 298°\text{K})$$

This equation is known as the Nernst equation. It quantitatively expresses the dependence of the maximum cell voltage upon the difference between the standard free energies of the reactants and products of the cell reaction and the activities of these species.

$\mathscr{E}_{\max}$ tends to be large when the reactant activities are large and the product activities are small. Often the symbol Q is used for the quotient of these activities.

$$Q \equiv \frac{a_C{}^c a_D{}^d}{a_A{}^a a_B{}^b}.$$

---

for other systems activities must be determined by experiment. The activity of a substance is often expressed as its "ideal activity" (its molality or its partial pressure, for example) multiplied by an "activity coefficient." In ideal systems activity coefficients are equal to 1. In non-ideal systems activity coefficients may be greater than or less than 1. The Debye-Hückel theory is a theory for estimating activity coefficients for strong electrolytes in dilute solutions.

* The units on n for the reaction $aA + bB = cC + dD$ are "faradays per a moles of A (or b moles of B) consumed."

With this abbreviation, the following equations may be written.

$$\Delta G_\sigma = \Delta G^0{}_\sigma + RT \ln Q$$

$$-n\mathscr{F}\mathscr{E}_{\max} = -n\mathscr{F}\mathscr{E}^0{}_{\max} + RT \ln Q$$

$$\mathscr{E}_{\max} = \mathscr{E}^0 - \frac{RT}{n\mathscr{F}} \ln Q.$$

The meaning of the Nernst equation can be illustrated by comparing the calculated and measured voltages for a galvanic cell based upon the spontaneous transfer of electrons from metallic zinc to hydrogen ions. Adequate to determine with reasonable accuracy the driving force behind the reaction

$$Zn(c) + \overset{2e}{\curvearrowright} 2H^+(aq) = Zn^{++}(aq) + H_2(gas)$$

is a cell made by simply connecting with wires a zinc strip, partially immersed in a zinc nitrate solution, through a voltmeter to a standard hydrogen electrode, Fig. 31.1. The circuit may be completed with a salt bridge.*

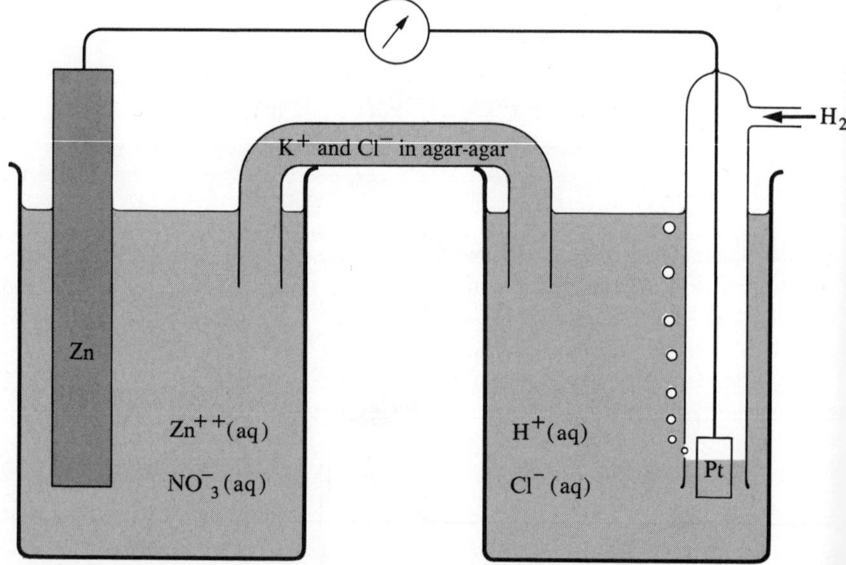

Fig. 31.1 Apparatus for measuring the driving force behind the spontaneous reaction of zinc with hydrochloric acid. Electrons released at the zinc electrode when zinc atoms go into solution as positive zinc ions flow through the voltmeter to the platinum strip, which is partially immersed in a solution of hydrochloric acid; here the electrons react with aquated hydrogen ions and produce molecular hydrogen.

* Introduced in this way are small liquid junction potentials, which are here neglected.

The reading on the voltmeter generally shows that electrons flow spontaneously from the zinc electrode to the hydrogen electrode, corresponding to the half-cell reactions

$$Zn(c) = Zn^{++}(aq) + 2e$$
$$2H^{+}(aq) + 2e = H_2(gas).$$

For this cell,

$$n = 2 \text{ faradays/mole Zn(c) consumed}$$
$$\Delta G^0_\sigma = G^0_{Zn^{++}(aq)} + G^0_{H_2(gas)} - G^0_{Zn(c)} - 2G^0_{H^+(aq)}$$
$$= -\ 35{,}184 \text{ cal/mole Zn(c) consumed} \quad (T = 298°K)$$

$$Q = \frac{a_{Zn^{++}(aq)} \cdot a_{H_2(gas)}}{a_{Zn(c)} \cdot a^2_{H^+(aq)}}$$
$$\approx \frac{(Zn^{++}) \cdot P_{H_2}}{(H^+)^2}.$$

Therefore, at 298°K

$$\mathscr{E}^0 = \frac{-(-35{,}184) \text{ cal}}{2 \times 96{,}500 \text{ coul.}} \times \frac{4.184 \text{ joules}}{1 \text{ cal}}$$
$$= 0.76 \text{ volts.}$$

By the Nernst equation, therefore,

$$\mathscr{E}_{max} \approx 0.76 - 0.0295 \log \frac{(Zn^{++}) \cdot P_{H_2}}{(H^+)^2}.$$

When the concentrations of the zinc ion about the zinc electrode and the hydrogen ion about the hydrogen electrode are each 1 molar, and when the partial pressure of the hydrogen gas at the hydrogen electrode is 1 atm., $Q \approx 1$ and the cell's maximum voltage, to which the measured voltage generally closely approximates (when the resistance of the voltmeter is large) is about 0.76 volts.

This reading can be altered in several ways. Decreasing the activity of the zinc ions—by adding excess ammonia to the zinc nitrate solution, for example (this ties up zinc ions as $Zn(NH_3)_6^{++}$)—increases the cell voltage. Lowering the zinc ion activity makes the Zn - Zn^{++} couple a more powerful electron donor. On the other hand, decreasing the activity of the hydrogen ion—by adding hydroxide ion to the hydrogen-electrode compartment, for example (this ties up the hydrogen ion as H_2O)—diminishes the cell voltage. Lowering the hydrogen ion activity makes the H_2 - H^+ couple a less powerful electron acceptor. A change in hydrogen ion concentration from 1 to 0.1 molar, for example, lowers the cell voltage approximately

$$0.0295 \log \frac{1}{(0.1)2} = 0.06 \text{ volts.}$$

That this is not an insignificant effect can be appreciated by considering what the theoretical cell voltage would be if, through the addition of excess hydroxide ion to the hydrogen-electrode compartment, the hydrogen ion concentration at the hydrogen electrode were reduced to 10^{-14} molar, a_{H_2} and $a_{Zn^{++}}$ remaining at unity. Under these conditions

$$\begin{aligned}\mathscr{E}_{\max} &\approx 0.76 - 0.0295 \log \frac{1 \cdot 1}{(10^{-14})^2} \\ &= 0.76 - 0.83 \\ &= -0.07 \text{ volts.}\end{aligned}$$

At the low hydrogen ion concentration of 10^{-14} molar, the H_2 - H^+ couple is an exceedingly poor electron acceptor, so poor, in fact, that no longer is it capable of accepting electrons spontaneously from the Zn - Zn^{++} couple. Indeed, the flow of electrons in the external circuit may now occur weakly in the opposite direction: from the hydrogen electrode to the zinc electrode. When this happens the half-cell reaction at the hydrogen electrode may be thought of as the former reaction reversed ($H_2 = 2H^+ + 2e$) followed by the reaction of the hydrogen ions with the excess hydroxide ions ($2H^+ + 2OH^- = 2H_2O$), to yield for the net half-cell reaction the reaction

$$H_2(\text{gas}) + 2OH^-(\text{aq}) = 2H_2O(\text{aq}) + 2e.$$

Together with the half-cell reaction at the zinc electrode

$$Zn^{++}(\text{aq}) + 2e = Zn(\text{c}),$$

this yields for the net cell reaction

$$H_2(\text{gas}) + 2OH^-(\text{aq}) + Zn^{++}(\text{aq}) = 2H_2O(\text{aq}) + Zn(\text{c}).$$

For this reaction $n = 2$ and

$$\begin{aligned}\Delta G_\sigma{}^0 &= [2G^0{}_{H_2O(\text{aq})} + G^0{}_{Zn(\text{c})}] - [G^0{}_{H_2(\text{gas})} + 2G^0{}_{OH^-(\text{aq})} + G^0{}_{Zn^{++}(\text{aq})}] \\ &= -3000 \text{ cal/mole Zn(c) produced.}\end{aligned}$$

Consequently, as already noted,

$$\begin{aligned}\mathscr{E}^0 &= -\frac{-3000}{2 \times 96{,}500} \times \frac{4.184}{1} \text{ volts} \\ &= 0.07 \text{ volts.}\end{aligned}$$

If the zinc electrode is replaced by a copper electrode in contact with a solution of copper nitrate, electrons will flow spontaneously from the hydrogen electrode to the metal electrode even when the hydrogen ion concentration at the hydrogen electrode is 1 mole per liter. The half-cell reactions in this case are

$$H_2 = 2H^+(aq) + 2e$$
$$Cu^{++}(aq) + 2e = Cu.$$

The net cell reaction is

$$Cu^{++}(aq) + H_2(gas) = Cu(c) + 2H^+(aq).$$

The number of faradays per mole of cupric ion consumed is 2. And

$$\begin{aligned}\Delta G^0_\sigma &= [G^0_{Cu(c)} + 2G^0_{H^+(aq)}] - [G^0_{Cu^{++}(aq)} + G^0_{H_2(gas)}] \\ &= -15{,}530 \text{ cal/mole } Cu^{++} \text{ consumed.}\end{aligned}$$

Consequently,

$$\begin{aligned}\mathscr{E}^0 &= -\frac{-15{,}530}{2 \times 96{,}500} \times \frac{4.184}{1} \\ &= 0.34 \text{ volts.}\end{aligned}$$

These results are summarized below in tabular form.

Zn ———	Zn^{++}	+0.76
H_2 ———	H^+	0.00
Cu ———	Cu^{++}	−0.34

Opposite each couple, written with its reduced form on the left and its oxidized form on the right, is given the value of the standard voltage $\mathscr{E}^0$ for the cell formed from this couple and the hydrogen electrode. A positive value signifies that electrons flow spontaneously from the metal electrode to the hydrogen electrode. A negative value indicates that the spontaneous flow of electrons is in the opposite direction: from the hydrogen electrode to the metal electrode. Briefly stated, electrons fall spontaneously from upper levels to lower levels.

In tabulations of this kind, electron donors are listed on the left; the most powerful donor (here Zn) occurs at the top of the list, the weakest donor (Cu) at the bottom. Electron acceptors are listed on the right; here the most powerful agent is found at the bottom of the list, the weakest at the top. A slightly longer tabulation of "redox" couples is given below. Underlined on the left are elements often found in impure, unrefined copper.

Na ——— Na^+

Zn ——— Zn^{++}

Fe ——— Fe^{++}

H_2 ——— H^+

Cu ——— Cu^{++}

Ag ——— Ag^+

Au ——— Au^+

F^- ——— F_2

The impurities, zinc, iron, silver, and gold, can be removed from copper by forcing electrons from the impure copper sample to a strip of pure copper dipped, along with the impure copper, into a solution of copper sulfate. First to release electrons in the impure copper is the best electron donor present, zinc; it loses two electrons per atom and enters solution as zinc ions.

$$Zn(c) = Zn^{++}(aq) + 2e.$$

Next to release electrons is iron.

$$Fe(c) = Fe^{++}(aq) + 2e.$$

When the electrode surface has been depleted of both zinc and iron, copper begins to dissolve.

$$Cu(c) = Cu^{++}(aq) + 2e.$$

With copper always present at the electrode surface, the poorer electron donors, silver and gold, have no opportunity to lose electrons. Unoxidized, they fall to the bottom of the cell. The only ions produced in solution therefore, are those circled on the right: Zn^{++}, Fe^{++}, and Cu^{++}. Of these the first to come out at the electron-rich, pure-copper electrode is the bes electron acceptor, Cu^{++}. The poorer electron acceptors Fe^{++} and Zn^{++} dc not plate out as long as there exists at the electrode surface an abundan supply of copper ions. The net effect of the process is threefold: the selectiv transfer of copper ions from impure copper to pure copper; the gradua replacement of copper ions in solution by zinc ions and iron ions, whic must occasionally be removed; and the production of an "anode sludge" rich in silver and gold, which helps pay for the process.

Problems for Chapters 29, 30, and 31

1. It was found in Chapter 29 that the maximum work available from the reaction H_2(gas) + (1/2)O_2(gas) = H_2O(liq) is less than the decrease in energy of the chemical system and the atmosphere, i.e. for the reaction cited $W_{max} < -\Delta H_\sigma$. Under what condition is $W_{max} > -\Delta H_\sigma$?
2. The strength of a donor-acceptor couple M - M^+ as a donor or acceptor of electrons depends upon the concentrations of the donor and acceptor species: donor strength increases as the donor concentration increases and the acceptor concentration decreases; acceptor strength increases as the acceptor concentration increases and the donor concentration decreases.

   Of the two couples

   $$H_2 \text{ (1 atm.)} \text{ ——— } H^+ \text{ (0.1 molar)}$$
   $$H_2 \text{ (1 atm.)} \text{ ——— } H^+ \text{ (1.0 molar)}$$

   (a) Which is the better electron donor? The better electron acceptor?
   (b) What are the half-cell reactions for a cell composed of these two couples?
   (c) What is the net cell reaction?
   (d) What is the maximum cell voltage?
   (e) What is the driving force behind the cell reaction?
   (f) What part of the universe provides the energy that appears as electrical work?
3. What is the value of the "activity quotient" Q in the expressions

   $$\Delta G_\sigma = \Delta G^0{}_\sigma + RT \ln Q$$

   and
   $$\mathscr{E}_{max} = \mathscr{E}^0{}_{max} - (RT/n\mathscr{F}) \ln Q$$

   when
   $$\Delta G_\sigma (= -n\mathscr{F}\mathscr{E}_{max}) = 0?$$

4. List several ways of increasing the maximum voltage available from a cell based upon the reaction

   $$H_2\text{(gas)} + Cu^{++}\text{(aq)} = 2H^+\text{(aq)} + Cu\text{(c)}.$$

   Under what conditions is $\mathscr{E}_{max} = 0$? How might these conditions be achieved?
5. What are the half-cell reactions and what is the value of $\mathscr{E}^0$ for a cell that has for one electrode a standard hydrogen gas - hydrogen ion electrode

and for the other electrode a strip of silver dipped into a solution of silver nitrate?

6. What happens to the voltage of the cell of problem 5 if the silver nitrate solution surrounding the silver electrode is replaced by a paste of silver chloride saturated with 1 molar HCl?

7. From the data in Appendix I, calculate $\mathscr{E}^0$ values for cells where one half-cell reaction is

$$H^+(aq) + e = \tfrac{1}{2}H_2(gas)$$

and the other half-cell reaction is

(a) $Cu = Cu^+ + e, \quad Cu^+ = Cu^{++} + e, \quad Cu = Cu^{++} + 2e$

(b) $Hg = \tfrac{1}{2}Hg_2^{++} + e, \quad \tfrac{1}{2}Hg_2^{++} = Hg^{++} + e, \quad Hg = Hg^{++} + 2e$

(c) $Fe = Fe^{++} + 2e, \quad Fe^{++} = Fe^{+++} + e, \quad Fe = Fe^{+++} + 3e.$

Summarize the results for each element in tabular form. Are the intermediate oxidation states Cu^+, Hg_2^{++}, and Fe^{++} stable with respect to the disproportionation reactions

$2Cu^+ = Cu + Cu^{++}$, $Hg_2^{++} = Hg + Hg^{++}$, and $3Fe^{++} = Fe + 2Fe^{+++}$?

Is there any relation between the $\mathscr{E}^\circ$ value for the couple Fe - Fe^{+++} and the $\mathscr{E}^0$ values for the couples Fe - Fe^{++} and Fe^{++} - Fe^{+++}?

8. Show that the effect of temperature upon cell voltage is given by the relation

$$\frac{d\mathscr{E}}{dT} = \frac{\Delta S}{n\mathscr{F}} \cdot \qquad \text{(Pressure and cell composition constant)}$$

9. For the cell whose half-cell reactions are

$$Ag(c) + Cl^-(aq) = AgCl(c) + e$$

$$\tfrac{1}{2}Cl_2(gas) + e = Cl^-(aq)$$

and whose net cell reaction, therefore, is simply the formation of silver chloride from its elements

$$Ag(c) + \tfrac{1}{2}Cl_2(gas) = AgCl(c),$$

the experimental values for $\mathscr{E}$ and $d\mathscr{E}/dT$ at 298°K and 1 atm. are 1.136 volts and -0.000595 volts/deg, respectively. Calculate from these data the values of ΔG^0_σ, ΔS^0_σ, and ΔH^0_σ. Compare these values with those obtained from the data in Appendix I.

Postscript to Chapters 31, 23, 14, and 7

The use of the symbol Q to represent thermal energy in transit in Chapter 7, heats of reaction in Chapter 14, partition functions in Chapter 23, and activity quotients in the present chapter may remind the reader of this celebrated dialogue:

"There's glory for you!"

"I don't know what you mean by 'glory,' " Alice said.

Humpty Dumpty smiled contemptuously. "Of course you don't—till I tell you. I meant 'there's a nice knockdown argument for you!' "

"But 'glory' doesn't mean 'a nice knockdown argument,' " Alice objected.

"When *I* use a word," Humpty Dumpty said, in rather a scornful tone, "it means just what I choose it to mean—neither more nor less."

"The question is," said Alice, "whether you *can* make words mean so many different things."

"The question is," said Humpty Dumpty, "which is to be master—that's all."*

* Lewis Carroll, *Through the Looking Glass*, Chapter VI.

32

Proton-Transfer Reactions and pH

Bronsted Acid-Base Theory

Of all gases, one of the most soluble in water is hydrogen chloride. Over three hundred liters of this gas will dissolve in one liter of water at 25°C.

This enormous solubility arises from the fact that hydrogen chloride gas and water react vigorously with each other. Per mole of hydrogen chloride consumed in the reaction, approximately eighteen thousand calories of thermal energy are liberated to the surroundings. The reaction product—a liquid under ordinary conditions—is called "hydrochloric acid." It has some interesting properties.

Unlike either reactant, "hydrochloric acid" is an excellent conductor of electricity. On freezing, ice precipitates at a temperature that suggests the presence in solution of two solute particles for each added molecule of hydrogen chloride. With silver nitrate "hydrochloric acid" gives an immediate precipitate of silver chloride. Spectroscopic examination yields no evidence for the attachment by covalent bonds of hydrogen atoms to chlorine atoms; instead, the data suggest—for the monohydrate at least—that all of the hydrogen atoms are covalently bound to oxygen atoms.

These facts can be accounted for by a simple model. It is supposed that when an HCl molecule collides with a water molecule in solution the proton in the HCl molecule migrates from its initial location in one of the electron pairs in the valence-shell of the chlorine atom of the HCl molecule into one of the unoccupied electron pairs in the valence-shell of the oxygen atom of the water molecule. This spontaneous transfer of a proton from HCl to H_2O produces in solution two charged species: a deprotonated chlorine atom, Cl^-, called the chloride ion, and a triply protonated oxygen atom, H_3O^+, called the hydronium ion. In chemical shorthand,

$$HCl + H_2O \overset{H^+}{\frown} \quad HCl + H_2O = H_3O^+ + Cl^- .$$

Proton-transfer reactions of this kind are among the simplest, fastest, and commonest reactions in chemistry. After centuries of study a terminology has developed for describing these reactions. Following Bronsted, proton donors are called acids, proton acceptors bases.

In the example above the HCl molecule is acting as an acid, the H_2O molecule as a base. Because water is not a strong base, its basic properties are often overlooked. To bring forth water's basic properties requires a powerful acid, such as HCl, HNO_3, or H_2SO_4.

When an acid loses its acidic proton, what remains is a potential proton acceptor—in a word, a base, called by Bronsted the acid's conjugate base. The conjugate base of HCl is Cl^-.*

Similarly, when a base accepts a proton, what is formed is a potential proton donor—i.e. an acid, called by Bronsted the base's conjugate acid. The conjugate acid of Cl^- is HCl, of H_2O, H_3O^+.

The hydronium ion is distinguished by the fact that it is the strongest acid that can exist in quantity in water. Any acid stronger than H_3O^+ reacts with water, always present in excess in aqueous solution, to form H_3O^+. Three common acids stronger than H_3O^+ are the mineral acids HCl, HNO_3, and H_2SO_4. In water the "Big Three" are leveled to the strength of H_3O^+.

$$\begin{array}{lcl} HCl & & Cl^- \\ HNO_3 & + H_2O = H_3O^+ + & NO_3^- \\ H_2SO_4 & & HSO_4^- \end{array}$$

To develop a protonating medium more powerful than aqueous H_3O^+ requires a solvent less basic than water, e.g. liquid hydrogen fluoride, or liquid sulfuric acid.

One characteristic feature of acid-base terminology should be noted. The statement "HCl is a strong acid" and the statement "Cl^- is a weak base" are conjugate statements. To say that an electron cloud loses protons easily is the same thing as saying that the deprotonated cloud accepts protons poorly.

The escaping tendency of protons from one electron cloud may be compared with the escaping tendency of protons from another electron cloud by a simple experiment. Take, for example, the chloride-ion electron cloud—written HCl when protonated, Cl^- when unprotonated—and the water-molecule electron cloud—written when protonated H_3O^+, when unprotonated H_2O. To determine which of these two electron clouds is the stronger proton donor, place the protonated HCl - Cl^- cloud in contact with the unprotonated H_3O^+ - H_2O cloud; i.e. add HCl to H_2O. If protons migrate spontaneously from the occupied cloud HCl - Cl^- to the vacant cloud

* That Cl^- can act as a base is shown by the common laboratory preparation of hydrogen chloride gas from solid sodium chloride and concentrated sulfuric acid.

$$H_2SO_4(\text{liq}) + Cl^-(\text{in } Na^+Cl^-) = HCl(\text{gas}) + HSO_4^-(\text{in } Na^+HSO_4^-)$$

H_3O^+ - H_2O, as in fact they do, one may say that the cloud HCl - Cl^- is a better proton donor in water than the cloud H_3O^+ - H_2O. This information can be displayed in a suggestive, tabular form.

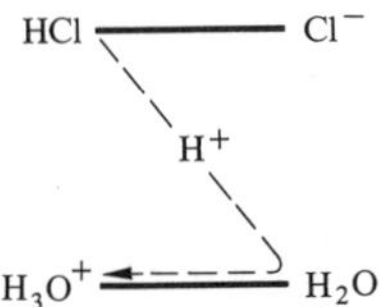

The dotted arrow indicates that protons tend to fall spontaneously from the upper level to the lower level. By "level" is meant an electron cloud which when occupied is a proton donor, when vacant a proton acceptor.

H_3O^+—the strongest acid water can tolerate in appreciable quantities—is capable of protonating many substances. The odor of vinegar produced when sodium acetate is added to "hydrochloric," "nitric," or "sulfuric acid" proclaims to the nose, if not to the eyes, that the following proton transfer has occurred.

$$\overset{H^+}{\overbrace{H_3O^+ + Ac^-}} = HAc + H_2O$$

The acetic acid formed in this reaction is in its own right a modest protonating agent—not as powerful as H_3O^+, much less HCl, but sufficiently powerful to protonate ammonia, carbonate ions, and hydroxide ions.

$$HAc + \begin{matrix} NH_3 \\ CO_3^{--} \\ OH^- \end{matrix} = \begin{matrix} NH_4^+ \\ HCO_3^- \\ H_2O \end{matrix} + Ac^-$$

The water formed in the last reaction is also a potential proton donor—not as strong as acetic acid, or even as strong as the mild acids NH_4^+ and HCO_3^-, but sufficiently strong to protonate the powerful bases O^{--}, NH_2^-, H^-, and CH_3^-.

$$H_2O + \begin{matrix} O^{--} \text{ (in } Na_2O) \\ NH_2^- \text{ (in } NaNH_2) \\ \\ H^- \text{ (in NaH)} \\ CH_3^- \text{ (in } NaCH_3) \end{matrix} = \begin{matrix} OH^- \\ NH_3 \\ \\ H_2 \\ CH_4 \end{matrix} + OH^-$$

The common product of these reactions, the hydroxide ion, is the stronges base that can exist in quantity in water. Any base stronger than OH^- i immediately protonated by water molecules. To develop a proton-accepting medium more powerful than aqueous OH^- requires a solvent less acidic than water, e.g. liquid ammonia, or dioxane, or a molten salt such as sodium chloride.

The facts cited above are summarized in Fig. 32.1. Listed on the left are Bronsted acids, on the right Bronsted bases. Acid strength increases upward, base strength downward. Acids above H_3O^+ and bases below OH^- are too powerful to exist in quantity in water. The direction of massive proton migration in spontaneous proton-transfer reactions is from upper occupied levels to lower vacant levels, i.e. from upper left to lower right.

It is impossible, however, to completely drain a level of all its protons, no matter how much strong base is added. Even in concentrated sodium hydroxide solutions some protons find their way to the H_3O^+ - H_2O level.

Proton donors Level occupied Bronsted acids		Proton acceptors Level vacant Bronsted bases
H_2SO_4	——	HSO_4^-
HCl	——	Cl^-
HNO_3	——	NO_3^-
H_3O^+	——	H_2O
HAc	——	Ac^-
NH_4^+	——	NH_3
HCO_3^-	——	$CO_3^=$
H_2O	——	OH^-
OH^-	——	$O^=$
NH_3	——	NH_2^-
CH_4	——	CH_3^-
H_2	——	H^-

Fig. 32.1

It is equally impossible to completely saturate a level with protons, regardless of how much strong acid is added. Even in concentrated "hydrochloric acid" some vacancies occur in the H_2O - OH^- level.

One might ask what is the cause of nature's reluctance to saturate a level, or to leave a level entirely empty? In sodium hydroxide solutions the protons could be distributed one and two per oxygen atom, but they are not; a few oxygen atoms carry three protons. In hydrochloric acid the protons could be distributed two and three per oxygen atom, but they are not; a few oxygen atoms carry only one proton. Why? The answer, in a word, is Entropy. Even the distribution of protons in pure water is influenced by it.

In 1894 the purest water in the world was placed between platinum electrodes and discovered to be a weak conductor of electricity. Evidently in water an internal process acts to produce charge carriers. That process is believed to be the auto-protolysis reaction

$$H_2O \overset{H^+}{\curvearrowright} H_2O = H_3O^+ + OH^- .$$

This is a striking conjecture. For when "hydrochloric acid" is added to sodium hydroxide, protons migrate spontaneously in the opposite direction. This is understandable. The reaction $H_3O^+ + OH^- = 2H_2O$ is strongly exothermic. For each mole of H_3O^+ and OH^- consumed in the reaction, 13,360 calories of energy are liberated to the thermal surroundings. At 25°C this increases the entropy of the surroundings

$$\frac{13{,}360 \text{ cal}}{298°\text{K}} = 44.82 \text{ cal/deg-mole } H_3O^+ \text{ consumed.}$$

Conversely, during the opposite reaction, $2H_2O = H_3O^+ + OH^-$, the entropy of the surroundings must decrease—by 44.82 cal/deg per mole of H_3O^+ and OH^- produced. This being so,

Why does water ionize?
For the same reason that water evaporates.

Evaporation, like self-ionization, is an endothermic process. Yet water will evaporate—decreasing simultaneously the entropy of its surroundings—*if the concentration of water vapor is not too great.*

For the same reason, water will ionize *if the concentration of the ions is not too great.*

Both processes increase the entropy of the universe.

Evaporation of water occurs at low water vapor concentrations because the vapor's large entropy compensates for both the entropy of the liquid consumed and for the entropy lost by the thermal surroundings.

Similarly, water ionizes spontaneously at low concentrations of H_3O^+ and/or OH^- (one at least must be small) because under these conditions the ions' large entropy compensates for both the entropy of the water molecules consumed in the reaction and for the entropy lost by the thermal surroundings.

These conclusions can be rendered more quantitative by using the data of Appendix I. At 25°C and 1 atm.

$$S_{H_3O^+} = 16.716 - 4.5757 \log (H_3O^+) \qquad (*)$$

$$S_{OH^-} = -2.519 - 4.5757 \log (OH^-)$$

$$S_{H_2O} = 16.716 - 4.5757 \log N_{H_2O}$$

$$\approx 16.716. \qquad (N_{H_2O} \approx 1)$$

The logarithmic term in the expressions for $S_{H_3O^+}$ and S_{OH^-} becomes important at low concentrations. For a solute concentration 10^{-10} molar its value is 45.757 cal/deg per mole. This large value reflects the fact that in dilute solutions a large number of distinguishable positions are available to the solute particles. At low concentrations the solute's configurational entropy is large. For this reason water can ionize, a bit, without violating the Second Law. Indeed, it would be absurd if it did not.

Imagine a beaker of absolutely pure H_2O—one containing no ions whatsoever. This pristine state is hardly likely to persist forever. By cooling slightly and utilizing the energy so obtained to promote protons from the H_2O - OH^- level to the H_3O^+ - H_2O level, the system can spontaneously increase its total entropy. For at low concentrations the reaction products, H_3O^+ and OH^-, have nearly infinite configurational entropies. Ionization under these conditions increases the system's configurational entropy more

* The thermodynamic properties of individual ions cannot be examined with the same freedom that the thermodynamic properties of neutral molecules can be examined. In experiments on macroscopic samples of matter ions of one charge are always accompanied by counter-ions of the opposite charge. The thermodynamic properties of cations and anions separately cannot be determined. Only their values relative to one another are known. By convention the thermodynamic properties of the aqueous proton, written "H^+," are arbitrarily set equal to zero. (Extra-thermodynamic arguments suggest that for the ion's entropy this choice may be approximately correct.) For the self-ionization of water written, noncommittally, as $H_2O(aq) = H^+(aq) + OH^-(aq)$, one finds from the data of Appendix I that

$\Delta S^0_\sigma = (S^0_{H^+(aq)} + S^0_{OH^-(aq)} - S^0_{H_2O(liq)}) = O + (-2.519) - 16.716 = -19.235$ cal/deg per mole of $H^+(aq)$ and $OH^-(aq)$ produced.

If the same value for the experimentally measurable quantity ΔS^0_σ is to be obtained from tabulated data when the self-ionization reaction is written in terms of H_3O^+, a procedure that introduces an additional molecule of water on the opposite side of the equation, the thermodynamic properties assigned to H_3O^+ in its standard state must be the same as those assigned to pure liquid water. Briefly stated, with the assignment $X^0_{H^+} = 0.0000$ ($X^0 = S^0$, G^0, H^0, C^0_P), $X^0_{H_3O^+} = X^0_{H_2O(liq)}$.

rapidly than it decreases the system's thermal entropy. It will continue until the concentrations of H_3O^+ and OH^- are such that

$$\begin{aligned}\Delta S_{\text{total}} &= \Delta S_\sigma + \Delta S_\theta \\ &= (S_{H_3O^+} + S_{OH^-} - 2S_{H_2O}) + \frac{\Delta E_\theta}{T} \\ &= -19.235 - R\ln(H_3O^+)(OH^-) - 44.82 \qquad (T = 298°K) \\ &= 0;\end{aligned}$$

that is, until

$$\begin{aligned}(H_3O^+)(OH^-) &= 10^{-\frac{19.235+44.82}{2.3R}} \\ &= 10^{-14}.\end{aligned}$$

A proton-transfer reaction such as

$$H_2O + H_2O = H_3O^+ + OH^- \quad (\overset{H^+}{\curvearrowright}\text{ from first } H_2O \text{ to second})$$

may be thought of as occurring in two steps: evaporation of protons from the H_2O - OH^- level

$$H_2O = OH^- + H^+$$

followed by condensation of protons in the H_3O^+ - H_2O level

$$H_2O + H^+ = H_3O^+.$$

The sum of these two "half-cell" reactions is the net reaction given above.

In a similar way, the reverse reaction, the neutralization of H_3O^+ by OH^-, or of OH^- by H_3O^+, may be thought of as evaporation of protons from the H_3O^+ - H_2O level followed by their condensation in the H_2O - OH^- level.

The direction of net proton migration between two levels depends upon the escaping tendency of protons from each level. If the proton escaping tendency from the upper level is low, owing to the absence (or near absence) of protons in that level, as in hypothetical pure H_2O (H_3O^+ concentration zero), a few protons will migrate spontaneously upward from a lower occupied level (e.g. H_2O - OH^-, an occupied level always present in aqueous solutions) to the sparsely populated upper level.

On the other hand, if the proton escaping tendency from an upper level is large, owing to the presence in that level of many protons (the H_3O^+ - H_2O level in 1 molar hydrochloric acid, for example), while elsewhere the proton escaping tendency from a lower level is small, owing to the presence in the lower level of many proton vacancies (the H_2O - OH^- level in 1 molar sodium hydroxide, for example), protons, given a chance, will migrate spontaneously from the upper level to the lower level, decreasing their escaping tendency from the upper level and increasing their escaping tendency from the lower level, until finally (very quickly, in fact) the proton escaping tendencies from both levels are the same. At that point the system is at equilibrium with respect to the distribution of protons between the two levels.

The escaping tendency of protons from a proton level is proportional to the activity of that level's proton-donor and inversely proportional to the activity of the level's proton-acceptor. The larger the former and the smaller the latter, the greater is the chance protons may escape from the level. Symbolically,

$$\text{Proton escaping tendency} \propto \frac{P\text{-donor activity}}{P\text{-acceptor activity}}.$$

This conclusion can be presented in a more analytic fashion as follows. Equilibrium with respect to the "half-cell reaction"

$$P\text{-donor} = P\text{-acceptor} + \text{H}^+$$

implies that

$$\frac{a_{\text{H}^+} \cdot a_{P\text{-acceptor}}}{a_{P\text{-donor}}} = K.$$

Hence,

$$a_{\text{H}^+} = K \frac{a^{P\text{-donor}}}{a_{P\text{-acceptor}}}.$$

Expressions of this form may be written down for each proton level in solution. For those two levels always present in aqueous solution, H_3O^+ - H_2O and H_2O - OH^-, and for the proton level of a typical weak acid - weak base pair, HAc - Ac^-, the following expressions are obtained.

$$a_{\text{H}^+}{}^{\text{H}_3\text{O}^+-\text{H}_2\text{O}} = K_{\text{H}_3\text{O}^+-\text{H}_2\text{O}} \frac{a_{\text{H}_3\text{O}^+}}{a_{\text{H}_2\text{O}}}$$

$$= K_{\text{H}_3\text{O}^+-\text{H}_2\text{O}} \frac{(\text{H}_3\text{O}^+)}{1} \qquad (N_{\text{H}_2\text{O}} \approx 1)$$

$$a_{\text{H}^+}{}^{\text{HAc}-\text{Ac}^-} = K_{\text{HAc}-\text{Ac}^-} \frac{a_{\text{HAc}}}{a_{\text{Ac}^-}}$$

$$= K_{\text{HAc}-\text{Ac}^-} \frac{(\text{HAc})}{(\text{Ac}^-)}$$

$$a_{\text{H}^+}{}^{\text{H}_2\text{O}-\text{OH}^-} = K_{\text{H}_2\text{O}-\text{OH}^-} \frac{a_{\text{H}_2\text{O}}}{a_{\text{OH}^-}}$$

$$= K_{\text{H}_2\text{O}-\text{OH}^-} \frac{1}{(\text{OH}^-)} \qquad (N_{\text{H}_2\text{O}} \approx 1)$$

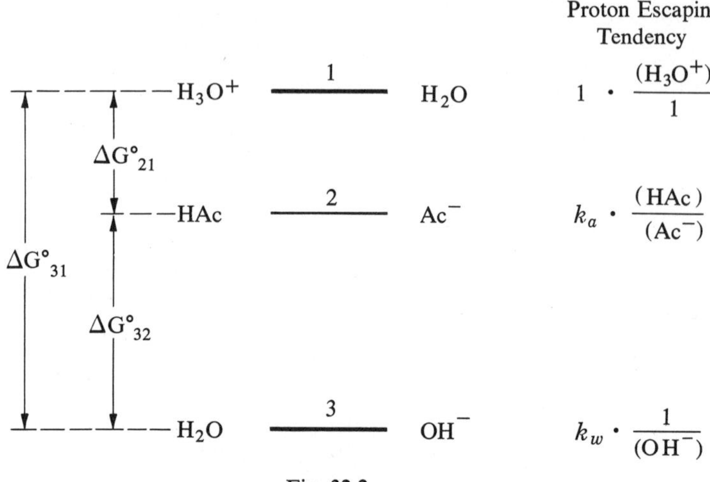

Fig. 32.2

The values of the proportionality constants $K_{H_3O^+-H_2O}$, K_{HAc-Ac^-}, and $K_{H_2O-OH^-}$ depend, among other things, upon the strengths of the corresponding proton-acceptor bonds. While the absolute values of these constants have not been determined (the vapor pressure of bare protons over aqueous solutions is too low to measure experimentally*), their relative values are known. Equating the expressions for $a_{H^+}{}^{H_3O^+-H_2O}$ and $a_{H^+}{}^{H_2O-OH^-}$, for example, one finds that when equilibrium prevails with respect to the distribution of protons between levels H_3O^+ - H_2O and H_2O - OH^-,

$$(H_3O^+)(OH^-) = \frac{K_{H_2O-OH^-}}{K_{H_3O^+-H_2O}}.$$

Evidently

$$\frac{K_{H_2O-OH^-}}{K_{H_3O^+-H_2O}} = K_w ,$$

where at 25°C

$$K_w = 10^{-14} .$$

The auto-protolysis constant K_w tends to be large when the solvent is simultaneously a strong acid ($K_{H_2O-OH^-}$ large) and a strong base ($K_{H_3O^+-H_2O}$ small). If arbitrarily we set $K_{H_3O^+-H_2O} = 1$, it follows that the constant $K_{H_2O-OH^-}$ must be identified with K_w. Similarly, by equating the expressions for $a_{H^+}{}^{H_3O^+-H_2O}$ and $a_{H^+}{}^{HAc-Ac^-}$, one finds that with the choice $K_{H_3O^+-H_2O} = 1$, K_{HAc-Ac^-} must be identified with the ionization constant K_a, 1.8×10^{-5} at 25°C. These results are summarized in Fig. 32.2.

* The value of the proportionality constant $K_{H_3O^+-H_2O}$ has been estimated to be $\sim 10^{-150}$.

By equating proton escaping tendencies in pairs, one finds that equilibrium exists with respect to proton distribution between levels 1 and 2 when

$$\frac{(H_3O^+)(Ac^-)}{(HAc)} = K_a \equiv K_{21} ;$$

between levels 1 and 3 when

$$(H_3O^+)(OH^-) = K_w \equiv K_{31} ;$$

and between levels 2 and 3 when

$$\frac{(HAc)(OH^-)}{(Ac^-)} = \frac{K_w}{K_a} \equiv K_{32} .$$

The ratio K_w/K_a is sometimes called the "hydrolysis constant," K_h. Evidently $K_{31}/K_{21} = K_{32}$, i.e.

$$K_{31} = K_{32} \cdot K_{21} .$$

This is a general result. For any electron cloud the product of the "hydrolysis constant" K_{32} and the "acid ionization constant" K_{21} is the solvent's auto-protolysis constant:

$$K_w = K_h \cdot K_a .$$

The same result may be obtained by noting that the standard free energy change for the promotion of protons from level 3 directly to level 1 is equal (Fig. 32.2) to the standard free energy change for the promotion of protons from level 3 to the intermediate level 2 plus the standard free energy change for the promotion of protons from level 2 to level 1.

$$\Delta G^0{}_{31} = \Delta G^0{}_{32} + \Delta G^0{}_{21}$$

Since $$\Delta G^0{}_{ij} = -RT \ln K_{ij} , \qquad (ij = 31, 32, 21)$$

it follows, as stated above, that $K_{31} = K_{32} \cdot K_{21}$.

When the escaping tendency of protons throughout a solution is uniform—a condition quickly reached in water—its value from any level, say level 1, or the logarithm of its value, or the logarithm of its reciprocal, the "*p*H," may be used as a measure of the solution's acidity.

Problems for Chapter 32

1. What are the following substances? $NaHSO_4$, NaCl, $NaNO_3$, NaAc, Na_2CO_3, NaOH, Na_2O, $NaNH_2$, $NaCH_3$, NaH.
2. What is "hydrobromic acid"?

3. The electron-level diagrams in the previous chapter and the proton-level diagrams in this chapter have what features in common?
4. In the analogy between Bronsted acid-base theory and oxidation-reduction theory, an acid is to a base as a what is to a what?
5. Using donor-acceptor terminology, define the following.

From acid-base theory	*From oxidation-reduction theory*
Neutralization	Reduction
Ionization	Oxidation
Hydrolysis	Strong oxidizing agent
Buffer	Strong reducing agent
Indicator	Electrolysis

6. Using the data of Appendix I, calculate the value of the acid ionization constant of NH_4^+ at 25°C. What is the numerical value of the equilibrium constant for the following reaction?

$$NH_3(aq) + H_2O = NH_4^+ + OH^-$$

7. Estimate the hydrogen ion concentration in these solutions: water with 0.5 moles per liter of dissolved (a) hydrochloric acid, (b) sodium oxide, (c) propane, (d) sodium chloride, (e) acetic acid, (f) sodium acetate, (g) acetic acid and sodium acetate, (h) sodium bicarbonate.

33

The Clausius-Clapeyron Equation

$$\left(\frac{\delta P}{\delta T}\right)_{G_X^{\text{liq}} = G_X^{\text{gas}}} = \frac{\Delta S}{\Delta V}$$

When liquid water is in equilibrium with its vapor according to the equation

$$H_2O(\text{liq}) = H_2O(\text{gas}),$$

the free energy of the water in the liquid phase must be equal to its free energy in the gaseous phase,

$$G_{H_2O}^{\text{liq}} = G_{H_2O}^{\text{gas}},$$

otherwise evaporation or condensation of a small amount of water would produce a change in the total entropy of the universe. At equilibrium the temperature and pressure of the liquid water and of the water vapor must therefore be such that

$$E_{H_2O}^{\text{liq}} - TS_{H_2O}^{\text{liq}} + PV_{H_2O}^{\text{liq}} = E_{H_2O}^{\text{gas}} - TS_{H_2O}^{\text{gas}} + PV_{H_2O}^{\text{gas}}.$$

The molar free energy of water, in either the liquid or the gaseous phase, is a function of the temperature and the pressure—not only because T and P occur explicitly in the definition of G, but, also, because E, S, and V depend, in general, upon the temperature and the pressure. In symbols,

$$\begin{aligned} G &\equiv E - TS + PV \\ &= E(T, P) - T \cdot S(T, P) + P \cdot V(T, P) \\ &= G(T, P). \end{aligned}$$

For every value of the absolute temperature T less than water's critical temperature there exists a value of the pressure P, called the vapor pressure, which is such that

$$G_{H_2O}^{\text{liq}}(T, P) = G_{H_2O}^{\text{gas}}(T, P).$$

If the temperature at an equilibrium point (T, P) is increased a small amount δT without altering the pressure, the water, we know, will evaporate. Now, for a process to be spontaneous, the free energy of the reactant(s) (here liquid water) must be greater than the free energy of the product(s) (water vapor); only then does the process produce a net increase in the entropy of the universe. It follows from this that at the non-equilibrium point $(T + \delta T, P)$,

$$G^{\text{liq}}_{\text{H}_2\text{O}}(T + \delta T, P) > G^{\text{gas}}_{\text{H}_2\text{O}}(T + \delta T, P).$$

In thermodynamic terms this inequality arises from the fact that an increase in temperature generally decreases a substance's contribution to the free energy of a phase, the more so the larger its contribution to the entropy of that phase. At first glance this may seem surprising; for E and PV generally increase as T increases (P constant); however, the increase in the TS product, which in the expression for the free energy is preceded by a minus sign, is always greater than the sum of the changes in E and PV. It will be shown below that the decrease in a substance's molar free energy $G = E - TS + PV$ is, in fact, proportional to the instantaneous value of its molar entropy; it is as if over small temperature intervals E, PV, and S did not change and (hence) that G were a linear function of T, decreasing as T increases at the rate of S cal/deg/mole. Since

$$S^{\text{gas}}_{\text{H}_2\text{O}} > S^{\text{liq}}_{\text{H}_2\text{O}},$$

the drop in $G^{\text{gas}}_{\text{H}_2\text{O}}$ produced by an increase δT in T exceeds the corresponding drop produced in $G^{\text{liq}}_{\text{H}_2\text{O}}$ and, hence, at temperature $T + \delta T$, and the original pressure, $G^{\text{gas}}_{\text{H}_2\text{O}} < G^{\text{liq}}_{\text{H}_2\text{O}}$, and the liquid evaporates.

The equality that originally existed between the molar free energies can be re-established by increasing the pressure. Experimentally, this corresponds to saying that the vapor pressure increases as the temperature increases. The reason for this, thermodynamically speaking, is that a substance's molar free energy increases with pressure at a rate proportional to its molar volume (on this, more later).* Since

$$V^{\text{gas}}_{\text{H}_2\text{O}} > V^{\text{liq}}_{\text{H}_2\text{O}},$$

a change δP in the pressure produces in $G^{\text{gas}}_{\text{H}_2\text{O}}$, the lesser of the two free energies, an increase that exceeds the corresponding increase in $G^{\text{liq}}_{\text{H}_2\text{O}}$. Thus, for a given value of δT there will exist a value for δP which is such that

$$G^{\text{liq}}_{\text{H}_2\text{O}}(T + \delta T, P + \delta P) = G^{\text{gas}}_{\text{H}_2\text{O}}(T + \delta T, P + \delta P).$$

* For an ideal gas (i.g.) this increase in free energy with pressure is entirely an entropy effect. As P increases (T constant) $E_{\text{i.g.}}$ and $(PV)_{\text{i.g.}}$ do not change; but $S_{\text{i.g.}}$ decreases (becomes less positive); therefore, the free energy increases.

In fact, for large δP the increase produced in $G^{\text{gas}}_{\text{H}_2\text{O}}$ may make the molar free energy of the vapor larger than the molar free energy of the liquid, and the vapor will condense.

The greater the difference between molar entropies, $\Delta S = S^{\text{gas}}_{\text{H}_2\text{O}} - S^{\text{liq}}_{\text{H}_2\text{O}}$, the greater the disturbance produced in the equilibrium condition $G^{\text{gas}}_{\text{H}_2\text{O}} = G^{\text{liq}}_{\text{H}_2\text{O}}$ by an isobaric temperature change δT. Likewise, the greater the difference between molar volumes, $\Delta V = V^{\text{gas}}_{\text{H}_2\text{O}} - V^{\text{liq}}_{\text{H}_2\text{O}}$, the greater the disturbance produced in the equilibrium condition $G^{\text{gas}}_{\text{H}_2\text{O}} = G^{\text{liq}}_{\text{H}_2\text{O}}$ by an isothermal pressure change δP. Hence, for given δT the larger ΔS and the smaller ΔV, the larger must be δP to maintain $G^{\text{gas}}_{\text{H}_2\text{O}} = G^{\text{liq}}_{\text{H}_2\text{O}}$. Mathematically speaking (see later),

$$\left(\frac{\delta P}{\delta T}\right)_{G^{\text{gas}}_{\text{H}_2\text{O}} = G^{\text{liq}}_{\text{H}_2\text{O}}} = \frac{\Delta S}{\Delta V}.$$

The change in vapor pressure with a change in temperature tends to be large when ΔS is large and ΔV is small.

A plot of the equilibrium points (T, P), $(T + \delta T, P + \delta P)$ ..., yields the curve shown in Fig. 33.1. Temperature-pressure points to the right of the curve represent states where the gas is more stable than the liquid. For points to the left of the curve the gas is less stable than the liquid. For points on the curve the gas and the liquid are equally stable. A proof that along the curve the value of $\delta P/\delta T$ is equal to the value of $\Delta S/\Delta V$ is given in the following section.

In a reversible process the changes that occur in the entropy, energy, and

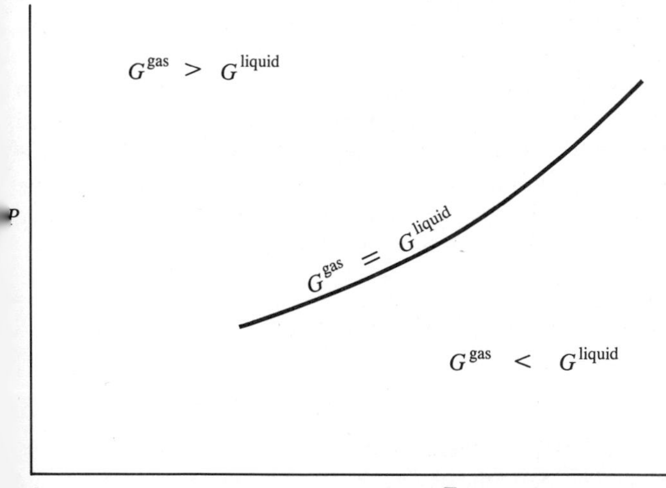

Fig. 33.1

volume of a chemical system are related to each other in the following manner.

$$\delta S - \frac{\delta E + P\delta V}{T} = 0. \qquad \text{(mass constant)}$$

The first term represents the entropy change in the chemical system, the second (with the minus sign) the entropy change in the system's thermal surroundings. The sum represents the total entropy change in the universe. For a reversible change this sum vanishes. Alternately, for a reversible change one may write that

$$\delta E = T\delta S - P\delta V \qquad (*)$$

where, evidently,

$$T = \left(\frac{\delta E}{\delta S}\right)_{\substack{V \text{ constant} \\ (\delta V = 0)}}$$

and

$$-P = \left(\frac{\delta E}{\delta V}\right)_{\substack{S \text{ constant} \\ (\delta S = 0)}}.$$

In words, the equation $\delta E = T\delta S - P\delta V$ says: the change in energy of a substance that is in internal equilibrium is equal to the change in its entropy times the rate at which its energy changes as its entropy alone changes, plus the change in its volume times the rate at which its energy changes as its volume alone changes.

From the fact that for small changes $\delta E = T\delta S - P\delta V$, and from the defining equation for the free energy, it follows that when the temperature and pressure of a substance change by small amounts δT and δP, the change produced in the substance's free energy is $-S\delta T + V\delta P$.

To see this, suppose that when the temperature and pressure change from their initial values T and P to the neighboring, final values $T + \delta T$ and $P + \delta P$, the values of the substance's energy, entropy, volume, and free energy change from their initial values E, S, V, and G to the neighboring, final values $E + \delta E$, $S + \delta S$, $V + \delta V$, and $G + \delta G$. Then, by the defining equation $G \equiv E - TS + PV$, which holds for both initial and final states,

$$\begin{aligned}\delta G &\equiv G_{\text{final}} - G_{\text{initial}} \\ &= [(E + \delta E) - (T + \delta T)(S + \delta S) + (P + \delta P)(V + \delta V)] \\ &\qquad\qquad - [E - TS + PV] \\ &= (\delta E - T\delta S + P\delta V) + (-S\delta T + V\delta P) - (\delta T \cdot \delta S - \delta P \cdot \delta V).\end{aligned}$$

The terms within the first parentheses add to zero for reversible changes. The terms within the last parentheses are negligible compared to the terms

* This is another way of saying that for a reversible change $\delta S_{\text{total}} = 0$.

within the second parentheses, if δT, δP, δV, and δS are small.* Thus, for reversible changes in temperature and pressure (or in energy and volume, from which follow changes in temperature and pressure)—but excluding changes in chemical composition,

$$\delta G = -S\delta T + V\delta P .$$

This equation summarizes two previous remarks: one, that the decrease in a substance's free energy for a given change in temperature, pressure constant, is proportional to the substance's entropy; and, two, that the increase in a substance's free energy for a given change in pressure, temperature constant, is proportional to the substance's volume.

Evidently

$$-S = \left(\frac{\delta G}{\delta T}\right)_{\substack{P \text{ constant} \\ (\delta P=0)}}$$

and

$$V = \left(\frac{\delta G}{\delta P}\right)_{\substack{T \text{ constant} \\ (\delta T=0)}} .$$

In words, the equation $\delta G = -S\delta T + V\delta P$ says: the change in free energy of a substance that is in internal equilibrium (and whose chemical composition or mass does not change) is equal to the change in its temperature times the rate at which its free energy changes as its temperature alone changes, plus the change in its pressure times the rate at which its free energy changes as its pressure alone changes.

An equation of the form $\delta G = -S\delta T + V\delta P$ holds for both liquid water and water vapor.

$$\delta G^{\text{liq}}_{\text{H}_2\text{O}} = -S^{\text{liq}}_{\text{H}_2\text{O}}\delta T + V^{\text{liq}}_{\text{H}_2\text{O}}\delta P$$

$$\delta G^{\text{gas}}_{\text{H}_2\text{O}} = -S^{\text{gas}}_{\text{H}_2\text{O}}\delta T + V^{\text{gas}}_{\text{H}_2\text{O}}\delta P \qquad (\dagger)$$

Now, it has been seen that for every point on the equilibrium curve illustrated in Fig. 33.1, $G^{\text{liq}}_{\text{H}_2\text{O}} = G^{\text{gas}}_{\text{H}_2\text{O}}$. This means that as one moves along the curve, T and P change in such a way that the change produced in $G^{\text{liq}}_{\text{H}_2\text{O}}$ is always equal to the change produced in $G^{\text{gas}}_{\text{H}_2\text{O}}$. In other words, for values of δT and δP that do not take one off the curve,

$$\delta G^{\text{liq}}_{\text{H}_2\text{O}} = \delta G^{\text{gas}}_{\text{H}_2\text{O}} .$$

* If in their respective units δP and δV are 10^{-10}, for example, and $V = 10$, the term $\delta P \cdot \delta V$ is one-hundred-billion times smaller than the term $V \cdot \delta P$.

† In writing these equations, it has been assumed that $T^{\text{liq}}_{\text{H}_2\text{O}} = T^{\text{gas}}_{\text{H}_2\text{O}} = T$ and that $P^{\text{liq}}_{\text{H}_2\text{O}} = P^{\text{gas}}_{\text{H}_2\text{O}} = P$.

In thermodynamic terms, this implies that the values of δT and δP must be such that

$$-S^{\text{liq}}_{\text{H}_2\text{O}}\delta T + V^{\text{liq}}_{\text{H}_2\text{O}}\delta P = -S^{\text{gas}}_{\text{H}_2\text{O}}\delta T + V^{\text{gas}}_{\text{H}_2\text{O}}\delta P.$$

Solving for the value of the ratio of δP to δT that maintains $G^{\text{liq}}_{\text{H}_2\text{O}} = G^{\text{gas}}_{\text{H}_2\text{O}}$, one finds that

$$\left(\frac{\delta P}{\delta T}\right)_{\substack{\text{along the}\\ \text{equil. curve}}} = \frac{S^{\text{gas}}_{\text{H}_2\text{O}} - S^{\text{liq}}_{\text{H}_2\text{O}}}{V^{\text{gas}}_{\text{H}_2\text{O}} - V^{\text{liq}}_{\text{H}_2\text{O}}}.$$

This may be written in a slightly different form. Since along the equilibrium curve $G^{\text{liq}}_{\text{H}_2\text{O}} = G^{\text{gas}}_{\text{H}_2\text{O}}$, i.e. since

$$E^{\text{liq}}_{\text{H}_2\text{O}} - TS^{\text{liq}}_{\text{H}_2\text{O}} + PV^{\text{liq}}_{\text{H}_2\text{O}} = E^{\text{gas}}_{\text{H}_2\text{O}} - TS^{\text{gas}}_{\text{H}_2\text{O}} + PV^{\text{gas}}_{\text{H}_2\text{O}},$$

it follows that when liquid and gaseous water are in equilibrium with each other

$$S^{\text{gas}}_{\text{H}_2\text{O}} - S^{\text{liq}}_{\text{H}_2\text{O}} = \frac{(E+PV)^{\text{gas}}_{\text{H}_2\text{O}} - (E+PV)^{\text{liq}}_{\text{H}_2\text{O}}}{T}$$

$$= \frac{H^{\text{gas}}_{\text{H}_2\text{O}} - H^{\text{liq}}_{\text{H}_2\text{O}}}{T}$$

$$\equiv \frac{\Delta H}{T}.$$

Introducing the term "VapPres" for P, to express explicitly the equilibrium condition written variously above as "along the equilibrium curve" or "$G^{\text{liq}}_{\text{H}_2\text{O}} = G^{\text{gas}}_{\text{H}_2\text{O}}$," one has

$$\frac{\delta(\text{VapPres})}{\delta T} = \frac{\Delta H}{T(V^{\text{gas}}_{\text{H}_2\text{O}} - V^{\text{liq}}_{\text{OH}_2})}.$$

This is known as the Clausius-Clapeyron equation.

For water at its normal boiling point, $\Delta H = 540$ cal/gm $= 9717$ cal/mole $= 4.01 \times 10^5$ cc-atm./mole, $T = 373.2°\text{K}$, $V^{\text{liq}}_{\text{H}_2\text{O}} \approx 1$ cc/gm $= 18$ cc/mole, and $V^{\text{gas}}_{\text{H}_2\text{O}} \approx RT/P = (82.06$ cc-atm./mole/deg$)(373.2°\text{K})/(1$ atm.$) = 30{,}600$ cc/mole. Hence,

$$\left.\frac{\delta(\text{VapPres})}{\delta T}\right|_{\substack{\text{H}_2\text{O at}\\ 100°\text{C}}} \approx \frac{4.01 \times 10^5 \text{ cc-atm./mole}}{373.2°\text{K}\,(30{,}600 - 18)\text{ cc/mole}}$$

$$= 0.0351 \text{ atm./deg}$$

$$= 26.6 \text{ mm Hg/deg.}$$

The experimental values of the vapor pressure of water at 99 and 101°C, respectively, are $(760 - 26.8)$ and $(760 + 27.6)$ mm Hg.

34

Configurational Entropy and Colligative Properties

$$-\Delta S \cdot \delta T + \Delta V \cdot \delta P + RT \cdot \delta N_2 = 0$$

If sugar, salt, or some other solute is added to liquid water, the contribution that the water makes to the entropy of the liquid phase increases and, therefore, the molar free energy of the water in the liquid phase decreases. Hence, if T and P represent a temperature and pressure at which pure liquid water is in equilibrium with its vapor, so that at this temperature and pressure

$$G^{\text{liq}}_{\text{H}_2\text{O}}(T, P, N^{\text{liq}}_{\text{H}_2\text{O}} = 1) = G^{\text{gas}}_{\text{H}_2\text{O}}(T, P),$$

then, assuming that the solute is non-volatile,*

$$G^{\text{liq}}_{\text{H}_2\text{O}}(T, P, N^{\text{liq}}_{\text{H}_2\text{O}} < 1) < G^{\text{gas}}_{\text{H}_2\text{O}}(T, P).$$

One sees from this inequality that water vapor at the original temperature T and pressure P should condense on the liquid phase. This can be prevented in two ways: one, by decreasing the pressure of the water vapor (this increases the entropy of the water vapor and decreases its free energy) or, two, by increasing the temperature of the system (this decreases the free energies of both the liquid water and the water vapor, but the latter more rapidly than the former). The first alternative corresponds to the statement that the addition of a non-volatile solute to a solvent decreases the solvent's vapor pressure. The second alternative corresponds to the statement that the addition of a non-volatile solute to a solvent increases the solvent's boiling point. These remarks can be made more precise by taking into account quantitatively the effect a solute has upon the configurational entropy of the solvent.

* A non-volatile solute in the liquid phase has no effect on the thermodynamic properties of the vapor phase.

It was seen in Chapter 25 that the solvent's contribution to the configurational entropy of a mixture of molecules of approximately the same size is $-R \ln N_1$. As the solvent's concentration changes, its configurational entropy per mole changes by an amount $-R\delta(\ln N_1)$ and, in the absence of thermal effects, its total molar entropy changes also by this amount. Correspondingly, the solvent's molar free energy changes by the amount $-T[-R\delta(\ln N_1)] = RT\delta(\ln N_1)$. This concentration-dependent change in free energy occurs independently of and in addition to changes that may occur in the solvent's free energy owing to changes in its temperature and its pressure; these latter changes, it was seen in the previous chapter, are equal, respectively, to $-S_1^{\text{liq}}\delta T$ and $+V_1^{\text{liq}}\delta P$.* Thus, for small changes in temperature and pressure and composition,

$$\delta G_1^{\text{liq}} = -S_1^{\text{liq}}\delta T + V_1^{\text{liq}}\delta P + RT\delta(\ln N_1).$$

For liquid and vapor to coexist, the changes δT, δP, and $\delta(\ln N_1)$ in the equilibrium values of T, P, and N_1 must be such that

$$-S_1^{\text{liq}}\delta T + V_1^{\text{liq}}\delta P + RT\delta(\ln N_1) = -S_1^{\text{gas}}\delta T + V_1^{\text{gas}}\delta P.$$

In this equation P represents the partial pressure of the solvent vapor. If the solute is non-volatile and there are no other gases present, P is equal to the total pressure of the gas phase.

When $\delta(\ln N_1) = 0$, this equation reduces to the expression from which the Clausius-Clapeyron equation was previously obtained.

When the solvent vapor behaves as an ideal gas,

$$V_1^{\text{gas}}\delta P = \frac{RT}{P}\,\delta P = RT\delta(\ln P).$$

When also $V_1^{\text{gas}} \gg V_1^{\text{liq}}$, the term $V_1^{\text{liq}}\delta P$ is unimportant and can be dropped. Under these conditions, to maintain equilibrium between the liquid solvent and solvent vapor δT, $\delta(\ln P)$ and $\delta(\ln N_1)$ must be such that

$$\Delta S \cdot \delta T - RT\delta(\ln P) + RT\delta(\ln N_1) = 0.$$

In this equation $\Delta S = S_1^{\text{gas}} - S_1^{\text{liq}} > 0$.

From this last expression it is seen that in order to maintain liquid-vapor equilibrium during an isothermal† change in concentration, the pressure of the gas must change in such a way that

$$-RT\delta(\ln P) + RT\delta(\ln N_1) = 0.$$

* The subscript 1 refers to the solvent, the subscript 2 to the solute.
† Isothermal means $T =$ constant, $\delta T = 0$.

In other words,

$$\delta(\ln P) - \delta(\ln N_1) = \delta(\ln P - \ln N_1) = \delta[\ln (P/N_1)] = 0.$$

That is to say, as N_1 changes P must change in such a way that there is no change in the value of the logarithm of the ratio P/N_1. The ratio itself must therefore be a constant:

$$\frac{P}{N_1} = \text{constant.}$$

This is Raoult's Law. The value of the constant is P^0, the equilibrium pressure of the vapor when $N_1 = 1$.

To maintain liquid-vapor equilibrium during an isobaric* change in concentration, the temperature of the system must change in such a way that

$$\Delta S \cdot \delta T + RT\delta(\ln N_1) = 0. \qquad \text{(VapPres constant)}$$

Now, $\ln N_1 = \ln (1 - N_2)$. For dilute solutions $N_2 \approx 0$ and $\ln (1 - N_2) \approx -N_2$. Hence, for dilute solutions

$$\Delta S \cdot \delta T - RT\delta N_2 = 0. \qquad \text{(VapPres constant)}$$

This equation shows that if there is to be no change in the vapor pressure of the solvent as the solute concentration increases, the temperature of the system must increase. The increase δT in T that maintains the solvent's vapor pressure constant, say at one atmosphere, when the solute concentration increases from $N_2 = 0$ to $N_2 = N_2$, N_2 a small number, so that one may take $\delta N_2 = N_2$, is given by the following expression:

$$\delta T = \frac{RT}{\Delta S} \cdot N_2,$$

or

$$\delta T = k_B \cdot N_2,$$

where

$$k_B = \frac{RT}{\Delta S}.$$

k_B is called "the boiling point constant on the mole fraction scale." For many non-associated liquids at their normal boiling points, $\Delta S\,(= S_1^{\text{gas}} - S_1^{\text{liq}}) \approx 20$ cal/deg-mole (Trouton's Rule). In these cases,

$$k_B \approx \frac{2T_{nbp}}{20} = \frac{T_{nbp}}{10}.$$

* Isobaric means P = constant, $\delta P = 0$. P here is the pressure of the solvent vapor.

For non-associated liquids the boiling point constant on the mole fraction scale is approximately one-tenth the normal boiling temperature of the solvent on the Kelvin scale (Table 34.1).

TABLE 34.1

Comparison of Normal Boiling Temperatures of Pure Substances with their Boiling Point Constants on the Mole Fraction Scale

Compound	$\frac{T_{nbp}(°\text{K})}{10}$	$k_B(°\text{K})$
Chlorine	23.85	23.3
Phosgene	28.14	29.0
Ethyl ether	30.76	29.2
Methyl sulfide	30.93	30.4
Dichloromethane	31.31	31.0
Carbon disulfide	31.94	31.0
Bromine	33.19	33.0
Hexane	34.21	34.0
Ethyl nitrate	38.79	37.0
Chlorobenzene	40.52	42.0
Butyric acid	43.66	44.8

Since at the normal boiling point

$$\Delta S = \frac{\Delta H}{T_{nbp}},$$

the equation $\delta T = (RT/\Delta S)N_2$ may also be written (bp = boiling point, nbp = normal boiling point)

$$\delta T_{bp} = \frac{RT^2_{nbp}}{\Delta H} \cdot N_2 .$$

This equation predicts that the boiling point of a one molal aqueous sucrose solution will be higher than the boiling point of the pure solvent by approximately the amount

$$\delta T_{bp} = \frac{(1.987 \text{ cal/°K/mole})(373.2°\text{K})^2}{9717 \text{ cal/mole}} \cdot \frac{1}{1 + \frac{1000}{18.0}}$$

$$= 0.50°\text{K}.$$

The equation $\delta T = (RT/\Delta S)N_2$ is also applicable to changes in freezing temperature. T in this case represents the solvent's normal freezing point

δT the change in its freezing point produced by an addition of solute to a concentration N_2, and $\Delta S = S_1^{\text{solid}} - S_1^{\text{liq}}$. Since $S_1^{\text{liq}} > S_1^{\text{solid}}$, ΔS is in this case a negative number. Thus, whereas the addition of a non-volatile solute elevates the solvent's boiling point, the presence of a solute insoluble in the solid solvent depresses the solvent's freezing point. The equation predicts that the freezing point of a one molal aqueous sucrose solution should differ from the freezing point of the pure solvent by approximately the amount

$$\delta T_{fp} = \frac{(1.987\ \text{cal/°K/mole})(273.2\text{°K})}{(-5.28\ \text{cal/°K/mole})} \cdot \frac{1}{56.5}$$

$$= -1.82\text{°K}.$$

The expression $\delta T/\delta N_2 \approx RT/\Delta S_\sigma$ has a simple geometrical proof, with which we conclude this section.

Fig. 34.1 is an expanded view of the free energy curves for the solid, pure liquid, and solution phases of a substance in the neighborhood of the solid-liquid crossover point, B.

Over small temperature intervals the three curves may be represented by straight lines. For dilute solutions G^{solution} runs nearly parallel to $G^{\text{pure liquid}}$. Its downward displacement from the latter is equal to the change in the solvent's configurational entropy in going from pure solvent ($N_2 = 0$) to a dilute solution ($N_2 = \delta N_2$), multiplied by the absolute temperature; i.e.

$$AB\ (\approx DE) = T[-R \ln (N_1)] = T[-R \ln (1 - N_2)] \approx RT\delta N_2 .$$

Now from Fig. 34.1

$$DE = CE - CD.$$

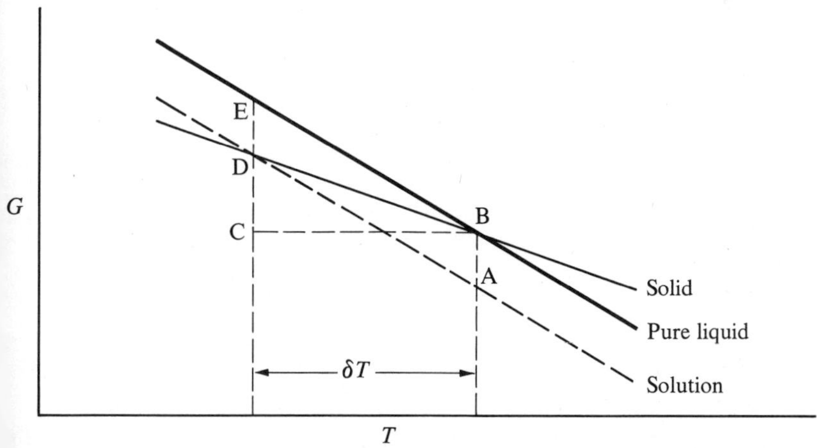

Fig. 34.1

Since $\delta G/\delta T = -S$, it follows that

$$\frac{CE}{\delta T} = -S^{\text{pure liquid}} \quad \text{and} \quad \frac{CD}{\delta T} = -S^{\text{solid}}.$$

Therefore,

$$DE = -S^{\text{pure liquid}}\,\delta T + S^{\text{solid}}\,\delta T$$
$$= \Delta S_\sigma \delta T.$$

In other words,

$$RT\delta N_2 \approx \Delta S_\sigma \delta T.$$

Qualitative Résumé of Chapters 33 and 34

The principal conclusions of Chapters 33 and 34 can be summarized in this remark: Maintenance of equilibrium with respect to the exchange of matter between two different phases involves manipulating the temperature, the pressure, and the solute concentration in such ways as to make the entropy changes that would occur in different parts of the universe for a small change of state add up to zero.

In one sense, this is obvious. When a chemical system and its thermal surroundings are at equilibrium with respect to an event, their joint entropy does not change for a small change in the degree of advancement of the event. Hence, any change within one part of the universe in the rate of change of the local entropy with respect to the degree of advancement of the event must at equilibrium be balanced in some other part of the universe by an opposite change in the rate of change of the local entropy with respect to the degree of advancement of the event.

More specifically, when a chemical system σ and its thermal surroundings θ are at equilibrium with respect to the change

$$\text{liquid} = \text{gas, or solid},$$

$$\Delta S_\sigma + \Delta S_\theta = 0,$$

where

$$\Delta S_\sigma = S_1^{\text{gas, or solid}} - S_1^{\text{liq}},$$

$$\Delta S_\theta = \frac{\Delta E_\theta}{T} = -\frac{\Delta H_\sigma}{T}$$

$$= -\frac{H_1^{\text{gas, or solid}} - H_1^{\text{liq}}}{T},$$

and where the entropy contribution S_1^{liq} of the solvent to the entropy of th[e] liquid phase is composed of two parts: its contribution $(S_{th}^{\text{liq}})_1$ to the liquid' thermal entropy and its contribution $(S_{cf}^{\text{liq}})_1$ to the liquid's configurational entropy.

$$S_1^{\text{liq}} = (S_{th}^{\text{liq}})_1 + (S_{cf}^{\text{liq}})_1.$$

A small addition of a second component to the liquid phase has generally a relatively small effect on the molar enthalpy of the solvent and no effect whatsoever on the molar enthalpy and entropy of the gas if the added substance is non-volatile; the molar enthalpy and entropy of the solid, also, are unaffected if the solute is insoluble in the crystalline solid. The configurational entropy of the solvent, however, is increased. Thus, of the six quantities H_1^{liq}, H_1^{gas}, S_1^{gas}, H_1^{solid}, S_1^{solid}, and S_1^{liq}, only the last one is seriously affected by the addition to the liquid phase of a non-volatile, solid-insoluble solute. As the configurational entropy of a solvent increases during an isothermal, isobaric increase in solute concentration, the difference between the molar entropy of the solvent vapor and the molar entropy of the liquid solvent, with which the vapor was perhaps in equilibrium, decreases. The original value of the difference $(S_1^{\text{gas}} - S_1^{\text{liq}})$ can be restored—and equilibrium with respect to vaporization of the liquid or condensation of its vapor thereby maintained—if the molar entropy of the vapor is increased. This can be accomplished by decreasing the vapor's pressure. One concludes from this, in agreement with Raoult's Law, that the vapor pressure of a solvent decreases as the concentration of the solute increases.

Alternatively, as the configurational entropy of a solvent initially in equilibrium with its vapor increases owing to an increase in solute concentration and, consequently, as the formerly vanishing sum $\Delta S_\sigma + \Delta S_\theta$ for the vaporization process ceases to vanish because the increased value of S_1^{liq} makes

$$|\Delta S_\sigma| < |\Delta S_\theta|,$$

one may try to restore the equilibrium condition $\Delta S_\sigma + \Delta S_\theta = 0$ by diminishing $|\Delta S_\theta|$. There would then be no need to decrease the pressure of the vapor. An easy way to diminish $|\Delta S_\theta|$ is to increase the temperature. Thus, as the concentration of a non-volatile solute increases, the temperature at which the vapor pressure remains constant, say at one atmosphere, increases; that is to say, as the solute concentration increases, the solution's boiling temperature increases.

On the other hand, as the configurational entropy of a solvent initially in equilibrium with its pure solid increases owing to an increase in the solute concentration in the liquid phase, the formerly vanishing sum $\Delta S_\sigma + \Delta S_\theta$ for the freezing process ceases to vanish because the increased value of S_1^{liq}, which already was larger than S_1^{solid}, makes

$$|\Delta S_\sigma| > |\Delta S_\theta|.$$

To restore the equilibrium condition, $|\Delta S_\theta|$ must be increased. Probably the easiest way to do this is to decrease the temperature. Thus, as the concentration of the solute in the liquid phase increases, the temperature at which the solvent is in equilibrium with its pure solid decreases; that is to say, as the solute concentration increases, the solution's freezing point decreases.

If the configurational entropy of a solvent initially in equilibrium with its vapor is left unchanged—i.e. if the solute concentration is not altered—but if the temperature of the system is increased, the formerly vanishing sum $\Delta S_\sigma + \Delta S_\theta$ for the vaporization process fails to vanish because the increased value of T makes

$$|\Delta S_\theta| < |\Delta S_\sigma| = S_1^{\text{gas}} - S_1^{\text{liq}}.$$

To restore the equilibrium condition, ΔS_σ must be decreased. Two ways to do this are: one, increase S_1^{liq} by adding some soluble impurity to the liquid phase or, two, decrease S_1^{gas} by increasing the pressure. (Increasing the pressure usually decreases S_1^{liq}, also; this is a relatively small, and usually negligible, effect, except near the critical point.) Thus, as the system's temperature increases without any change in the solute concentration, the pressure at which the solvent remains in equilibrium with its vapor increases; that is to say, as the temperature increases, the solvent's vapor pressure increases.

Problems for Chapters 33 and 34

1. When liquid water is in equilibrium with its pure solid, $G_{H_2O}^{\text{liq}}$ is equal to ———(a). Addition of an impurity to the liquid phase ———(b) the ———(c) entropy of the liquid water and causes $G_{H_2O}^{\text{liq}}$ to ———(d). If, as is usually the case, the impurity is ———(e) in the ———(f) phase, there will be no change in the thermodynamic properties of the (g)——— Hence, $G_{H_2O}^{\text{solid}}$ will now be ———(h) than ———(i), and the ———(j) should ———(k). This can be prevented by ———(l) the temperature; this causes both ———(m) and ———(n) to ———(o), but ———(p) more than ———(q) because ———(r) is ———(s) than ———(t). Eventually, therefore, there will be restored the equilibrium condition ———(u). Thus, as the solute concentration in the liquid phase ———(v), the ———(w) at which the liquid solvent remains in ———(x) with its ———(y) ———(z).
2. At the triple point of water, what relations exist between the molar free energies of the solid, liquid, and gaseous water? How many independent conditions does this place upon the temperature and pressure of the system? Consequently, how many degrees of freedom are there in a

system that contains simultaneously pure solid, liquid, and gaseous water?

3. Show that the difference δT between the *fp* or *bp* of a dilute solution and the *fp* or *bp* of the pure solvent is given by the equation

$$\delta T = \frac{RT^2_{ntp}}{\Delta H} \cdot \frac{M_1}{1000} \cdot m,$$

where T_{ntp} is the normal transition temperature, M_1 is the molecular weight of the solvent, m is the molality of the solute, and $\Delta H = H_1^{\text{solid, or gas}} - H_1^{\text{liq}}$. What determines the sign of δT?

4. Show that because the heat of sublimation of ice is greater than the heat of vaporization of liquid water, the vapor pressure curve of ice in the neighborhood of 0°C must increase more rapidly with temperature than does the corresponding vapor pressure curve of liquid water. (If this were not so, ice might melt on being cooled and water might freeze on being heated.)

5. (a) Using Raoult's Law, estimate the vapor pressure of a one molal aqueous sucrose solution at 100°C. (b) Using Raoult's Law and the Clausius-Clapeyron equation, estimate the vapor pressure of a one molal aqueous sucrose solution at 101°C. (c) Using the previous results, estimate the normal boiling point of a one molal aqueous sucrose solution.

6. Show that for substances not subject to spontaneous, irreversible, entropy-producing changes,

$$\left(\frac{\delta S}{\delta V}\right)_E = \frac{P}{T}.$$

Argue, hence, that for an ideal gas

$$\delta S = nR\delta(\ln V). \qquad (T \text{ constant})$$

From this last equation, by integration (or summation), is obtained the useful result $S(V_2) - S(V_1) = nR \ln (V_2/V_1)$.

7. When liquid water is in equilibrium with its solid, ———(a) is equal to ———(b). Increasing the pressure on the system causes both ———(c) and ———(d) to ———(e), but ———(f) more than ———(g) because ———(h) is ———(i) than ———(j). Hence, ———(k) will now be ———(l) than ———(m), and the ———(n) should ———(o). This can be prevented by ———(p) the ———(q); this causes both ———(r) and ———(s) to ———(t), but ———(u) more than ———(v) because ———(w) is ———(x) than ———(y). Eventually, therefore, the equilibrium condition ———(z) will be restored. Thus, as the pressure on

the system ———(a′), the ———(b′) at which the liquid remains in ———(c′) with the ———(d′) ———(e′).

8. According to the Clausius-Clapeyron equation coexisting water and ice will continue to coexist when a change δP is made in the pressure of the system if concomitantly there is a temperature change δT which is such that

$$\frac{\delta P}{\delta T} = \frac{\Delta S}{\Delta V}. \qquad (N_2 \text{ constant})$$

By approximately how much is the water-ice coexistence temperature altered by a change in pressure from 1 to 11 atm.?

THE GIBBS-POYNTING EFFECT

9. When a liquid solvent is in equilibrium with its vapor, ———(a) is equal to ———(b). If, now, the pressure on the liquid phase is increased by the amount δP^{liq}, the molar free energy of the liquid is ———(c) by the amount ———(d). To maintain the equilibrium condition ———(e) without changing either the temperature or the composition of the system, the pressure of the vapor must be ———(f) by the amount ———(g). In short, squeezing on a liquid ———(h) its ——— (i); however, since generally the molar volume of the liquid is very much ———(j) than the ———(k) of the ———(l), the change in ———(m) is generally very much ———(n) than the change in ———(o). At 25°C, for example, increasing the pressure on liquid water from 1 to 11 atm. increases its vapor pressure only ———(p).

THE OSMOTIC PRESSURE EFFECT

10. The configurational entropy of the solvent in a solution that contains a solute at a low concentration N_2 is ———(a) than the configurational entropy of the pure solvent (the latter is ———(b)) by approximately the amount ———(c). Therefore, the molar free energy of the solvent in the solution phase is ———(d) than the molar free energy of the ———(e) by approximately the amount ———(f). If the solution is physically separated from the pure solvent by a membrane permeable to the solvent but not to the solute, ———(g) molecules, given time, wil diffuse spontaneously from ———(h) into ———(i). This spontaneou diffusion of molecules from a region where their concentration is ——— (j) to one where their concentration is ———(k) can be prevented if th molar free energy of the solvent in the solution phase is ———(l). On way to do this is to ———(m) the pressure on the ———(n). (Decreasin

the ———(o) on the ———(p) would help to make equal the ———(q), but there are obvious practical limitations to this method.) If the solution is essentially incompressible, and if the molar volume of the solvent in the solution phase is essentially equal to the molar volume of the pure solvent, a pressure increase π on the solution phase will ———(r) the molar free energy of the solvent in the solution phase by the amount ———(s). Equilibrium between the solvent in the solution phase and solvent in the pure solvent phase will exist if $\pi =$ ———(t). Now, by definition, $N_2 =$ ———(u). For dilute solutions, therefore, $N_2 \approx$ ——— (v). But the product ———(w) is essentially just the total ———(x) of the ———(y) phase. Furthermore, by definition the ratio ———(z) is the solute concentration in the solution phase expressed in ———(a′). Hence, for dilute solutions, $\pi \approx$ ———(b′). π is called the solution's osmotic pressure. Like the solution's density, refractive index, vapor pressure, freezing point, and boiling point, the osmotic pressure is an ———(c′) property of the solution. Because the molar volume of a condensed phase is relatively small, a relatively large change in pressure is required to balance in its effect on a solvent's molar free energy the effect of a modest change in concentration. For example, at 25°C the osmotic pressure of a 1 molar aqueous sucrose solution is ———(d′) atm. The osmotic pressure of a solution is a very sensitive measure of the solute concentration.

11. What is the osmotic pressure at 25°C of an aqueous solution whose freezing point is −0.00186°C?
12. For a system to be at equilibrium with respect to the migration of thermal energy from one part of the system to another, the ———(a) must be ———(b) throughout the system. For a system to be at equilibrium with respect to the migration of a chemical from one part of the system to another, the ———(c) must be ———(d) throughout the system.

HE IDEAL SOLUBILITY EQUATION

Vhen a solution is saturated with respect to a solid, i.e. when the solute in olution is in equilibrium with pure solid solute,

$$(S_2^{\text{soln}} - S_2^{\text{solid}}) - \frac{(H_2^{\text{soln}} - H_2^{\text{solid}})}{T} = 0.$$

 this equation S_2^{soln} and H_2^{soln} are the contributions the solute (component 2) akes to the entropy and enthalpy of the solution. According to the lattice odel of solutions,

$$S_2^{\text{soln}} = (S_{th}^{\text{soln}})_2 - R \ln N_2 .$$

Hence, at saturation

$$-R \ln N_2 = \frac{(H_2^{\text{soln}} - H_2^{\text{solid}})}{T} - [(S_{th}^{\text{soln}})_2 - S_2^{\text{solid}}].$$

If this equation holds for all values of N_2, the solution is said to be an ideal solution.

When $N_2 = 1$, the solution corresponds to pure liquid component 2. The temperature at this point corresponds to the normal melting point of component 2. Calling this temperature T_{mp}, one finds that

$$[(S_{th}^{\text{soln}})_2 - S_2^{\text{solid}}] = \frac{(H_2^{\text{soln}} - H_2^{\text{solid}})}{T_{mp}}.$$

In effect what this equation says is that an ideal solution is one in which the contributions a component makes to the enthalpy and the thermal entropy of the solution are the same as the contributions the component makes to the corresponding properties of its own pure liquid. In such cases as this, the solubility equation becomes

$$-R \ln N_2 = \Delta H \left(\frac{1}{T} - \frac{1}{T_{mp}}\right),$$

where

$$\Delta H = H_2^{\text{pure liquid}} - H_2^{\text{solid}}.$$

According to this equation, the solubility of a solute in a solvent with which it forms an ideal solution is independent of the properties of the solvent. The solute's ideal solubility depends only upon its heat of fusion, its melting point, and the temperature.

13. Calculate the ideal solubility of naphthalene at 25°C. The heat of fusion of naphthalene is 4,440 cal/mole; its melting point is 80.05°C.
14. Which would you expect to be more soluble in benzene at 25°C, phenanthrene (*mp* 100°C) or anthracene (*mp* 217°C)?
15. If ΔH is independent of the temperature, a plot of $-\ln N_2$ at saturation against the reciprocal of the absolute temperature yields for ideal solutions a ———(a) whose slope is ———(b) and whose intercept at $1/T = 0$ is ———(c). This straight line can be determined from any of the following three pairs of data: (i) the melting temperature and the solubility at one other temperature, (ii) ———(d), (iii) ———(e).

LATTICE VACANCIES

Illustrated below schematically is a crystal lattice that has produced on its surface through an atomic displacement an adsorbed hole that subsequently becomes through a second atomic displacement an absorbed hole. Through further atomic displacements this hole can wander throughout the crystal lattice.

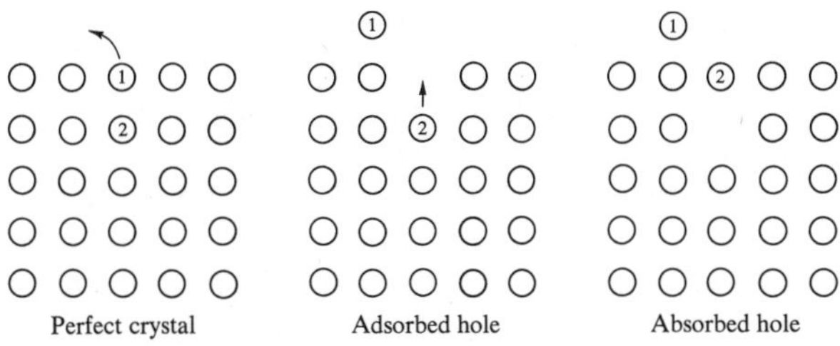

Perfect crystal Adsorbed hole Absorbed hole

Formation of lattice vacancies at the expense of the thermal energy of the surroundings diminishes the entropy of the surroundings; simultaneously, however, the configurational entropy of the crystal is increased, for there are many ways the vacancies formed might be distributed over the available lattice sites. In this respect, a crystal that contains N atoms and n vacancies is like a solution that contains N solvent molecules and n solute molecules. Referring to the discussion of configurational entropy in Chapter 25, one sees that formation of one additional defect will increase the configurational entropy of a crystal by the amount $k \ln (N + n)/n$ when $n \gg 1$; simultaneously, if the thermal energy required is ε cal/vacancy, the entropy of the thermal surroundings will change by the amount $-\varepsilon/T$. Hence, if formation of lattice vacancies does not seriously affect the thermal entropy of the crystal,

$$\Delta S_\sigma + \Delta S_\theta \approx k \ln \left(\frac{N + n}{n}\right) - \frac{\varepsilon}{T}.$$

If the crystal already is saturated with vacancies, $\Delta S_{\text{total}} = \Delta S_\sigma + \Delta S_\theta = 0$. Therefore, at saturation

$$\frac{n}{N + n} \approx e^{-\varepsilon/kT}.$$

At 0°K the equilibrium value of n is zero. With rising temperature the equilibrium number of vacancies rapidly increases.

6. For many solids ε has a value of approximately 1 electron volt per vacancy; this is approximately 23,000 calories per mole of vacancies. Using this value for the "heat of formation of vacancies," estimate for such solids the equilibrium mole fraction of lattice vacancies at 1000°K.

Quantitative Résumé of Chapters 33 and 34

The quantitative conclusions of Chapters 33 and 34 are summarized in the equation

$$-\Delta S \cdot \delta T + \Delta V \cdot \delta P + RT \cdot \delta N_2 = 0.$$

This equation is of the form

$$a \cdot x + b \cdot y + c \cdot z = 0,$$

where, for example,

$$\begin{aligned} x &= \delta T \\ y &= \delta P \\ z &= \delta N_2 \end{aligned} \qquad \text{and} \qquad \begin{aligned} a &= -\Delta S \\ b &= \Delta V \\ c &= RT. \end{aligned}$$

For small values of the variables x, y, z, the coefficients a, b, c can be treated as constants.

The equation $ax + by + cz = 0$ has two essentially different solutions. More precisely (see below), it has two linearly independent solutions. It permits, so to speak, two degrees of freedom. Any pair of variables may be considered as independent variables and values for this pair may be selected quite arbitrarily; but when this has been done, the value of the third variable is determined by the equation. If the values of z and y are given, the value of x can be calculated. For example, one might take (I) $z = 0$ and $y = -a$; then $x = b$. Or, one might take (II) $z = -a$ and $y = 0$; then $x = c$. These two solutions are tabulated below. A third solution with a vanishing x component is listed also.

	(I)	(II)	(III)
x	b	c	0
y	$-a$	0	$-c$
z	0	$-a$	b

The components of solution (III) can be expressed as a linear combination of the corresponding components of solutions (I) and (II). Thus,

$$x_{\text{III}} = \frac{c}{a}\, x_{\text{I}} - \frac{b}{a}\, x_{\text{II}}$$

$$y_{\text{III}} = \frac{c}{a}\, y_{\text{I}} - \frac{b}{a}\, y_{\text{II}}$$

$$z_{\text{III}} = \frac{c}{a}\, z_{\text{I}} - \frac{b}{a}\, z_{\text{II}}\,.$$

These equations illustrate the meaning of the remark that the equation $ax + by + cz = 0$ has only two linearly independent solutions.

From solutions I, II, and III, one finds that

$$\left(\frac{y}{x}\right)_{z=0} = -\frac{a}{b} \qquad \left(\frac{x}{z}\right)_{y=0} = -\frac{c}{a} \qquad \left(\frac{y}{z}\right)_{x=0} = -\frac{c}{b}.$$

In terms of the variables δT, δP, δN_2, and the coefficients $-\Delta S$, ΔV, RT, these equations read

$$\left(\frac{\delta P}{\delta T}\right)_{\delta N_2=0} = \frac{\Delta S}{\Delta V} \qquad \left(\frac{\delta T}{\delta N_2}\right)_{\delta P=0} = \frac{RT}{\Delta S} \qquad \left(\frac{\delta P}{\delta N_2}\right)_{\delta T=0} = -\frac{RT}{\Delta V}.$$

The first equation is the Clausius-Clapeyron equation. The second equation is the freezing-point-lowering or the boiling-point-rise equation. The third equation is equivalent to Raoult's Law.* Usually the subscript conditions $\delta N_2 = 0$, $\delta P = 0$, and $\delta T = 0$ are replaced by the symbols N_2, P, T. Used as subscripts, these symbols stand for the phrases "N_2 constant," "P constant," and "T constant." With this notation, the equations read

$$\left(\frac{\delta P}{\delta T}\right)_{N_2} = \frac{\Delta S}{\Delta V} \qquad \left(\frac{\delta T}{\delta N_2}\right)_{P} = \frac{RT}{\Delta S} \qquad \left(\frac{\delta P}{\delta N_2}\right)_{T} = -\frac{RT}{\Delta V}.$$

All three equations should also carry the subscript S_{total}, meaning "S_{total} = constant," or "$\Delta S_{\text{total}} = 0$."

It may be noted from the last equations that

$$\left(\frac{\delta T}{\delta N_2}\right)_{P} = -\left(\frac{\delta P}{\delta N_2}\right)_{T} \Big/ \left(\frac{\delta P}{\delta T}\right)_{N_2}. \qquad (S_{\text{total}} \text{ constant})$$

This relation shows that the quantity on the left, the boiling-point-rise constant, can be calculated if the two quantities on the right, the change in solvent vapor pressure with composition and the change in solvent vapor pressure with temperature, are known. A numerical example of the interdependence of the three effects—the *bp*-rise effect, Raoult's Law effect, and the Clausius-Clapeyron effect—was given in the fifth problem for Chapter 34.

A Concluding Remark on Raoult's Law

The discussion in Chapter 34 of the colligative properties of solutions has been based upon the idea that the addition of a solute to a solvent increases

* Let δN_2 represent a solute concentration change from $N_2 = 0$ to $N_2 = N_2$, N_2 small. Then

$$\delta P \equiv P_1 - P_1^0 = -\frac{RT}{\Delta V}\cdot N_2 \approx -\frac{RT}{(RT/P_1^0)}\cdot N_2 = -P_1^0(1 - N_1);$$

that is, $P_1 = P_1^0 N_1$, which is Raoult's Law.

the solvent's contribution to the entropy of the liquid phase and decreases, therefore, the solvent's molar free energy in this phase, by an amount, according to the lattice model of solutions described in Chapter 25, $RT\delta(\ln N_1)$; from this result Raoult's Law was obtained. Briefly stated, the order of the discussion has been:

$$\text{Lattice Model} \xrightarrow{S=k\ln\Omega} \delta G_1^{\text{liq}} \overset{T,P\text{ constant}}{=} RT\delta(\ln N_1) \xrightarrow{G_1^{\text{liq}}=G_1^{\text{gas}}} \text{Raoult's Law.}$$

It is our purpose here to point out that the last step in this sequence may be taken in the opposite sense: Given Raoult's Law, it follows that

$$\delta G_1^{\text{liq}}(T, P \text{ constant}) = RT\delta(\ln N_1).$$

For if the condition $G_1^{\text{liq}} = G_1^{\text{gas}}$ holds for all states that are at equilibrium with respect to the evaporation or condensation of solvent, concentration changes of solvent molecules in the liquid and gaseous phases must be such that

$$\delta G_1^{\text{liq}} = \delta G_1^{\text{gas}}.$$

Now earlier in this chapter it was found that at constant temperature

$$\delta G_1^{\text{gas}} = V_1^{\text{gas}}\delta P_1^{\text{gas}}. \qquad (T\text{ constant})$$

V_1^{gas} is the molar volume of the gas. If the gas behaves as an ideal gas, $V_1^{\text{gas}} = RT/P_1^{\text{gas}}$ and

$$\begin{aligned}\delta G_1^{\text{gas}} &= \frac{RT}{P_1^{\text{gas}}}\,\delta P_1^{\text{gas}} \qquad (T\text{ constant})\\ &= RT\delta(\ln P_1^{\text{gas}}).\end{aligned}$$

Combining this last result with the equation $\delta G_1^{\text{liq}} = \delta G_1^{\text{gas}}$, one finds that

$$\delta G_1^{\text{liq}} = RT\delta(\ln P_1^{\text{gas}}). \qquad (T\text{ constant})$$

But by Raoult's Law

$$P_1^{\text{gas}} = P_1{}^0 N_1.$$

$P_1{}^0$ is the vapor pressure of the pure solvent, i.e. the equilibrium value of P_1^{gas} when $N_1 = 1$. For solutions that obey Raoult's Law, therefore,

$$\begin{aligned}\delta G_1^{\text{liq}} &= RT\delta(\ln P_1{}^0 N_1)\\ &= RT\delta(\ln N_1 + \ln P_1{}^0) \qquad (T\text{ constant})\\ &= RT\delta(\ln N_1) + RT\delta(\ln P_1{}^0).\end{aligned}$$

When $\delta P_1^{\text{liq}} = 0$, the vapor pressure of the pure solvent does not change at constant temperature. Under these conditions $\delta(\ln P_1{}^0) = 0$ and, as stated earlier,

$$\delta G_1^{\text{liq}} = RT\delta(\ln N_1). \qquad (T, P\text{ constant})$$

35

Concluding Comments

Physical Significance of Free Energy

Since by the additivity of entropy $\Delta S_{\text{total}} = \Delta S_\sigma + \Delta S_\theta$, and since by the definition of absolute temperature $\Delta S_\theta = \Delta E_\theta / T$, and since by the First Law $\Delta E_\theta = -(\Delta H_\sigma + W)$, it follows, as written many times previously, that

$$\Delta S_{\text{total}}(W = 0) = \Delta S_\sigma - \frac{\Delta H_\sigma}{T} = -\frac{\Delta G_\sigma}{T}.$$

This equation suggests a physical interpretation for the free energies of chemical compounds. Consider the formation of a compound A from its elements in their standard states. For the reaction

$$\text{Elements in their standard states} = \text{Compound } A$$

$$\Delta S_\sigma = S_A - \sum_{\text{reactants}} S_i^0$$

$$\Delta H_\sigma = H_A - \sum_{\text{reactants}} H_i^0 .$$

By convention the enthalpies of elements in their standard states are equal to zero. For the above reaction, therefore,

$$\Delta H_\sigma = H_A .$$

Hence,

$$\Delta S_{\text{total}}\,(W = 0) = \left(S_A - \frac{H_A}{T}\right) - \sum_{\text{reactants}} S_i^0$$

$$= -\frac{G_A}{T} - \sum_{\text{reactants}} S_i^0 .$$

This equation states that

> A compound's free energy, divided by $-T$ and diminished by the entropies of its constituent elements, is equal to the increase in entropy of the universe when the compound is synthesized from its elements without the production or consumption of mechanical work.

The more negative a compound's free energy, the larger tends to be the entropy increase in the universe when the compound is synthesized from its elements.

It is instructive to consider these results from the more detailed point of view of statistical thermodynamics. What is the change in entropy of the universe when a particle is synthesized in its ground level, for example, or in an excited state?

For convenience let compound A be an ensemble of loosely coupled localized particles. By Boltzmann's relation, $S_A = k \ln (N!/n_0!n_1!n_2! \ldots)$, where $N = n_0 + n_1 + n_2 \ldots.$ Synthesis of an additional particle of A in the ground level ($n_0 \gg 1$) increases the entropy of A by the amount $k \ln (N/n_0)$ per particle or, per mole, $R \ln (N/n_0)$. Simultaneously the chemical system loses the entropy associated with the reactants. For the synthesis of one mole of A in the ground level, therefore,

$$\Delta S_\sigma = R \ln (N/n_0)_A - \sum_{\text{reactants}} S_i^0 .$$

The entropy change in the thermal surroundings is $-1/T$ times the amount by which the product's enthalpy exceeds the enthalpy of the reactants. Let $\varepsilon_A{}^0$ be the energy of the ground level for a particle in compound A measured with respect to the enthalpies of the elements in their standard states. With respect to these enthalpies the enthalpy of a mole of particles in A's ground level is $N_0\varepsilon_A{}^0 + PV_A$. Therefore,

$$\Delta S_\theta = -\frac{N_0\varepsilon_A{}^0 + PV_A}{T} .$$

Combining the expressions for ΔS_σ and ΔS_θ, one finds that for the synthesis from the elements of a mole of particles of compound A in their ground level

$$\Delta S_{\text{total}} (W = 0) = R \ln (N/n_0)_A - \frac{N_0\varepsilon_A{}^0 + PV_A}{T} - \sum_{\text{reactants}} S_i^0 .$$

The same result would be obtained if one supposed that the particles of A were synthesized in an excited state. To see that this is so, consider that the synthesis is executed in two steps: synthesis of the particles in the ground level (as above) followed by their thermal promotion to an excited state. The latter step, however, does not alter the total entropy of the universe if A is in thermal equilibrium with its surroundings; it therefore makes no contribution to ΔS_{total}.

Comparison of the two expressions for ΔS_{total} shows that for an ensemble of loosely coupled localized particles

$$G_A = -RT \ln (N/n_0)_A + N_0 \varepsilon_A{}^0 + PV_A \,.$$

Substitution of the entropy expression derived in Chapter 23

$$S = R \ln Q + \frac{E_{th}}{T}$$

into the definition of the free energy

$$G = E + PV - TS$$

yields the same result

$$G_A = -RT \ln Q_A + (E - E_{th})_A + PV_A$$

when it is noted that

$$\frac{N}{n_0} = Q \qquad \text{(Chapter 23)}$$

and that $(E - E_{th})$—the difference between an object's internal energy and its thermal energy—is just the energy of its ground level per mole.

$$\begin{aligned}(E - E_{th}) &= [n_0\varepsilon_0 + n_1\varepsilon_1 + n_2\varepsilon_2 + \cdots] - [n_1(\varepsilon_1 - \varepsilon_0) + n_2(\varepsilon_2 - \varepsilon_0) \\ &\qquad + \cdots] \\ &= n_0\varepsilon_0 + n_1\varepsilon_0 + n_2\varepsilon_0 + \cdots \\ &= N\varepsilon_0 \,.\end{aligned}$$

Insertion of the definition of the partition function

$$Q = 1 + e^{-\varepsilon_1/kT} + e^{-\varepsilon_2/kT} + \cdots$$

into these results yields the following expression for the free energy of an ensemble of loosely coupled localized particles.

$$G_A = N_0\varepsilon_A{}^0 + PV_A - RT \ln (1 + e^{-\varepsilon_1/kT} + e^{-\varepsilon_2/kT} + \cdots) \,.$$

This relation expresses quantitatively two broad generalizations about chemical stability.

Forms of matter stable at low temperatures have low-lying energy levels.

Forms of matter stable at high temperatures have closely spaced energy levels.

The accompanying figure illustrates schematically these two generalizations.

HIGH T FORM
Large ϵ_0, small spacing

LOW T FORM
Small ϵ_0, large spacing

Low T form	High T form
H_2O (solid)	H_2O (liquid)
H_2O (liquid)	H_2O (gas)
H_2O (gas)	H_2 (gas) + O_2 (gas)
H_2 (gas)	H (gas)
H (gas)	H (gas) + e^- (gas)

At low temperatures the low-lying, widely spaced energy levels on the left are populated in preference to the more closely spaced, higher-lying levels on the right. This is an entropy effect, an entropy effect in the thermal surroundings, whose entropy would rapidly diminish if the energy necessary to populate the energy levels on the right were removed from it at low temperatures.

At high temperatures, where the same loss in energy by the thermal surroundings produces a smaller loss in its entropy, the closely spaced energy levels on the right are populated in preference to the more widely spaced levels on the left. This, again, is an entropy effect, an entropy effect in the chemical system—whose entropy increases as the spacing between its energy levels decreases.

In summary, the two factors that determine chemical stability are: (1) location of the ground level; by the First Law this determines the energy and, hence, the entropy of the thermal surroundings; and (2) location of the upper levels with respect to the ground level; by Boltzmann's relation this determines the entropy of the chemical system. Together these two factors determine the entropy of the universe.

At any temperature the preferred set of energy levels is the set with the smallest value of $H - TS$. The first term in this expression reflects the location of the ground level; the second term reflects the spacing of the upper levels with respect to the ground level.

At intermediate temperatures both sets of levels may be populated, if in this way the entropy of the universe is made as large as possible. Under these conditions the two sets of energy levels have the same value of $H - TS$.

Problems for Chapter 35

1. As a measure of chemical stability Planck used the function

$$\Psi = S - \frac{H}{T}\cdot$$

What is the relation between Planck's function Ψ and the free energy? Between Ψ and the entropy of the universe? Is chemical stability associated with large or small values of Ψ? Under what conditions is Ψ large?

V

The Role of Mathematics in Thermodynamics

36

Uses of T

For reversible exchanges of energy and volume between a chemical system σ and its thermal surroundings θ

$$dS_{\text{total}} = dS_\sigma + dS_\theta = dS_\sigma - \frac{dE_\sigma + PdV_\sigma}{T} = 0.$$

Now changes in the thermodynamic properties of a chemical system are determined solely by the initial and final states of the system; only indirectly are dE_σ, dV_σ, and dS_σ affected by what happens in the chemical system's surroundings. The equation $dS_\sigma - (dE_\sigma + PdV_\sigma)/T = 0$ has therefore a broader use than the word "reversible" in the first sentence might imply. It is applicable to all systems that are in internal equilibrium—provided their masses do not change, provided $W = 0$, and provided there are no changes in "other variables" (see Chapter 37). It is often written, without the subscript σ, in the form

$$dE = TdS - PdV.$$

From this it is seen that

$$T = \left(\frac{\partial E}{\partial S}\right)_V \quad \text{and} \quad P = -\left(\frac{\partial E}{\partial V}\right)_S .$$

The relation $T = (\partial E/\partial S)_V$ has many uses. It has been used in previous chapters to calculate heat engine efficiencies, Third Law entropies, and changes in thermal entropies. It has been used to introduce the parameter T into the equations for Einstein's solid. And it has been used in the derivation of Boltzmann's factor. It will now be used to identify the constant k in Boltzmann's relation $S = k \ln \Omega$, and to derive Maxwell's relations.

Boltzmann's Constant. Consider an ensemble of non-localized, independent, point particles—briefly, an ideal gas—and suppose that without altering the ensemble's internal energy its volume is doubled. What happens to Ω? Each particle of the ensemble will have accessible to it twice as many locations in space as previously it had. If the ensemble contains N particles, the value of Ω will increase by the factor 2^N. Trebling the volume would increase Ω by the factor 3^N. Halving the volume would decrease Ω by the factor $(1/2)^N$, and so forth. Symbolically,

$$\Omega(2V) = 2^N\Omega(V)$$
$$\Omega(3V) = 3^N\Omega(V)$$
$$\vdots$$
$$\Omega(1/2V) = (1/2)^N\Omega(V)$$
$$\Omega(1/3V) = (1/3)^N\Omega(V)\,.$$

More generally, for λ any rational number, E and N constant,

$$\Omega(\lambda V) = \lambda^N\Omega(V).$$

Setting $\lambda = 1/V$ one finds that

$$\Omega(V) = V^N \cdot \Omega(1).$$

By Boltzmann's relation, therefore,

$$S = Nk \ln V + k \ln \Omega(1).$$

$\Omega(1)$ is a function of E and N, but not of V. Differentiating the last expression with respect to V holding constant E and N, one finds that for an ideal gas

$$\left(\frac{\partial S}{\partial V}\right)_{E,\,N} = \frac{Nk}{V}\,.$$

Now for any substance (N constant)

$$\left(\frac{\partial S}{\partial V}\right)_E = -\left(\frac{\partial S}{\partial E}\right)_V\left(\frac{\partial E}{\partial V}\right)_S \qquad \text{Calculus}$$

$$= -\left(\frac{1}{T}\right)(-P) \qquad \text{Thermodynamic definition of } T, P$$

$$= \frac{P}{T}.$$

And for an ideal gas

$$\frac{P}{T'} = \frac{nR}{V}.$$

T has been primed in this last expression to distinguish momentarily between the temperature defined by an ideal gas and the temperature defined by the relation $T = (\partial E/\partial S)_V$. If these two temperatures are set equal to each other, it follows by the last two relations that for an ideal gas

$$\left(\frac{\partial S}{\partial V}\right)_{E,N} = \frac{nR}{V}\,.$$

Two expressions for the value of $(\partial S/\partial V)_{E,N}$ for an ideal gas have been obtained: Nk/V, from theory—$\Omega(V) = V^N\Omega(1)$—and Boltzmann's relation; and nR/V, from experiment—$PV = nRT$—and the thermodynamic definitions of T and P. Recalling that $N = nN_0$ (N_0 is Avogadro's number), one sees that for these two expressions to be equivalent, Boltzmann's constant must be equal to the gas constant per molecule.

$$k = \frac{nR}{N} = \frac{R}{N_0}\,.$$

The previous results may be used to show that the internal energy of an ideal gas does not depend upon its volume.

For an ideal gas

$$\frac{\partial^2 S}{\partial E \partial V} = 0.$$

This result is obtained by differentiating the expression $(\partial S/\partial V)_{E,N} = nR/V$ with respect to E keeping V and n (which is proportional to N) constant. Now for any system in internal equilibrium

$$\left(\frac{\partial E}{\partial V}\right)_T = -\left(\frac{\partial E}{\partial T}\right)_V \left(\frac{\partial T}{\partial V}\right)_E = -C_V \left[\frac{\partial(\partial E/\partial S)_V}{\partial V}\right]_E$$

$$= -C_V \left[\frac{\partial(\partial S/\partial E)_V^{-1}}{\partial V}\right]_E = C_V \left(\frac{\partial S}{\partial E}\right)_V^{-2} \frac{\partial^2 S}{\partial E \partial V}$$

$$= C_V T^2 \frac{\partial^2 S}{\partial E \partial V}\,.$$

Therefore, for an ideal gas

$$\left(\frac{\partial E}{\partial V}\right)_T = 0.$$

Maxwell's Relations. Using calculus and the definitions $T \equiv (\partial E/\partial S)_V$ and $P \equiv -(\partial E/\partial V)_S$, one may write

$$\begin{aligned}
\left(\frac{\partial S}{\partial V}\right)_T &= -\left(\frac{\partial S}{\partial T}\right)_V \left(\frac{\partial T}{\partial V}\right)_S \\
&= -\left(\frac{\partial S}{\partial T}\right)_V \left[\frac{\partial}{\partial V}\left(\frac{\partial E}{\partial S}\right)_V\right]_S \\
&= -\left(\frac{\partial S}{\partial T}\right)_V \left[\frac{\partial}{\partial S}\left(\frac{\partial E}{\partial V}\right)_S\right]_V \\
&= -\left(\frac{\partial S}{\partial T}\right)_V \left[\frac{\partial(-P)}{\partial S}\right]_V \\
&= \left(\frac{\partial P}{\partial T}\right)_V .
\end{aligned}$$

Alternatively, one may note that the function

$$A \equiv E - TS$$

has for its total differential the expression

$$\begin{aligned}
dA &= dE - TdS - SdT \\
&= (TdS - PdV) - TdS - SdT \\
&= -SdT - PdV.
\end{aligned}$$

The independent variables here are T and V, the same as those in the expression $(\partial S/\partial V)_T$. Evidently

$$-S = \left(\frac{\partial A}{\partial T}\right)_V \quad \text{and} \quad -P = \left(\frac{\partial A}{\partial V}\right)_T .$$

Since

$$\left[\frac{\partial}{\partial V}\left(\frac{\partial A}{\partial T}\right)_V\right]_T = \left[\frac{\partial}{\partial T}\left(\frac{\partial A}{\partial V}\right)_T\right]_V ,$$

it follows that

$$\left(\frac{\partial S}{\partial V}\right)_T = \left(\frac{\partial P}{\partial T}\right)_V .$$

Similarly, from the function

$$G \equiv E + PV - TS,$$

whose total differential is given by the expression

$$dG = -SdT + VdP,$$

one finds that

$$(\partial S/\partial P)_T = -(\partial V/\partial T)_P .$$

The relations

$$\left(\frac{\partial S}{\partial V}\right)_T = \left(\frac{\partial P}{\partial T}\right)_V \quad \text{and} \quad \left(\frac{\partial S}{\partial P}\right)_T = -\left(\frac{\partial V}{\partial T}\right)_P$$

are known as Maxwell's relations. Their form may be recalled by noting that S has the same dimensions as PV/T. Thus $(\partial S/\partial V)_T = \pm(\partial P/\partial T)_V$, and $(\partial S/\partial P)_T = \pm(\partial V/\partial T)_P$. The correct sign is selected by noting that generally $(\partial S/\partial V)_T$, $(\partial P/\partial T)_V$, and $(\partial V/\partial T)_P$ are positive, whereas generally $(\partial S/\partial P)_T$ is negative.

Maxwell's relations may be used to show that the internal energy of an ideal gas does not depend upon its volume. From the relation $dE = TdS - PdV$ it follows that

$$\left(\frac{\partial E}{\partial V}\right)_T = T\left(\frac{\partial S}{\partial V}\right)_T - P.$$

By Maxwell's relation, therefore,

$$\left(\frac{\partial E}{\partial V}\right)_T = T\left(\frac{\partial P}{\partial T}\right)_V - P.$$

This can be written

$$\left(\frac{\partial E}{\partial V}\right)_T = T^2\left[\frac{\partial(P/T)}{\partial T}\right]_V$$

$$= -\left[\frac{\partial(P/T)}{\partial(1/T)}\right]_V.$$

For an ideal gas

$$d\left(\frac{P}{T}\right)_V = d\left(\frac{nR}{V}\right)_V = 0.$$

37

Thermodynamics of Rubber Elasticity

Hold one end of the slip [of rubber] . . . between the thumb and forefinger of each hand; bring the middle of the piece into slight contact with the lips; . . . extend the slip suddenly; and you will immediately perceive a sensation of warmth in that part of the mouth which touches it. . . .

So wrote John Gough in 1806. The reader may easily verify that rapid extension of a rubber band does indeed increase the rubber band's temperature slightly. James Joule examined the effect with a thermocouple in 1849 and found that an adiabatic elongation of vulcanized rubber by 65 per cent produces a temperature rise of 0.1°C. This small effect can be easily magnified by alternating rapid extensions with free contractions. The free contraction of a rubber band—like the free expansion of an ideal gas—produces little change in the rubber band's temperature. Each cycle of operation—a forced extension followed by a free contraction—therefore increases the temperature of the working substance a few tenths of a degree.

Conversely, when a rubber band is subjected to a rapid, constrained contraction, its temperature drops.

Gough noted another interesting fact. A loaded rubber band when heated contracts. In his words:

If one end of a slip of Caoutchouc be fastened to a rod of metal or wood, and a weight be fixed at the other extremity . . . the thong will be found to become shorter with heat and longer with cold.

Gough's two observations embrace most of the important thermodynamic features of rubber elasticity. These observations, to repeat, are:

(I) A change in rubber temperature with a forced, rapid change in length.

(II) A change in rubber length with a change in temperature, tension constant.

Two closely related phenomena are:

(I′) A change in gas temperature with a forced, rapid change in volume.

(II′) A change in gas volume with a change in temperature, pressure constant.

These phenomena will be discussed in the order (I′), (I); (II′), (II).

I′. *Temperature change in a reversible adiabatic compression.* To say that a change is reversible ($\Delta S_{\text{total}} = 0$) and adiabatic ($\Delta S_\theta = 0$) is the same thing as saying that it is isentropic ($\Delta S_\sigma = 0$). The quantity of interest in the present instance is, therefore,

$$\left(\frac{\partial T}{\partial V}\right)_S . \tag{1}$$

In this expression the independent variables, V and S, are the same ones that appear on the right in the equation

$$dE = TdS - PdV. \tag{2}$$

Since

$$\left[\frac{\partial}{\partial V}\left(\frac{\partial E}{\partial S}\right)_V\right]_S = \left[\frac{\partial}{\partial S}\left(\frac{\partial E}{\partial V}\right)_S\right]_V, \tag{3}$$

it follows that

$$\left(\frac{\partial T}{\partial V}\right)_S = -\left(\frac{\partial P}{\partial S}\right)_V . \tag{4}$$

In attempting to interpret the derivative on the right, one is led to ask, how can P and S be varied when V is held constant? The answer—by changing T. This suggests writing

$$\left(\frac{\partial P}{\partial S}\right)_V = \left(\frac{\partial P}{\partial T}\right)_V \cdot \left(\frac{\partial T}{\partial S}\right)_V . \tag{5}$$

The derivative $(\partial T/\partial S)_V$ can be expanded in a similar fashion.

$$\left(\frac{\partial T}{\partial S}\right)_V = \left(\frac{\partial T}{\partial E}\right)_V \cdot \left(\frac{\partial E}{\partial S}\right)_V$$

$$= \frac{1}{C_V} \cdot T . \tag{6}$$

Substituting from (6) into (5) and from (5) into (4), one finds that

$$\left(\frac{\partial T}{\partial V}\right)_S = -\frac{T}{C_V}\left(\frac{\partial P}{\partial T}\right)_V . \tag{7}$$

This equation expresses $(\partial T/\partial V)_S$ in terms of the measurable quantities T, C_V, and $(\partial P/\partial T)_V$.

The same result can be obtained in another way. By calculus

$$\left(\frac{\partial T}{\partial V}\right)_S = -\left(\frac{\partial T}{\partial S}\right)_V \cdot \left(\frac{\partial S}{\partial V}\right)_T. \tag{8}$$

The first derivative on the right has already been evaluated (equation (6)). The independent variables in the second derivative are the same ones that appear on the right in the equation

$$d(E - TS) = -SdT - PdV. \tag{9}$$

Since

$$\left\{\frac{\partial}{\partial V}\left[\frac{\partial(E - TS)}{\partial T}\right]_V\right\}_T = \left\{\frac{\partial}{\partial T}\left[\frac{\partial(E - TS)}{\partial V}\right]_T\right\}_V, \tag{10}$$

it follows (Chapter 36) that

$$\left(\frac{\partial S}{\partial V}\right)_T = \left(\frac{\partial P}{\partial T}\right)_V. \tag{11}$$

Substitution from (11) and (6) into (8) yields (7).

T and C_V are positive for all substances. For most substances $(\partial P/\partial T)_V$ is positive also; this corresponds to saying, equation (11), that the substance's entropy increases in an isothermal expansion. For most substances, therefore, the temperature drops in a reversible adiabatic expansion.

I. *Temperature change of rubber in a reversible adiabatic extension.* The quantity of interest in the present instance is

$$\left(\frac{\partial T}{\partial \ell}\right)_S. \tag{1'}$$

The independent variables in this expression are the same ones that appear on the right in the equation

$$dE = TdS - PdV + \not{f} d\ell. \tag{2'}$$

In this expression dE is the energy change of the rubber; TdS is the energy change of the thermal surroundings; PdV is the energy change of the atmosphere; and the last term is the energy change of a weight that does the amount of work $\not{f} d\ell$ in extending the rubber by the amount $d\ell$ against an elastic restoring force $\not{f}$. Equation (2′) may be taken to be the defining equation for $\not{f}$. Thus, by (2′) $\not{f} = (\partial E/\partial \ell)_{S,V}$. In words, the elastic restoring force $\not{f}$ is the rate at which the energy of the rubber increases in an adiabatic $(\Delta E_\theta = 0)$, constant volume $(\Delta E_{\text{atm.}} = 0)$ extension. In an ordinary adiabatic extension of a rubber band, it is the pressure on the rubber band, not its volume, that remains constant. To be precise, the quantity of interest in the present instance is, therefore,

$$\left(\frac{\partial T}{\partial \ell}\right)_{S,P}. \tag{1''}$$

In this expression the independent variables ℓ, S, and P are the same ones that appear on the right in the equation

$$d(E + PV) \equiv dH = TdS + VdP + f d\ell \tag{2''}$$

Since

$$\left[\frac{\partial}{\partial \ell}\left(\frac{\partial H}{\partial S}\right)_{\ell,P}\right]_{S,P} = \left[\frac{\partial}{\partial S}\left(\frac{\partial H}{\partial \ell}\right)_{S,P}\right]_{\ell,P}, \tag{3'}$$

where $(\partial H/\partial S)_{\ell,P} = T$ and $(\partial H/\partial \ell)_{S,P} = f$, it follows that

$$\left(\frac{\partial T}{\partial \ell}\right)_{S,P} = \left(\frac{\partial f}{\partial S}\right)_{\ell,P}. \tag{4'}$$

The derivative on the right can be written

$$\left(\frac{\partial f}{\partial S}\right)_{\ell,P} = \left(\frac{\partial f}{\partial T}\right)_{\ell,P} \cdot \left(\frac{\partial T}{\partial S}\right)_{\ell,P}. \tag{5'}$$

In a similar fashion,

$$\left(\frac{\partial T}{\partial S}\right)_{\ell,P} = \left(\frac{\partial T}{\partial H}\right)_{\ell,P} \cdot \left(\frac{\partial H}{\partial S}\right)_{\ell,P}$$

$$= \frac{1}{C_{\ell,P}} \cdot T. \tag{6'}$$

Substituting from (6′) into (5′) and from (5′) into (4′), one finds that

$$\left(\frac{\partial T}{\partial \ell}\right)_{S,P} = \frac{T}{C_{\ell,P}} \left(\frac{\partial f}{\partial T}\right)_{\ell,P}. \tag{7'}$$

This equation expresses $(\partial T/\partial \ell)_{S,P}$ in terms of the measurable quantities T, $C_{\ell,P}$ (the heat capacity measured at constant extension and constant pressure), and $(\partial f/\partial T)_{\ell,P}$. The same result can be obtained in another way. By calculus

$$\left(\frac{\partial T}{\partial \ell}\right)_{S,P} = -\left(\frac{\partial T}{\partial S}\right)_{\ell,P} \left(\frac{\partial S}{\partial \ell}\right)_{T,P}. \tag{8'}$$

The first derivative on the right has already been evaluated (equation (6′)). The independent variables in the second derivative are the same ones that appear on the right in the equation

$$d(E + PV - TS) = -SdT + VdP + f d\ell. \tag{9'}$$

Since

$$\left\{\frac{\partial}{\partial \ell}\left[\frac{\partial(E + PV - TS)}{\partial T}\right]_{\ell,P}\right\}_{T,P} = \left\{\frac{\partial}{\partial T}\left[\frac{\partial(E + PV - TS)}{\partial \ell}\right]_{T,P}\right\}_{\ell,P}, \quad (10')$$

it follows that

$$\left(\frac{\partial S}{\partial \ell}\right)_{T,P} = -\left(\frac{\partial f}{\partial T}\right)_{\ell,P}. \quad (11')$$

Substitution from (11′) and (6′) into (8′) yields (7′).

As before, the absolute temperature and the heat capacity are positive quantities. By experiment it is found that $(\partial f/\partial T)_{\ell,P}$ is positive also. Consequently, equation (7′), rubber's temperature increases in a reversible, adiabatic extension.

To say that the elastic tension of rubber at constant extension and constant pressure increases as the temperature increases corresponds to saying, equation (11′), that rubber's entropy decreases in an isothermal, isobaric extension. This seems reasonable.

II′. *Thermal expansion under constant pressure.* The quantity of interest here is $(\partial V/\partial T)_P$. By calculus

$$\left(\frac{\partial V}{\partial T}\right)_P = -\left(\frac{\partial V}{\partial P}\right)_T \left(\frac{\partial P}{\partial T}\right)_V. \quad (12)$$

Since $(\partial V/\partial P)_T$ is always negative, while $(\partial P/\partial T)_V$ is generally positive, substances generally expand when heated.

II. *Thermal contraction of rubber under constant tension.* The quantity of interest here is $(\partial \ell/\partial T)_{f,P}$. By calculus

$$\left(\frac{\partial \ell}{\partial T}\right)_{f,P} = -\left(\frac{\partial \ell}{\partial f}\right)_{T,P} \left(\frac{\partial f}{\partial T}\right)_{\ell,P}. \quad (12')$$

Below the elastic limit both derivatives on the right are positive. Therefore $(\partial \ell/\partial T)_{f,P}$ is negative. Rubber, as Gough observed, contracts when heated.

Earlier it was mentioned that the temperature of a rubber band does not change appreciably in a free contraction. This fact will now be considered in more detail. As before, an analogous problem will be considered first.

III′. *Temperature change in an adiabatic free expansion.* The quantity of interest here is

$$\left(\frac{\partial T}{\partial V}\right)_E. \quad (13)^*$$

* To say that an expansion is adiabatic ($\Delta E_\theta = 0$) and free ($\Delta E_{\text{wt.}} = 0$) is the same thing as saying that it is isocaloric ($\Delta E_\sigma = 0$), since by the First Law

$$\Delta E_{\text{total}} = \Delta E_\sigma + \Delta E_\theta + \Delta E_{\text{wt.}} = 0.$$

The independent variables in this expression are the same ones that appear on the right in the equation

$$dS = \frac{1}{T}\,dE + \frac{P}{T}\,dV. \tag{14}$$

Since

$$\left[\frac{\partial}{\partial V}\left(\frac{\partial S}{\partial E}\right)_V\right]_E = \left[\frac{\partial}{\partial E}\left(\frac{\partial S}{\partial V}\right)_E\right]_V, \tag{15}$$

it follows that

$$\left[\frac{\partial(1/T)}{\partial V}\right]_E = \left[\frac{\partial(P/T)}{\partial E}\right]_V.$$

Or,

$$\left(\frac{\partial T}{\partial V}\right)_E = -T^2\left[\frac{\partial(P/T)}{\partial E}\right]_V. \tag{16}$$

The derivative on the right vanishes for an ideal gas. Heating such a substance at constant volume increases the pressure by the same factor it increases the absolute temperature; thus the ratio of P to T $(= nR/V)$ remains constant. For an ideal gas, therefore,

$$\left(\frac{\partial T}{\partial V}\right)_E = 0. \tag{17}$$

The temperature of an ideal gas does not change in a free expansion. Now by calculus

$$\left(\frac{\partial T}{\partial V}\right)_E = -\left(\frac{\partial T}{\partial E}\right)_V\left(\frac{\partial E}{\partial V}\right)_T. \tag{18}$$

Since

$$\left(\frac{\partial T}{\partial E}\right)_V \neq 0, \tag{19}$$

it follows that for an ideal gas

$$\left(\frac{\partial E}{\partial V}\right)_T = 0. \tag{20}$$

The internal energy of an ideal gas is independent of its volume.

III. *Temperature change of rubber in an adiabatic free contraction.* If the volume of rubber does not change significantly in a free contraction, the quantity of interest here may be taken to be

$$\left(\frac{\partial T}{\partial \ell}\right)_{E,V}. \tag{13'}$$

The independent variables in this expression are the same ones that appear on the right in the equation

$$dS = \frac{1}{T}dE - \frac{f}{T}d\ell + \frac{P}{T}dV. \tag{14'}$$

Since

$$\left[\frac{\partial}{\partial \ell}\left(\frac{\partial S}{\partial E}\right)_{\ell,V}\right]_{E,V} = \left[\frac{\partial}{\partial E}\left(\frac{\partial S}{\partial \ell}\right)_{E,V}\right]_{\ell,V}, \tag{15'}$$

it follows that

$$\left(\frac{\partial T}{\partial \ell}\right)_{E,V} = T^2\left[\frac{\partial(f/T)}{\partial E}\right]_{\ell,V}. \tag{16'}$$

The derivative on the right is almost zero. Heating rubber at constant elongation increases the elastic tension by almost the same factor it increases the absolute temperature; thus the ratio of f to T remains almost constant and, therefore,

$$\left(\frac{\partial T}{\partial \ell}\right)_{E,V} \approx 0. \tag{17'}$$

The temperature of rubber does not change appreciably in a free contraction. Now by calculus

$$\left(\frac{\partial T}{\partial \ell}\right)_{E,V} = -\left(\frac{\partial T}{\partial E}\right)_{\ell,V}\left(\frac{\partial E}{\partial \ell}\right)_{T,V}. \tag{18'}$$

Since

$$\left(\frac{\partial T}{\partial E}\right)_{\ell,V} \neq 0, \tag{19'}$$

it follows that for rubber

$$\left(\frac{\partial E}{\partial \ell}\right)_{T,V} \approx 0. \tag{20'}$$

The internal energy of a rubber sample is almost independent of its length.

IV. *Other Variables.* One further example of the application of thermodynamics to "other variables" may be cited. The change in energy of a system whose surface area changes by the amount $d\mathscr{A}$ may be written

$$dE = TdS - PdV + \gamma d\mathscr{A}.$$

This equation defines γ to be the rate at which the energy of the system changes in a reversible, adiabatic, constant volume change in surface area.

$$\gamma = \left(\frac{\partial E}{\partial \mathscr{A}}\right)_{S,V}.$$

Another expression for γ is obtained by noting that when surface effects are included

$$dG \equiv d(E - TS + PV) = -SdT + VdP + \gamma d\mathscr{A}.$$

Hence
$$\gamma = \left(\frac{\partial G}{\partial \mathscr{A}}\right)_{T,P}.$$

γ is called the "surface tension." It is the work required to increase the system's surface area by unit amount at constant temperature and constant pressure.

From the expression

$$d(E - TS) = -SdT - PdV + \gamma d\mathscr{A},$$

is obtained the Maxwell-type relation

$$\left(\frac{\partial S}{\partial \mathscr{A}}\right)_{T,V} = -\left(\frac{\partial \gamma}{\partial T}\right)_{\mathscr{A},V}.$$

Since $(\partial\gamma/\partial T)_{\mathscr{A},V}$ is negative (at the critical temperature γ vanishes), it follows that the entropy of a one-component system increases with an isothermal, constant volume increase in surface area. This seems reasonable.

38

Algebraic Method for Obtaining Thermodynamic Formulas

This chapter describes a method for obtaining thermodynamic formulas from fundamental and generally applicable considerations.* The method generates Bridgman's Tables of First Derivatives and extends their scope.

The energy of a stationary, internally stable, one-component system whose mass does not change may be expressed as a function of the entropy and volume of the system: $E = E(S, V)$. For such a system

$$dE = \left(\frac{\partial E}{\partial S}\right)_V dS + \left(\frac{\partial E}{\partial V}\right)_S dV.$$

Introducing the definitions

$$T \equiv \left(\frac{\partial E}{\partial S}\right)_V = T(S, V)$$

$$P \equiv -\left(\frac{\partial E}{\partial V}\right)_S = P(S, V),$$

one may write

$$dE - TdS + PdV = 0.$$

For small values of dE, dS, and dV, this last equation may be viewed as a linear, homogeneous equation in the variables dE, dS, dV with constant coefficients 1, $-T$, $+P$, respectively. Such a system of equations (one equation, three unknowns) has two linearly independent solutions.

Linearly independent solutions to a set of linear, homogeneous equations with two degrees of freedom (two more variables than equations) can be systematically set down by assigning one variable, say in this case dV, the value 0 and another, say dS, the value 1; the value(s) of the other variable(s) can then be solved for; in the present case this procedure yields for dE the

* H. A. Bent, "A Simplified Algebraic Method for Obtaining Thermodynamical Formulas," *J. Chem. Phys.* Vol. 21, p. 1408 (1953).

value T. Then reverse the first two assignments: take $dV = 1$ and $dS = 0$; dE must then have the value $-P$. These results are tabulated below.

	I	II
dE	T	$-P$
dS	1	0
dV	0	1

Any other solution to the equation $dE - TdS + PdV = 0$ can be expressed as a linear combination of solutions I and II. The solution with $dS = a$ and $dV = b$, for example, is equal to (a) I + (b) II. A solution with $dE = 0$ is (P) I + (T) II.

	P(I) + T(II)
dE	0
dS	P
dV	T

The absolute values of the components dE, dS, and dV in these solutions are not significant in the present context. Any solution to a linear, homogeneous equation can be multiplied by a constant factor—which, if one wishes, may be very small—without altering the solution's characteristic features: the values of the ratios of its components to each other.

The value of the ratio of dE to dS in solution I is T.

$$\frac{dE_V}{dS_V} = T$$

The subscript V in this expression indicates that the values of dE and dS have been taken from a solution in which $dV = 0$. That is to say, at constant volume the ratio of the change in E to the corresponding change in S is T. In conventional notation, $(\partial E/\partial S)_V = T$. Similarly, from II $(\partial E/\partial V)_S = -P$. From (P)I + (T)II $(\partial S/\partial V)_E = P/T$.

If the equations $T = T(S, V)$ and $P = P(S, V)$ can be solved for S and V—i.e. if the Jacobian $\partial(T, P)/\partial(S, V)$ does not vanish—dT and dP can be introduced into the discussion by adjoining to the equation $dE - TdS + PdV = 0$ the two equations

$$\begin{aligned} dS &= \left(\frac{\partial S}{\partial T}\right)_P dT + \left(\frac{\partial S}{\partial P}\right)_T dP \\ &= \left(\frac{C_P}{T}\right) dT - \left(\frac{\partial V}{\partial T}\right)_P dP \end{aligned}$$

$$dV = \left(\frac{\partial V}{\partial T}\right)_P dT + \left(\frac{\partial V}{\partial P}\right)_T dP.$$

Formed in this way is a set of three equations in five variables. The number of degrees of freedom is the same as before: two. Adjoining to this set the three additional equations $dH = TdS + VdP$, $dA = -SdT - PdV$, and $dG = -SdT + VdP$ yields a set of six equations in eight variables. These equations are set forth below in a form that makes relatively easy the calculation of dG, dA, dH, dE, dS, and dV once values for dT and dP have been selected.

$$\begin{aligned}
&dG && && + SdT && - VdP = 0.\\
&dA && + PdV && + SdT && = 0.\\
&dH - TdS && && && - VdP = 0.\\
&dE - TdS + PdV && && && = 0.\\
&dS && && - \left(\frac{C_P}{T}\right) dT && + \left(\frac{\partial V}{\partial T}\right)_P dP = 0.\\
&dV && && - \left(\frac{\partial V}{\partial T}\right)_P dT && - \left(\frac{\partial V}{\partial P}\right)_T dP = 0.
\end{aligned}$$

Again there are only two linearly independent solutions. Two such solutions, obtained by setting $dP = 0$ and $dT = 1$, and vice versa, are given below.

	I	II
dG	$-S$	V
dA	$-S - P(\partial V/\partial T)_P$	$-P(\partial V/\partial P)_T$
dH	C_P	$V - T(\partial V/\partial T)_P$
dE	$C_P - P(\partial V/\partial T)_P$	$-T(\partial V/\partial T)_P - P(\partial V/\partial P)_T$
dS	C_P/T	$-(\partial V/\partial T)_P$
dV	$(\partial V/\partial T)_P$	$(\partial V/\partial P)_T$
dT	1	0
dP	0	1

From solution I may be read off the familiar relations $(\partial G/\partial T)_P = -S$, $(\partial H/\partial T)_P = C_P$, and $(\partial S/\partial T)_P = C_P/T$. From solution II are obtained the relations $(\partial G/\partial P)_T = V$, $(\partial S/\partial P)_T = -(\partial V/\partial T)_P$, and $(\partial H/\partial P)_T = V - T\cdot(\partial V/\partial T)_P$. The last relation also may be obtained by dividing the equation $dH = TdS + VdP$ by dP, imposing the condition $T =$ constant, and making use of the relation $(\partial S/\partial P)_T = -(\partial V/\partial T)_P$.

The components of solution I are listed by Bridgman as $d(\ )_P$; those of II as $-d(\ )_T$. Any linear combination of I and II whose dV component is zero—e.g. $(\partial V/\partial P)_T$I $-$ $(\partial V/\partial T)_P$II—is equivalent—in this case equal—to Bridgman's values for $d(\ \ )_V$. More generally, Bridgman's set $d(\ \ )_X$ is equal to solution I multiplied by the X component of solution II less solution II multiplied by the X component of solution I. In symbols,

$$d(\ \ )_{X,\,\text{Bridgman}} = X_{\text{II}}(\text{I}) - X_{\text{I}}(\text{II}).$$

Bridgman found that it was possible to select values for dX_i, X_k constant, and dX_k, X_i constant, such that

$$d(X_i)_{X_k} = -d(X_k)_{X_i}\,.$$

The existence of a tabulation with this useful property stems from the fact that the solution space of a set of n linear, homogeneous equations in $n + 2$ variables is, as we have seen, spanned by two vectors of the form

	I	II
	$(X_{n+2} = 0)$	$(X_{n+1} = 0)$
X_1	a_1	b_1
X_2	a_2	b_2
$\cdot$	$\cdot$	$\cdot$
$\cdot$	$\cdot$	$\cdot$
$\cdot$	$\cdot$	$\cdot$
$\cdot$	$\cdot$	$\cdot$
X_i	a_i	b_i
$\cdot$	$\cdot$	$\cdot$
$\cdot$	$\cdot$	$\cdot$
$\cdot$	$\cdot$	$\cdot$
$\cdot$	$\cdot$	$\cdot$
X_k	a_k	b_k
$\cdot$	$\cdot$	$\cdot$
$\cdot$	$\cdot$	$\cdot$
$\cdot$	$\cdot$	$\cdot$
$\cdot$	$\cdot$	$\cdot$
X_{n+1}	1	0
X_{n+2}	0	-1

The second vector above corresponds to the previous solution II multiplied by -1. Linear combination of I and II in the manner previously indicated yields solutions of the following general type.

	III	IV
	$-b_i(\mathrm{I}) + a_i(\mathrm{II})$	$-b_k(\mathrm{I}) + a_k(\mathrm{II})$
	$(X_i = 0)$	$(X_k = 0)$
X_1	$-b_i a_1 + a_i b_1$	$-b_k a_1 + a_k b_1$
X_2	$-b_i a_2 + a_i b_2$	$-b_k a_2 + a_k b_2$
$\vdots$	$\vdots$	$\vdots$
X_i	0	$-b_k a_i + a_k b_i$
$\vdots$	$\vdots$	$\vdots$
X_k	$-b_i a_k + a_i b_k$	0
$\vdots$	$\vdots$	$\vdots$
X_{n+1}	$-b_i$	$-b_k$
X_{n+2}	$-a_i$	$-a_k$

The components of solutions I, II, III, and IV satisfy Bridgman's condition. Thus, as the reader may verify, for example,

$$(X_{n+1})_{X_{n+2}} = -(X_{n+2})_{X_{n+1}} \quad \text{and} \quad (X_k)_{X_i} = -(X_i)_{X_k}\,.$$

39

Le Chatelier's Principle

Discussion of Le Chatelier's principle is facilitated if use is made of the variable ξ, "the degree of advancement of a chemical reaction." The meaning of ξ is best conveyed by an example. Consider the reaction

$$N_2 + 3H_2 = 2NH_3.$$

Let the number of moles of ammonia produced in the reaction be 2ξ. This corresponds to the consumption of ξ moles of nitrogen and 3ξ moles of hydrogen. If initially the system contained $n^0{}_{NH_3}$ moles of ammonia, $n^0{}_{N_2}$ moles of nitrogen, and $n^0{}_{H_2}$ moles of hydrogen, production of 2ξ moles of ammonia makes

$$n_{NH_3} = n^0{}_{NH_3} + 2\xi$$
$$n_{N_2} = n^0{}_{N_2} - \xi$$
$$n_{H_2} = n^0{}_{H_2} - 3\xi.$$

These equations are valid for all permissible values of ξ. Differentiation of the mole numbers n_{NH_3}, n_{N_2}, n_{H_2} with respect to ξ gives the following expressions.

$$\frac{dn_{NH_3}}{d\xi} = 2$$

$$\frac{dn_{N_2}}{d\xi} = -1$$

$$\frac{dn_{H_2}}{d\xi} = -3$$

More generally, if ν_i is the coefficient of the formula of the ith chemical in the balanced chemical equation, ν_i being taken positive if the ith chemical is a product of the reaction and negative if it is a reactant, one may write

$$\frac{dn_i}{d\xi} = \nu_i\,.$$

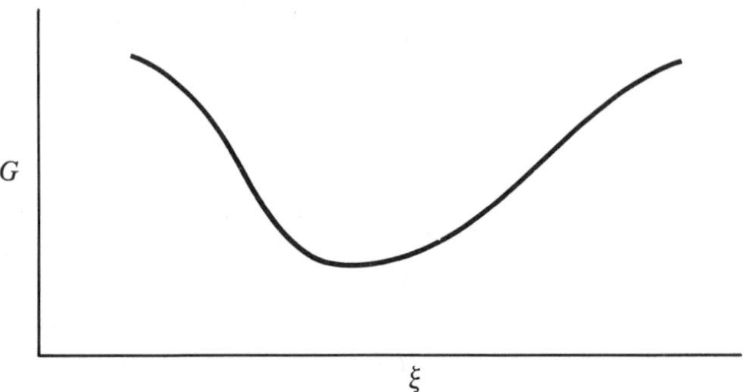

Fig. 39.1

In the present example ($i = NH_3$, N_2, or H_2), $\nu_{NH_3} = 2$, $\nu_{N_2} = -1$, and $\nu_{H_2} = -3$.

As a reaction advances under its own initiative, the free energy of the chemical system decreases. A plot of G against ξ therefore has the general shape shown in Fig. 39.1.

Two important features of this figure should be noted. When the free energy has reached its minimum value, i.e. when the system has reached equilibrium with respect to the reaction represented by ξ,

$$\frac{\partial G}{\partial \xi} = 0 \qquad (T, P \text{ constant})$$

and

$$\frac{\partial^2 G}{\partial \xi^2} > 0.$$

The statement $\partial G/\partial \xi = 0$ is equivalent to the statement $\Delta G = 0$. For at constant temperature and pressure

$$\begin{aligned} dG &= G_1 dn_1 + G_2 dn_2 + G_3 dn_3 + \cdots \\ &= \Sigma G_i dn_i, \end{aligned}$$

where

$$G_i = \left(\frac{\partial G}{\partial n_i}\right)_{T,P,n_{j \neq i}}.$$

The subscripts 1, 2, 3, ... stand for the reactant and product species in the chemical reaction to which ξ refers. Dividing the expression for dG by $d\xi$ holding constant T and P, and introducing for the ratios $dn_i/d\xi$, $i = 1, 2, 3, \ldots$, the symbols ν_i, one finds that

$$\left(\frac{\partial G}{\partial \xi}\right)_{T,P} = \Sigma \nu_i G_i\,.$$

Since ν_i is positive for product species and negative for reactant species, this last equation may be written

$$\left(\frac{\partial G}{\partial \xi}\right)_{T,P} = \sum_{\text{products}} |\nu_i| G_i - \sum_{\text{reactants}} |\nu_i| G_i$$
$$= \Delta G.$$

More generally, for any extensive property X—G, H, E, S, or V—

$$\left(\frac{\partial X}{\partial \xi}\right)_{T,P} = \Delta X.$$

The temperature, pressure, and degree of advancement variable ξ of a reaction mixture that is at equilibrium with respect to the reaction to which ξ refers can be altered without destroying the equilibrium condition $\partial G/\partial \xi = 0$ if the sum of the changes in $\partial G/\partial \xi$ produced by the separate changes in T, P, and ξ add algebraically to zero, i.e. if the values of dT, dP, and $d\xi$ are such that

$$d\left(\frac{\partial G}{\partial \xi}\right)_{T,P} = 0.$$

Written out fully, the condition on dT, dP, and $d\xi$ required to maintain $\partial G/\partial \xi = 0$ is

$$\left[\frac{\partial}{\partial T}\left(\frac{\partial G}{\partial \xi}\right)_{T,P}\right]_{P,\xi} dT + \left[\frac{\partial}{\partial P}\left(\frac{\partial G}{\partial \xi}\right)_{T,P}\right]_{T,\xi} dP +$$
$$+ \left[\frac{\partial}{\partial \xi}\left(\frac{\partial G}{\partial \xi}\right)_{T,P}\right]_{T,P} d\xi = 0.$$

The order of differentiation in the first two terms of this expression may be interchanged, to give

$$\left[\frac{\partial}{\partial \xi}\left(\frac{\partial G}{\partial T}\right)_{P,\xi}\right]_{T,P} dT + \left[\frac{\partial}{\partial \xi}\left(\frac{\partial G}{\partial P}\right)_{T,\xi}\right]_{T,P} dP + \left(\frac{\partial^2 G}{\partial \xi^2}\right)_{T,P} d\xi = 0.$$

Now $(\partial G/\partial T)_{P,\xi} = -S$ and $(\partial G/\partial P)_{T,\xi} = V$. Therefore, the previous equation becomes

$$\left[\frac{\partial}{\partial \xi}(-S)\right]_{T,P} dT + \left[\frac{\partial}{\partial \xi}(V)\right]_{T,P} dP + \left(\frac{\partial^2 G}{\partial \xi^2}\right)_{T,P} d\xi = 0.$$

Also, $(\partial S/\partial \xi)_{T,P} = \Delta S$ and $(\partial V/\partial \xi)_{T,P} = \Delta V$. Thus, to maintain $\partial G/\partial \xi = 0$, dT, dP, and $d\xi$ must be such that

$$-(\Delta S)dT + (\Delta V)dP + \left(\frac{\partial^2 G}{\partial \xi^2}\right)_{T,P} d\xi = 0.$$

This equation has several interesting implications.

1. Setting $dP = 0$, one finds that

$$\left(\frac{\partial \xi}{\partial T}\right)_{\substack{P\\ \Delta G=0}} = \frac{\Delta S}{\left(\dfrac{\partial^2 G}{\partial \xi^2}\right)_{T,P}} = \frac{\Delta H}{T\left(\dfrac{\partial^2 G}{\partial \xi^2}\right)_{T,P}}. \qquad (*)$$

Now when $\Delta G = 0$, $(\partial^2 G/\partial \xi^2)_{T,P}$ is positive. Also, T is always positive. Therefore, at equilibrium $(\partial \xi/\partial T)_P$ has the same sign as ΔH.

For endothermic reactions ΔH is positive. In these cases, therefore, $(\partial \xi/\partial T)_{\substack{P\\ \Delta G=0}}$ is positive. For endothermic reactions the degree-of-reaction-advancement at equilibrium increases as the temperature increases.

2. Setting $dT = 0$, one finds that

$$\left(\frac{\partial \xi}{\partial P}\right)_{\substack{T\\ \Delta G=0}} = -\frac{\Delta V}{\left(\dfrac{\partial^2 G}{\partial \xi^2}\right)_{T,P}}.$$

Again, the denominator is positive when $\Delta G = 0$. At equilibrium, therefore, $(\partial \xi/\partial P)_T$ and ΔV have opposite signs. When ΔV is negative—as in the reaction $N_2 + 3H_2 = 2NH_3$, or as in the reaction H_2O(gas) $= H_2O$(liq)—the degree-of-reaction-advancement at equilibrium increases as the pressure increases.

Finally,

3. Setting $d\xi = 0$, one finds that

$$\left(\frac{\partial P}{\partial T}\right)_{\substack{\xi\\ \Delta G=0}} = \frac{\Delta S}{\Delta V}.$$

This is a generalized Clausius-Clapeyron equation. When ΔS and ΔV a equilibrium are both positive—as in the reaction H_2O(liq) $= H_2O$(gas)—o both negative—as in the reaction $N_2 + 3H_2 = 2NH_3$—the pressure mus increase as the temperature increases if the degree-of-reaction-advancemen at equilibrium is to be maintained constant.

* When $\Delta G = 0$, $\Delta S = \Delta H/T$.

40

Homogeneous Functions

The energy of an internally stable, one-component system for which surface effects may be neglected is a function of its entropy, volume, and mole number n: $E = E(S, V, n)$. This functional relation has the property that doubling the values of S, V, and n doubles the value of E: $E(2S, 2V, 2n) = 2E(S, V, n)$. More generally, for any positive number λ

$$E(\lambda S, \lambda V, \lambda n) = \lambda E(S, V, n).$$

The function $E(S, V, n)$ is said to be a homogeneous function of the first degree in the variables S, V, and n. This property is a consequence of the additivity of E, S, V, and n, and the fact that the entropy of an internally stable system is as large as the system's energy, volume, and mass permit. Before considering a formal proof of this statement, it may be noted that when the system's mole number is treated as an independent variable on the same footing as the system's entropy and volume, "we shall have for the complete value of the differential of E"*

$$dE = TdS - PdV + \mu dn,$$

where

$$T \equiv \left(\frac{\partial E}{\partial S}\right)_{V,n}$$

$$P \equiv -\left(\frac{\partial E}{\partial V}\right)_{S,n}$$

$$\mu \equiv \left(\frac{\partial E}{\partial n}\right)_{S,V}.$$

For a one-component system the partial derivative μ is equal to the system's molar free energy—or, what is the same thing, to the derivative of the system's free energy with respect to the mole number n, T and P constant. For when n is a variable,

* Willard Gibbs, *Collected Works*, New Haven: Yale University Press, 1948, Vol. I, p. 63.

$$dG[\equiv d(E + PV - TS)] = -SdT + VdP + \mu dn.$$

Hence
$$\mu = \left(\frac{\partial G}{\partial n}\right)_{T,P}.$$

A system is internally stable if, and only if, the partial derivatives T, P, and μ are uniform throughout the system.

*Proof of the Homogeneity Condition.** A one-component system is stable when its entropy is as large as its energy, volume, and mole number allow. Change E, V, and n, and the maximum value of S will change; i.e. for stable states S is a function of E, V, and n.

$$S = S(E, V, n)$$

If $(\partial S/\partial E)_{V,n} \neq 0$, this relation can be solved for E in terms of S, V, and n.

$$E = E(S, V, n)$$

Now consider the system in two parts, a and b. Part a is internally stable when the value of S^a is as large as E^a, V^a, and n^a permit. These four values are related to each other by the formula

$$E^a = E(S^a, V^a, n^a).$$

Similarly, when part b is internally stable

$$E^b = E(S^b, V^b, n^b).$$

The total energy, E^{a+b}, cannot be calculated from a corresponding formula, however, unless the system as a whole is stable. But when

$$T^a = T^b$$
$$P^a = P^b$$
$$\mu^a = \mu^b,$$

S^{a+b} is as large as E^{a+b}, V^{a+b}, and n^{a+b} permit. Under these conditions

$$E^{a+b} = E(S^{a+b}, V^{a+b}, n^{a+b}).$$

Since by additivity
$$S^{a+b} = S^a + S^b$$
$$V^{a+b} = V^a + V^b$$
$$n^{a+b} = n^a + n^b$$
$$E^{a+b} = E^a + E^b,$$

it follows that when the composite system $a + b$ is internally stable

$$E(S^a + S^b, V^a + V^b, n^a + n^b) = E(S^a, V^a, n^a) + E(S^b, V^b, n^b).$$

* Alternative proofs are given in H. A. Bent, "Homogeneity Condition in Thermodynamics," *J. Chem. Phys.*, Vol. 23, p. 2199 (1955).

Taking $S^a = S^b = S$, $V^a = V^b = V$, $n^a = n^b = n$, one finds that

$$E(2S, 2V, 2n) = 2E(S, V, n),$$

reminiscent of the homogeneity condition.

These results may be stated as a theorem. If

$$f(x^a + x^b, y^a + y^b, z^a + z^b) = f(x^a, y^a, z^a) + f(x^b, y^b, z^b) \tag{1}$$

whenever

$$f_i(x^a, y^a, z^a) = f_i(x^b, y^b, z^b), \tag{2}$$

$$i = x, y, z \text{ and } f_x = (\partial f/\partial x)_{y,z} \text{ etc.,}$$

then

$$f(\lambda x, \lambda y, \lambda z) = \lambda f(x, y, z). \tag{3}$$

Proof. Equations (2) are satisfied by $x^a = x^b = x$, $y^a = y^b = y$, $z^a = z^b = z$. Therefore, by equation (1), $f(2x, 2y, 2z) = 2f(x, y, z)$. This equation is valid for all allowed values of x, y, and z. Differentiation with respect to x, y, and z yields $f_i(2x, 2y, 2z) = f_i(x, y, z)$; i.e. equations (2) are satisfied also by the choice of variables $x^a = 2x^b = 2x$, $y^a = 2y^b = 2y$, $z^a = 2z^b = 2z$, which in equation (1) gives $f(3x, 3y, 3z) = f(2x, 2y, 2z) + f(x, y, z) = 3f(x, y, z)$, and so forth, for λ any positive integer. To establish the theorem for fractional values of λ—and, hence, for λ any rational number—set x, y, and z in the relation $f(nx, ny, nz) = nf(x, y, z)$, n an integer, equal to x/n, y/n, and z/n. To establish the theorem for $\lambda = 0$, take in equation (1) $(x^a, y^a, z^a) = (x^b, y^b, z^b) = (0, 0, 0)$.

The equation $E(\lambda S, \lambda V, \lambda n) = \lambda E(S, V, n)$ has several interesting implications.

Intensive Properties. The substitution $\lambda = 1/n$ shows that the molar energy, E/n, can be expressed as a function of two variables: the molar entropy and the molar volume.

$$\frac{E(S, V, n)}{n} = E\left(\frac{S}{n}, \frac{V}{n}, 1\right).$$

Moreover, it is evident that the molar energy is an intensive property; it is homogeneous to the zeroth degree in S, V, and n. The same statement holds for the derivatives of E with respect to S, V, and n. To show this, differentiate the expression $E(\lambda S, \lambda V, \lambda n) = \lambda E(S, V, n)$ with respect to S holding constant V, n and λ. This yields

$$\frac{\partial E(\lambda S, \lambda V, \lambda n)}{\partial(\lambda S)} \cdot \frac{\partial(\lambda S)}{\partial S} = \lambda \frac{\partial E(S, V, n)}{\partial S}.$$

Now $\partial(\lambda S/\partial S) = \lambda$. Introducing the symbol T for the derivative of E with respect to its first argument (S or λS), one finds that

$$T(\lambda S, \lambda V, \lambda n) = \lambda^0 \cdot T(S, V, n).$$

Similar expressions may be written for P and μ.

Euler's Relation and the Gibbs-Duhem Equation. Differentiation of the expression $E(\lambda S, \lambda V, \lambda n) = \lambda E(S, V, n)$ with respect to λ holding constant S, V, and n yields Euler's relation.

$$\begin{aligned} E(S, V, n) &= \frac{\partial E(\lambda S, \lambda V, \lambda n)}{\partial(\lambda S)} \cdot \frac{\partial(\lambda S)}{\partial \lambda} + \frac{\partial E(\lambda S, \lambda V, \lambda n)}{\partial(\lambda V)} \cdot \frac{\partial(\lambda V)}{\partial \lambda} \\ &\qquad + \frac{\partial E(\lambda S, \lambda V, \lambda n)}{\partial(\lambda n)} \cdot \frac{\partial(\lambda n)}{\partial \lambda} \\ &= TS - PV + \mu n. \end{aligned}$$

μ, one sees, is equal to the system's free energy per mole.

$$\mu = \frac{E - TS + PV}{n}$$

A comparison of the expression obtained from Euler's relation for the total differential of E with the expression $dE = TdS - PdV + \mu dn$ yields the so-called Gibbs-Duhem equation.

$$-SdT + VdP + nd\mu = 0$$

Intercept Theorem. At constant temperature and pressure the volume of a two-component system is a homogeneous function of the first order in the mole numbers n_1 and n_2.

$$V(\lambda n_1, \lambda n_2) = \lambda V(n_1, n_2)$$

By Euler's relation, therefore,

$$V = V_1 n_1 + V_2 n_2 \,,$$

where

$$V_1 = \left(\frac{\partial V}{\partial n_1}\right)_{n_2,T,P}$$

$$V_2 = \left(\frac{\partial V}{\partial n_2}\right)_{n_1,T,P} .$$

The corresponding Gibbs-Duhem equation is

$$n_1 dV_1 + n_2 dV_2 = 0.$$

V_1 and V_2 are called "partial molal volumes." They give the rates of change of the corresponding chemical potentials with respect to pressur

Thus,

$$\left(\frac{\partial \mu_1}{\partial P}\right)_{n_1,n_2,T} = \left[\frac{\partial}{\partial P}\left(\frac{\partial G}{\partial n_1}\right)_{P,n_2,T}\right]_{n_1,n_2,T}$$

$$= \left[\frac{\partial}{\partial n_1}\left(\frac{\partial G}{\partial P}\right)_{n_1,n_2,T}\right]_{P,n_2,T}$$

$$= \left(\frac{\partial V}{\partial n_1}\right)_{P,n_2,T}$$

$$= V_1.$$

Similarly, $\qquad (\partial \mu_2/\partial P)_{n_1,n_2,T} = V_2.$

V_1 and V_2 are equal to the intercepts of the tangent line to the curve formed by plotting the volume per mole, $V/(n_1 + n_2)$, against the mole fraction of one of the components. To prove this, divide the Euler expression $V = V_1 n_1 + V_2 n_2$ by the sum $n_1 + n_2$, symbolizing by N_1 and N_2 the mole fractions $n_1/(n_1 + n_2)$ and $n_2/(n_1 + n_2)$.

$$\frac{V}{n_1 + n_2} = V_1 N_1 + V_2 N_2$$

Take the total differential of the resulting expression

$$d\left(\frac{V}{n_1 + n_2}\right) = V_1 dN_1 + V_2 dN_2 + N_1 dV_1 + N_2 dV_2$$

and simplify with the Gibbs-Duhem equation ($N_1 dV_1 + N_2 dV_2 = 0$) and the mole-fraction relation $dN_2 = -dN_1$. One finds that

$$d\left(\frac{V}{n_1 + n_2}\right) = (V_1 - V_2) dN_1 \ .$$

The calculation of the $N_1 = 1$ and $N_2 = 1$ intercepts is given below.

$$N_1 = 1 \text{ intercept} = \left(\frac{V}{n_1 + n_2}\right) + \frac{d[V/(n_1 + n_2)]}{dN_1} \cdot (1 - N_1)$$

$$= (V_1 N_1 + V_2 N_2) + (V_1 - V_2) N_2$$

$$= V_1$$

$$N_2 = 1 \text{ intercept} = \left(\frac{V}{n_1 + n_2}\right) - \frac{d[V/(n_1 + n_2)]}{dN_1} \cdot N_1$$

$$= (V_1 N_1 + V_2 N_2) - (V_1 - V_2) N_1$$

$$= V_2$$

Similar expressions may be written for partial molal enthalpies, entropies, free energies, and heat capacities.

Black-body Radiation (A case of two independent variables). The entropy of the radiation in a cavity is a function of the radiation's energy and volume: $S = S(E, V)$; or, solving for E, $E = E(S, V)$. The three quantities E, S, and V form in this case a complete extensive set; i.e. for black-body radiation

$$E(\lambda S, \lambda V) = \lambda E(S, V).$$

By Euler's relation, therefore,

$$E = TS - PV.$$

The corresponding Gibbs-Duhem equation is

$$-SdT + VdP = 0.$$

Hence, for black-body radiation,

$$\frac{dP}{dT} = \frac{S}{V}$$

$$= \frac{E + PV}{TV}.$$

Now the energy of black-body radiation is equal to its PV product multiplied by 3. This relation follows from an argument analogous to the argument that shows that the energy of an ideal gas is equal to its PV product multiplied by 3/2. The latter argument will be considered first.

1. *Energy of an ideal monatomic gas.* The translational energy of an ideal gas is equal to the sum over the translational quantum numbers n_x, n_y, n_z of the number of particles n_{n_x, n_y, n_z} in a quantum level (n_x, n_y, n_z) multiplied by the energy of that quantum level, $\varepsilon_{n_x,n_y,n_z}$. The latter is equal to $(n_x^2 + n_y^2 + n_z^2)h^2/8mV^{2/3}$. Thus

$$E = \sum_{n_x,n_y,n_z} n_{n_x,n_y,n_z} \frac{(n_x^2 + n_y^2 + n_z^2)h^2}{8mV^{2/3}}.$$

This expression shows that for an ideal gas the product $EV^{2/3}$ is a function of the population numbers n_{n_x,n_y,n_z}.

$$EV^{2/3} = f(n_{n_x,n_y,n_z})$$

At equilibrium the population numbers n_{n_x,n_y,n_z} determine also the entropy of the gas, through Boltzmann's relation.

Conversely, the maximum value of S under the prevailing conditions together with the value of the mole number n, determines the values of the population numbers n_{n_x,n_y,n_z}. In other words, for an ideal gas that is in internal equilibrium $EV^{2/3}$ is a function of S and n.

$$EV^{2/3} = f'(S, n)$$

Differentiation of the product $EV^{2/3}$ with respect to V holding constant S and n yields

$$E(2/3)V^{-1/3} + V^{2/3}\left(\frac{\partial E}{\partial V}\right)_{S,n} = 0.$$

Introducing the abbreviation $-P$ for $(\partial E/\partial V)_{S,n}$ and rearranging, one finds that for an ideal gas that is in internal equilibrium

$$E = \frac{3}{2}PV.$$

2. *Energy of a photon gas.* The factor 3/2 in the expression $E = (3/2)PV$ for the translational energy of an ideal particle gas arises from the exponent 2/3 in the expression $\varepsilon_{n_x,n_y,n_z} = (n_x^2 + n_y^2 + n_z^2)h^2/8mV^{2/3}$ for the translational energy levels of a particle in a cubical box of volume V, length $V^{1/3}$. The factor 2 in the numerator of the exponent 2/3 arises from the fact that a particle's kinetic energy is inversely proportional to the *second* power of its wave length: K.E. $= p^2/2m = (h/\lambda)^2/2m$. Now a photon's energy is inversely proportional to the *first* power of its wave length: $\varepsilon = h\nu = h(c/\lambda)$. By an argument analogous to the one given above for a particle gas, it follows that for a photon gas that is in internal equilibrium $EV^{1/3}$ is a function of S.

$$EV^{1/3} = g(S).$$

Differentiation of the product $EV^{1/3}$ with respect to V holding constant S shows that for a photon gas

$$E = 3PV.$$

The result $E = 3PV$ can be used to eliminate E from the expression $dP/dT = (E + PV)/TV$. The equation obtained

$$\frac{dP}{dT} = \frac{4P}{T}$$

may be written

$$\frac{dP}{P} - 4\frac{dT}{T} = d\ln P - 4d\ln T = d\ln\left(\frac{P}{T^4}\right) = 0.$$

This implies that

$$\frac{P}{T^4} = \text{constant}.$$

Since $P = (1/3)E/V$, the last equation is equivalent to the statement that the energy density of equilibrium radiation is proportional to the fourth power of the absolute temperature.

$$\frac{E}{V} = \text{constant} \times T^4.$$

This formula was discovered empirically by Stefan in 1879 and derived theoretically by Boltzmann five years later. H. A. Lorentz has called Boltzmann's thermodynamic derivation of Stefan's equation "a true pearl of theoretical physics."

41

Lagrange's Multipliers*

Stationary values and critical points. An important problem in statistical thermodynamics is the problem of finding the values of the population numbers $n_1, n_2, n_3, \ldots$ that maximize the function

$$S = k \ln \Omega(n_1, n_2, n_3 \ldots), \tag{1a}$$

subject to the constraints

$$\begin{aligned} \Sigma n_i \varepsilon_i &= \text{constant, } E \\ \Sigma n_i &= \text{constant, } N \\ V &= \text{constant.} \end{aligned} \tag{1b}$$

A simpler problem with analogous mathematical properties is the problem of finding the values of x and y that maximize the function

$$A = xy \tag{2a}$$

subject to the condition

$$2x + 2y = \text{constant, } P. \tag{2b}$$

A may be taken to represent the area, P the perimeter of a rectangle of length x, width y. The problem is to find values for x and y that make

$$dA = xdy + ydx = 0 \tag{3a}$$

for any values of dx and dy that satisfy the condition

$$dP = 2dx + 2dy = 0. \tag{3b}$$

If x increases, P constant, y must decrease, and vice versa. More precisely, by (3b) $dy = -dx$. Making this substitution in (3a), one finds that x and y must be such that for any value of dx

$$dA = (y - x)dx = 0. \tag{4}$$

Since dx need not be zero, y must be equal to x. Thus, A has a stationary value at the critical point

$$x = y = P/4. \tag{5}$$

* The material of this chapter is adapted from a series of seminars on the calculus of stationary states given by the author in the department of mathematics at the University of California in 1951.

A's value at this point is given by the expression

$$A = P^2/16. \tag{6}$$

This is A's maximum value. For by (4) when $dy = -dx$, $dA/dx = y - x$; hence

$$\frac{d^2A}{dx^2} = \frac{dy}{dx} - 1 = -2 < 0.$$

Of all rectangles the square has the largest area for a given perimeter.

More generally, a function $f(x_1, \ldots, x_n) = f(x)$ subject to the constraints $g^k(x) = C^k$, $k = 1, \ldots, m < n$, is said to have a stationary value at the critical point $X = (X_1, \ldots, X_n)$ if X satisfies all the constraints and if at the critical point X

$$df = f_1 dx_1 + \cdots + f_n dx_n = 0 \tag{7a}$$

for all $dx = dx_1, \ldots, dx_n$ that satisfy the equations

$$\begin{aligned} dg^1 &= g_1{}^1 dx_1 + \cdots + g_n{}^1 dx_n = 0 \\ &\;\;\vdots \\ dg^m &= g_1{}^m dx_1 + \cdots + g_n{}^m dx_n = 0. \end{aligned} \tag{7b}$$

*The Multiplier Rule.** The multiplier rule states that the function $f(x)$ with constraints $g^k(x) = C^k$ has the same critical points as the function

$$M(x, \lambda) = f(x) - \sum_{k=1}^{m} \lambda_k [g^k(x) - C^k] \tag{8}$$

with no constraints. The parameters $\lambda_1, \ldots, \lambda_m$ are called Lagrange's multipliers. The critical points of the multiplier expression $M(x, \lambda)$ are found by equating to zero the partial derivatives of $M(x, \lambda)$ with respect to $x_1, \ldots, x_n$ and $\lambda_1, \ldots, \lambda_m$. These equations may be written in the following form.

$$\begin{aligned} \frac{\partial M}{\partial x_1} = 0: &\qquad f_1 = \lambda_1 g_1{}^1 + \cdots + \lambda_m g_1{}^m \\ \vdots &\qquad \vdots \\ \frac{\partial M}{\partial x_n} = 0: &\qquad f_n = \lambda_1 g_n{}^1 + \cdots + \lambda_m g_n{}^m \end{aligned} \tag{9}$$

* This rule is derived on p. 289.

and

$$\frac{\partial M}{\partial \lambda_1} = 0: \qquad g^1 = C^1$$
$$\vdots \qquad\qquad \vdots \qquad (10)$$
$$\frac{\partial M}{\partial \lambda_m} = 0: \qquad g^m = C^m.$$

The multiplier expression for the area-perimeter problem is

$$M(x, y, \lambda) = xy - \lambda(2x + 2y - P). \qquad (8')$$

The equations corresponding to (9) and (10) are

$$\frac{\partial M}{\partial x} = 0: \qquad y = 2\lambda$$
$$\qquad (9')$$
$$\frac{\partial M}{\partial y} = 0: \qquad x = 2\lambda$$

$$\frac{\partial M}{\partial \lambda} = 0: \qquad 2x + 2y = P. \qquad (10')$$

As before, one finds that the critical point occurs at $x = y = P/4$. At this point

$$\lambda = P/8. \qquad (11)$$

The multiplier expression for the entropy problem (equations (1a) and (1b)) is

$$M(n_i; \lambda_E, \lambda_N, \lambda_V) =$$
$$k \ln \Omega - [\lambda_E (\Sigma n_i \varepsilon_i - E) + \lambda_N(\Sigma n_i - N) + \lambda_V(V(n_i) - V)]. \qquad (8'')$$

When $\Omega = N!/\Pi n_i!$, the equations corresponding to (9) are

$$\frac{\partial M}{\partial n_i} = 0: \; -k \ln \frac{n_i}{N} = \lambda_E \varepsilon_i + \lambda_N + \lambda_V \frac{\partial V}{\partial n_i}. \qquad (9'')$$

Evidently at a stationary point

$$n_i = Ne^{-\frac{1}{k}\left(\lambda_N + \lambda_V \frac{\partial V}{\partial n_i}\right)} e^{-\frac{\lambda_E \varepsilon_i}{k}}. \qquad (12)$$

This expression is equivalent to the expression $n_i = (N/Q)e^{-\varepsilon_i/kT}$ derived in Chapter 23 if the factor $e^{-\frac{1}{k}\left(\lambda_N + \lambda_V \frac{\partial V}{\partial n_i}\right)}$ is equal to Q^{-1} and if $\lambda_E = 1/T$. In these expressions the leading term in Q is $e^{-\varepsilon_0/kT}$.

Identification of Lagrange's Multipliers. At a critical point the values of the variables $x_1, \ldots, x_n$ in the function $f(x_1, \ldots, x_n)$ are dependent upon the values of the constants C^k in the constraining conditions $g^k(x) = C^k$. (In the area-perimeter problem x and y at the critical point were equal to $P/4$.) Change the values of C^k and the critical-point values of $x_1, \ldots, x_n$ will change; so, also, will the critical-point value of $f(x)$. Indirectly, therefore, the critical-point value of $f(x)$ is a function of the parameters $C^k = g^k(x)$.

$$f(x) = F[g^1(x), \ldots, g^m(x)] \tag{13}$$

(At the critical point of the area-perimeter problem $A = P^2/16$.) From (13) are obtained by differentiation*

$$\begin{aligned} f_1 &= F_1 g_1{}^1 + \cdots + F_m g_1{}^m \\ &\cdot \\ &\cdot \\ &\cdot \\ f_n &= F_1 g_n{}^1 + \cdots + F_m g_n{}^m , \end{aligned} \tag{14}$$

where

$$F_k = (\partial f/\partial g^k)_{g^i \neq k} .$$

Comparison of equations (14) with equations (9) shows—on the assumption the functions $g^k(x)$ are functionally independent—that Lagrange's multipliers are the derivatives of the steadied function f with respect to the constraining functions g^k.

$$\boxed{\lambda_k = \left(\frac{\partial f}{\partial g^k}\right)_{g^i \neq k}} \tag{15}$$

This means that in the area-perimeter problem

$$\lambda = \frac{dA}{dP} = \frac{d(P^2/16)}{dP} = \frac{P}{8},$$

in agreement with equation (11).

In the entropy maximum problem, as suggested earlier,

$$\lambda_E = \left(\frac{\partial S}{\partial E}\right)_{V,N} = \frac{1}{T}. \tag{16}$$

Also,

$$\begin{aligned} \lambda_N &= \left(\frac{\partial S}{\partial N}\right)_{E,V} \\ &= \left(\frac{\partial S}{\partial n}\right)_{E,V} \cdot \left(\frac{\partial n}{\partial N}\right)_{E,V} \\ &= -\frac{\mu}{T} \cdot \frac{1}{N_0}. \end{aligned} \tag{17a†}$$

* It can be shown that this procedure is valid although only $n - m$ of the n variables $x_1, \ldots, x_n$ are independent.

† $n = N/N_0$.

And
$$\lambda_V = \left(\frac{\partial S}{\partial V}\right)_{E,N}$$
$$= \frac{P}{T}. \tag{17b}$$

Substitution into equation (12) from equations (16), (17a), and (17b) and replacing $\partial V/\partial n_i$ by V/N_0, the volume per particle, one finds that when Ω is equal to, or replaceable by, $N!/\Pi ni!$,
$$n_i = Ne^{(\mu - PV)/RT}\, e^{-\varepsilon_i/kT}. \tag{18}$$

Since $\Sigma n_i = N$,
$$e^{(\mu - PV)/RT} = (e^{-\varepsilon_0/kT} + e^{-\varepsilon_1/kT} + \cdots)^{-1}$$
$$= Q^{-1} \qquad \text{(Leading term in } Q = e^{-\varepsilon_0/kT})$$
$$= (e^{-\varepsilon_0/kT} \cdot Q)^{-1} \qquad \text{(Leading term in } Q = 1).$$

Hence,
$$n_i = (N/Q)e^{-\varepsilon_i/kT}.$$

Also, as found in Chapter 35,
$$\mu\ (= G \text{ per mole}) = N_0\varepsilon_0 + PV - RT \ln Q.$$

Reciprocity of Stationary States. If a function $f(x)$ with constraints $g^1(x) = C^1, g^2(x) = C^2, \ldots, g^m(x) = C^m$ has a maximum value C at a critical point X, the function $g^1(x)$ with constraints $f(x) = C, g^2(x) = C^2, \ldots, g^m(x) = C^m$ has a minimum value at X if the derivative $(\partial f/\partial g^1)_{g^2,\ldots,g^m}$ is positive; if this derivative is negative, $g^1(x)$ has a maximum value at X.

In the area-perimeter problem the derivative dA/dP is positive. Therefore, by the statement above the function $P = 2x + 2y$ with the constraint $xy = P^2/16$, P a constant, has a minimum value at the point $x = y = P/4$. This conclusion may be verified as follows. From the condition $d(xy) = 0$, $dy = -(y/x)dx$. Hence, $dP = (2 - 2y/x)dx$. dP vanishes when $x = y$. Since $xy = P^2/16$, the critical point occurs at $x = y = P/4$. Also, since $(dP/dx)_{xy\ \text{constant}} = 2 - 2y/x$,
$$(d^2P/dx^2)_{xy\ \text{constant}} = -2\left(x\frac{dy}{dx} - y\right)\Big/x^2 = 4y/x^2 = 4/x = 16/P > 0.$$

In other words, the value of P at the critical point $x = y = P/4$ is a minimum value for P. Of all rectangles the square has the smallest perimeter for a given area.

In the entropy-energy-volume-mass problem the derivative $(\partial S/\partial E)_{V,n}$ is always positive. Accordingly, since S has a maximum when E, V, and n are held constant, E has a minimum when S, V, and n are held constant.

The existence of this minimum may be verified as follows. Consider a composite system composed of two objects that are initially in thermal equilibrium with each other. The bulk transfer of thermal energy from one object to the other decreases the system's entropy. Since the derivative $(\partial S/\partial E)_{V,n}$ is positive, the system's entropy may be restored to its initial value by increasing E. In this way the energy of a system can be increased without limit while holding constant the system's entropy, volume, and mass. Under the conditions S, V, n constant no upper bound to E exists. A lower bound exists, however, since the entropy of an internally stable system vanishes if all its thermal energy is removed.

The previous results can be generalized. Let ξ represent the degree of advancement of an event—for example, a chemical reaction, or the exchange between two parts of a system of thermal energy, volume, or mass. A plot of S against ξ holding constant E, V, and n will have the general shape shown in Fig. 41.1. At the equilibrium point ξ^0

$$\left(\frac{\partial S}{\partial \xi}\right)_{E,V,n} = 0 \tag{19a}$$

and

$$\left(\frac{\partial^2 S}{\partial \xi^2}\right)_{E,V,n} < 0. \tag{19b}$$

By calculus, V and n constant,

$$\left(\frac{\partial E}{\partial \xi}\right)_S = -\left(\frac{\partial E}{\partial S}\right)_\xi \left(\frac{\partial S}{\partial \xi}\right)_E. \tag{20}$$

Therefore, (19a),

$$\left(\frac{\partial E}{\partial \xi}\right)_{\substack{S,V,n\\ \xi=\xi^0}} = 0. \tag{21}$$

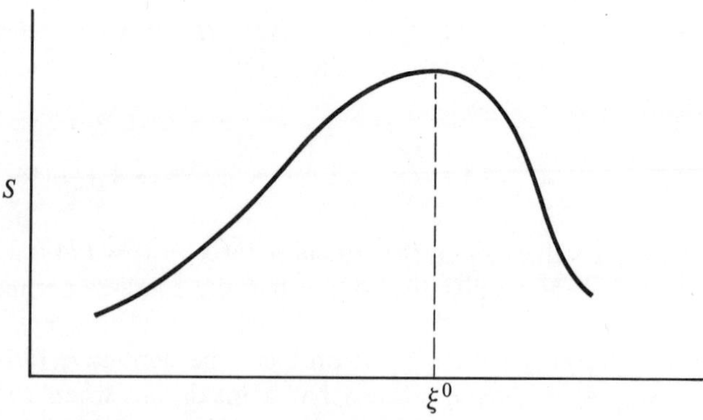

Fig. 41.1

E, like S, has a stationary value at ξ^0. To determine the character of this stationary value it is necessary to determine the sign at ξ^0 of $(\partial^2 E/\partial\xi^2)_{S,V,n}$. From (20), V and n constant,

$$\left(\frac{\partial^2 E}{\partial \xi^2}\right)_S = \left\{\frac{\partial}{\partial \xi}\left[-\left(\frac{\partial E}{\partial S}\right)_\xi \left(\frac{\partial S}{\partial \xi}\right)_E\right]\right\}_S$$

$$= -\left(\frac{\partial S}{\partial \xi}\right)_E \left[\frac{\partial}{\partial \xi}\left(\frac{\partial E}{\partial S}\right)_\xi\right]_S - \left(\frac{\partial E}{\partial S}\right)_\xi \left[\frac{\partial}{\partial \xi}\left(\frac{\partial S}{\partial \xi}\right)_E\right]_S. \tag{22}$$

The first term in (22) is zero at ξ^0, by (19a). To evaluate the second term, it may be noted that

$$d\left(\frac{\partial S}{\partial \xi}\right)_E = \left[\frac{\partial}{\partial \xi}\left(\frac{\partial S}{\partial \xi}\right)_E\right]_E d\xi + \left[\frac{\partial}{\partial E}\left(\frac{\partial S}{\partial \xi}\right)_E\right]_\xi dE.$$

Therefore,

$$\left[\frac{\partial \left(\frac{\partial S}{\partial \xi}\right)_E}{\partial \xi}\right]_S = \left(\frac{\partial^2 S}{\partial \xi^2}\right)_E + \left[\frac{\partial}{\partial E}\left(\frac{\partial S}{\partial \xi}\right)_E\right]_\xi \left(\frac{\partial E}{\partial \xi}\right)_S. \tag{23}$$

The second term in (23) is zero at ξ^0, by (21). Thus, at ξ^0

$$\left[\frac{\partial}{\partial \xi}\left(\frac{\partial S}{\partial \xi}\right)_E\right]_S = \left(\frac{\partial^2 S}{\partial \xi^2}\right)_E. \tag{24}$$

Substitution from (24) into (22) yields

$$\left(\frac{\partial^2 E}{\partial \xi^2}\right)_{\substack{S,V,n\\ \xi=\xi^0}} = -\frac{1}{T}\left(\frac{\partial^2 S}{\partial \xi^2}\right)_{\substack{E,V,n\\ \xi=\xi^0}}. \tag{25}$$

Since $\partial^2 S/\partial\xi^2$ is negative at ξ^0 and T is always positive,

$$\left(\frac{\partial^2 E}{\partial \xi^2}\right)_{\substack{S,V,n\\ \xi=\xi^0}} > 0. \tag{26}$$

The stationary value E has at ξ^0 when S, V, and n are held constant represents a minimum value for E.

A Self-Evident Principle. If a function $M(x)$ has a stationary value at X and if $G^k(X) = C^k$, $M(x)$ remains stationary at X if to such constraints as may already exist are added the further constraints $G^k(x) = C^k$.

This principle may be used to show that the multiplier expression (8) has the same critical points as the function $f(x)$ with constraints $g^k(x) = C^k$. For suppose $M(x, \lambda)$ has a stationary value at X. By equations (10),

$g^k(X) = C^k$. Now annex as constraints to $M(x, \lambda)$ the relations $g^k(x) = C^k$. By the self-evident principle $M(x, \lambda)$ retains its stationary value at X. But now $M(x, \lambda) = f(x)$.

The Legendre Transformation. The multiplier expression for finding the stationary value of E subject to the conditions $S = S^0$, $V = V^0$, $n = n^0$ is $M(S, V, n; \lambda_S, \lambda_V, \lambda_n) = E - \lambda_E(S - S^0) - \lambda_V(V - V^0) - \lambda_n(n - n^0)$. The constants S^0, V^0, and n^0 are the values of S, V, and n at the critical point. The corresponding values of λ_E, λ_V, and λ_n may be designated $\lambda_E{}^0$, $\lambda_V{}^0$, and $\lambda_n{}^0$. If one annexes as constraints the three relations

$$\lambda_E = \lambda_E{}^0$$
$$\lambda_V = \lambda_V{}^0$$
$$n = n^0,$$

the multiplier expression becomes

$$E - \lambda_E{}^0(S - S^0) - \lambda_V{}^0(V - V^0).$$

Dropping from this expression the constant terms $\lambda_E{}^0 S^0$ and $\lambda_V{}^0 V^0$ and introducing the multiplier identification, (equation (15)), one finds that when T, P, and n are held constant the function that has a stationary value is the free energy

$$E - TS + PV.$$

Answers to the Problems

Chapter 1

1. $(E_{\text{total}})_{\substack{\text{final}\\\text{state}}} = (E_{\text{total}})_{\substack{\text{initial}\\\text{state}}}$.

   $E_{\text{total}} = \text{constant}.$

   $E_{\text{universe}} = \text{constant}.$

   $E_{\text{isolated system}} = \text{constant}.$

   Energy is conserved.

   Energy can be neither created nor destroyed.

   Any process whose net effect is equivalent to the raising or the lowering of a weight is impossible.

$$(E_{\text{total}})_{\substack{\text{final}\\\text{state}}} - (E_{\text{total}})_{\substack{\text{initial}\\\text{state}}} \equiv \Delta E_{\text{total}} = 0.$$

2. Probably not. The earth is not an isolated system. From outer space it receives such things as radiant energy, cosmic rays, dust, and meteorites, and to outer space it loses such things as radiant energy, atmospheric gases, and space vehicles. It seems unlikely that its total energy remains strictly constant.
3. If energy cannot be created, the total energy in the final state must be less than or at most equal to the total energy in the initial state.

$$(E_{\text{total}})_{\substack{\text{final}\\\text{state}}} \leq (E_{\text{total}})_{\substack{\text{initial}\\\text{state}}}$$

On the other hand, if energy cannot be destroyed, the total energy in the final state must be at least as great as and perhaps greater than the total energy in the initial state.

$$(E_{\text{total}})_{\substack{\text{final}\\\text{state}}} \geq (E_{\text{total}})_{\substack{\text{initial}\\\text{state}}}$$

Both statements are true if, and only if,

$$(E_{\text{total}})_{\substack{\text{final}\\\text{state}}} = (E_{\text{total}})_{\substack{\text{initial}\\\text{state}}}.$$

4. Particularly to be noted in bookkeeping energy changes for eventual use with the Second Law are possible changes in the thermal energies of objects.
   (a) The aqueous phase and the solid ammonium nitrate phase; also, because the solution process is not athermal, the immediate thermal surroundings of these two phases. Neglected here is a small change that occurs in the energy of the atmosphere if the solution process occurs in an open beaker and is accompanied by a net change in the combined volume of the condensed phases; for pressures up to several atmospheres energy changes of this kind are normally unimportant.
   (b) The contents of the bomb and, because the reaction is exothermic, the bomb itself and the bomb's immediate thermal surroundings.
   (c) The liquid phases and their thermal surroundings.
   (d) The zinc phase, the solution phase, the hydrogen-enriched atmosphere, and the thermal surroundings of these phases.
   (e) The weight and the rubber band—but not their thermal surroundings, although in fact during a rapid extension the rubber band does become warmer; however, if the extension is *rapid*, the system hasn't time during the extension to exchange thermal energy with its surroundings.
   (f) The weight, the rubber band, and its thermal surroundings. Isothermal extension of some elastomers produces an increase in the thermal energy of the surroundings almost equal to the decrease in the potential energy of the weight. (What does this imply? Answer: That the energies of these elastomers are almost independent of their length.)
   (g) The weight (and piston head) and the gas. During a rapid compression the gas becomes warmer, but it hasn't time to exchange thermal energy with its surroundings. A process in which the thermal energy of the surroundings does not change is called adiabatic.
   (h) The weight, the gas, and the thermal surroundings of the gas. Isothermal compression of most gases produces an increase in the thermal energy of the surroundings almost equal to the decrease in the potential energy of the confining weight. (What does this imply?)
   (i) The thermal surroundings of the glass and the glass itself, both for its diminished potential energy and for energy stored in newly created internal strains and in the freshly created glass-air interface. The last term could be particularly important if the glass were to shatter to a fine powder.
   (j) Your mother-in-law, whose altitude will change slightly as her mass comes to bear on the scale's platform; the scale's platform; the

balance spring, owing to an increase in length, and, if extended beyond its elastic limit, to an altered state of internal strain (How could this be determined?); and the spring's thermal surroundings, if the spring is one whose temperature changes during an adiabatic extension. (Adiabatic extension of a spring alters the spring temperature if the spring's tension at fixed extension is a function of the spring temperature.)

(k) The ice tray and its contents; the room (energy extracted from the freezing water is expelled, eventually, through cooling coils, to the room in which the refrigerator stands); the power plant that provides electrical energy to the refrigerator machinery whose function it is to keep the cold box cold; and the thermal surroundings of the wires that connect the refrigerator machinery to the power plant (because of the *IR* drop along these wires); but, note, not the refrigerator itself, which, if a long-lasting refrigerator, survives essentially unchanged the process of pumping thermal energy from the freezing water to the warmer room. (Question: What happens to the energy delivered by the power plant to the refrigerator? Answer: In the end it appears as thermal energy in the room (Chapter 9).)

5. (a) $3 \times E_{H_2O}^{solid}$. $3 \times E_{H_2O}^{liq}$.

   (b) $n_{H_2O}^{solid} \times E_{H_2O}^{solid} + n_{H_2O}^{liq} \times E_{H_2O}^{liq}$.

   (c) $n_{H_2O}^{solid} - 1$. $n_{H_2O}^{liq} + 1$. $(n_{H_2O}^{solid} - 1) \times E_{H_2O}^{solid} + (n_{H_2O}^{liq} + 1) \times E_{H_2O}^{liq}$.

6. No. The energy liberated is not produced from nothing. The energy of two protons, an oxygen nucleus, and ten electrons arranged as indicated by the symbols $H_2 + (1/2)O_2$ is substantially greater than the energy of the same particles in the arrangement HOH (Chapter 2).
7. Energy is not destroyed. The energy of liquid water is greater than the energy of the same mass of ice.
8. The mechanical energy spent winding the wire appears eventually as thermal energy, some at the time of winding, some later during the reaction with hydrochloric acid. The process of solution in hydrochloric acid of coiled spring wire is slightly more exothermic than is solution of the same length of uncoiled, unstrained wire. The difference between the two heats of solution constitutes, in fact, a quantitative measure of the strain energy stored in the coiled wire.
9. Contrary to popular belief, the open-refrigerator-door policy makes kitchens hotter—unless the working parts of the refrigerator, the cold-box excepted, are mounted outside the kitchen (Chapter 9).
10. Some physicists questioned the accuracy of the experiments.

    Niels Bohr conjectured that energy might not always be conserved after all. (Currently it is conjectured that the energy conservation law may indeed be violated, but generally for only very short intervals of time of the order of $h/\Delta E$ seconds, where h is Planck's constant and ΔE is the extent to which energy is not conserved.)

Other physicists believed energy might be carried away undetected by tiny, uncharged particles, christened by Fermi, father of beta-decay theory, neutrinos.

Neutrinos also help to conserve angular momentum during beta-decay. At least four different kinds of neutrinos are believed to exist. They have zero, or nearly zero, rest mass and interact so weakly with matter they can typically pass through millions of planets the size and composition of the earth before suffering a collision. Virtually all the neutrinos ever produced in the history of the universe—through beta decay and nuclear reactions in stellar interiors—are estimated to be still at large. They represent a vast sink into which the universe continually pours energy and from which, evidently, energy rarely returns in other forms.

Chapter 2

1. -1440 cal/mole.
2. The chemical and thermochemical equations are

$$H_2 + 103 \text{ kcal} = 2H$$

$$E_{H_2} + 103 \text{ kcal/mole} = 2E_H .$$

Therefore, if one sets $E_{H_2} = 0$, $E_H = +51.5$ kcal/mole.

3. $$(E_\sigma)_{\substack{\text{initial}\\\text{state}}} = E_{\text{reactants}} = E_{H_2}^{\text{gas}} + (1/2)E_{O_2}^{\text{gas}} .$$

$$(E_\sigma)_{\substack{\text{final}\\\text{state}}} = E_{\text{products}} = E_{H_2O}^{\text{liq}} .$$

4. (a) exothermic. (b) negative ($E_{\text{products}} < E_{\text{reactants}}$). An example is the reaction H_2(gas) + (1/2)O_2(gas) = H_2O(liq). Here

$$\Delta E_\theta = +67{,}400 \text{ cal/mole } H_2O\text{(liq) formed.}$$

$$\Delta E_\sigma = -67{,}400 \text{ " " " " "}$$

5. (a) endothermic. (b) positive ($E_{\text{products}} > E_{\text{reactants}}$). An example is th change H_2O(solid) = H_2O(liq). Here

$$\Delta E_\theta = -1440 \text{ cal/mole of } H_2O\text{(liq) formed.}$$

$$\Delta E_\sigma = +1440 \text{ " " " " " "}$$

6. V_σ would generally be considered to be the volume of the bomb's interior. If the bomb is perfectly rigid, $(V_\sigma)_{\substack{\text{final}\\\text{state}}} = (V_\sigma)_{\substack{\text{initial}\\\text{state}}}$. Therefore, $\Delta V_\sigma \equiv (V_\sigma)_{\substack{\text{final}\\\text{state}}} - (V_\sigma)_{\substack{\text{initial}\\\text{state}}} = 0$.
7. The flame is so hot that, contrary to one of the assumptions of the calculation, reaction of the hydrogen and the oxygen is not complete. To put it another way, at the high temperature of the flame water vapor is partially dissociated.
8. No. The net changes in the universe are not the same in the two cases. In the spontaneous dissociation of liquid water into hydrogen and oxygen, 68 kcal/mole of H_2O dissociated has to be absorbed from the surroundings (and at 25°C this does not happen). In the electrolytic dissociation of water, only 11.7 kcal/mole is absorbed from the surroundings; the remainder of the energy necessary to make up the difference between $E_{H_2O}^{liq}$ and $E_{H_2}^{gas} + (1/2)E_{O_2}^{gas}$ is provided electrochemically.
9. The energy of sublimation does not vanish at absolute zero; in fact, for most substances the energy of sublimation is larger at absolute zero than at any other temperature. What faulty assumption in the statement leads to the totally erroneous conclusion?
10. Reactions spontaneous at low temperatures are exothermic reactions. Reactions spontaneous at high temperatures are endothermic reactions.
11. According to Dr. Black's data the energy required to melt a given mass of ice is equal to the energy required to increase the temperature of the same mass of liquid water about 140 Fahrenheit degrees, or

$$140°\text{F} \times \frac{100°\text{C}}{(212 - 32)°\text{F}} = 78°\text{C}.$$

Now, the energy required to increase the temperature of one gram of liquid water one centigrade degree is 1 calorie. Therefore, the energy required to increase the temperature of one gram of liquid water 78 centigrade degrees, or to melt one gram of ice, is 78 calories. In other words, according to Dr. Black's data the heat of melting of ice is about 78 cal/gm.

Chapter 3

1. The First Law is not necessarily violated.
   They are all highly improbable processes.
   The reverse of each process is a possible process.
2. No. The *net effect* of a complete cycle would be the production of mechanical work solely at the expense of the thermal energy of the

surroundings. It would correspond in Fig. 3.1 to passage from State 2 back to State 1.

Evidently a catalyst cannot alter the composition of a system that is at equilibrium with respect to the change catalyzed by the catalyst. For if it could, it would be possible, as in the nitrogen-hydrogen-ammonia machine, to raise weights solely at the expense of the thermal energy of their surroundings. Therefore, in a system at equilibrium a catalyst must affect the rates of the forward and backward reactions by exactly equal amounts.

Machines whose net effect in operation would be equivalent to the raising of a weight solely at the expense of the thermal energy of their surroundings are called perpetual motion machines of the second kind. Patent applications on machines of this kind are no longer entertained in France, England, and most other nations of the world.

3. Yes, from the thermodynamic point of view the two statements are equivalent. Like bankers mainly interested in cash balances and not

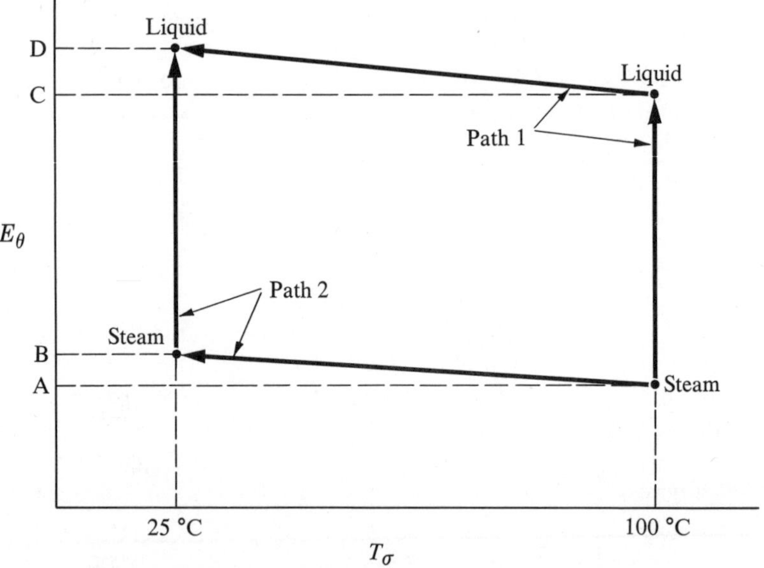

Fig. 3.2 Energy liberated to the thermal surroundings when steam condenses.
AC = Heat of condensation of steam at 100°C.
CD = Energy liberated when liquid water cools from 100° to 25°C.
AB = Energy liberated when steam cools from 100° to 25°C.
BD = Heat of condensation of steam at 25°C.

Because the heat capacity of steam (8.02 cal/deg-mole) is less than the heat capacity of liquid water (18.0 cal/deg-mole), AB $(= C_P^{\text{steam}} \times \Delta T) <$ CD $(= C_P^{\text{liq water}} \times \Delta T)$. Therefore, BD $>$ AC: the heat of condensation of steam increases with decreasing temperature.

whether drafts and deposits arrive by mail or messenger, thermodynamics is concerned with final effects and not with the manner in which these effects are achieved. Processes whose final effects are identical are regarded as identical regardless of whether the effects are in one instance achieved "spontaneously" and in another through the intervention of special agents that in the end, like true catalysts, are recovered unchanged. In thermodynamics it is the ends, not the means, that are important.

The quantity of energy liberated to the thermal surroundings during the condensation of steam initially at 100°C to liquid water finally at 25°C, for example, is the same regardless of whether the water first condenses at 100°C and then cools as the liquid to 25°C or cools first to 25°C as a supercooled vapor and then condenses (Fig. 3.2).

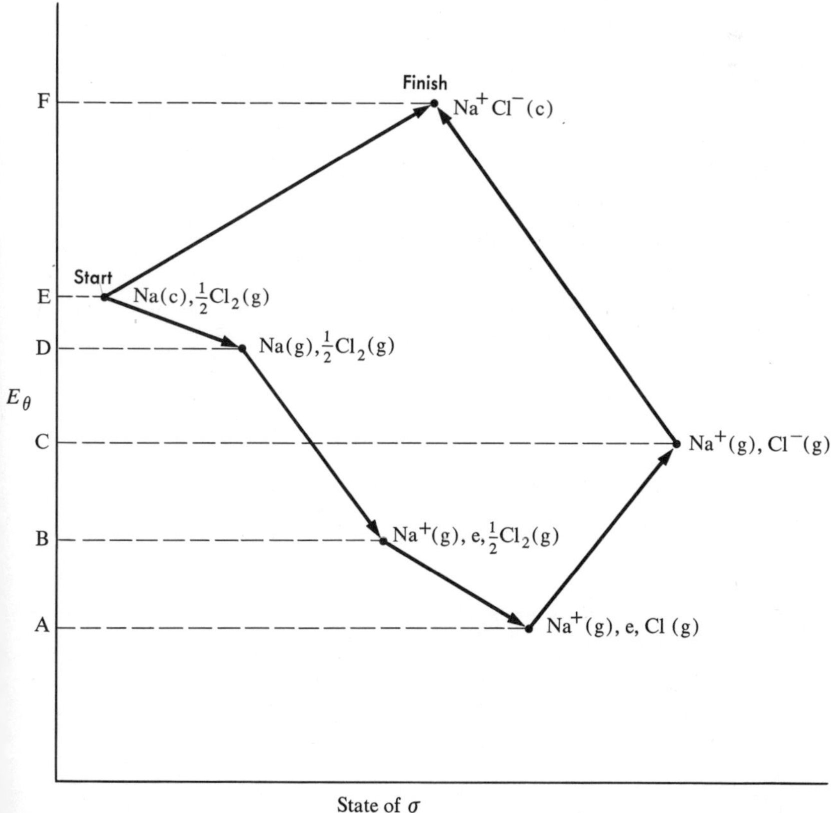

ig. 3.3 The Born-Haber cycle. The standard heat of formation of sodium chloride (EF) equal to the heat of sublimation of sodium (DE) plus sodium's first ionization potential 3D) plus the bond dissociation energy of chlorine (AB) less chlorine's electron affinity C), all this (CE) subtracted from sodium chloride's lattice energy (CF).

Similarly, if metallic sodium is vaporized to atomic sodium and each atom so formed is ionized to sodium ions and electrons, and if simultaneously gaseous chlorine is dissociated to atomic chlorine and to each atom so formed is added an electron to form chloride ions, and if, then, the sodium ions and the chloride ions are allowed to condense to form crystalline sodium chloride, the net energy liberated to the thermal surroundings is the same as the energy liberated to the surroundings when metallic sodium reacts directly with chlorine gas to form sodium chloride (Fig. 3.3).

Chapter 4

1. The disorder in a universe tends to increase.*
   The entropy of a universe tends to increase.
   Entropy is conserved only in reversible processes; in all spontaneous, irreversible processes it increases.
   Any spontaneous process is a change from a less to a more disordered state.
   Any change from a more to a less disordered state is impossible.
   Any process that decreases the entropy of a universe is impossible.
   Any process whose net effect is equivalent to the raising of a weight solely at the expense of the thermal energy of a body is impossible.
   Any process whose sole effect is equivalent to the transfer of heat from a cold to a warmer body is impossible.

   $$(S_{\text{total}})_{\substack{\text{final}\\\text{state}}} \geq (S_{\text{total}})_{\substack{\text{initial}\\\text{state}}} .$$

   $$[(S_{\text{total}})_{\substack{\text{final}\\\text{state}}} - (S_{\text{total}})_{\substack{\text{initial}\\\text{state}}}] \equiv \Delta S_{\text{total}} \geq 0.$$

   $$\Delta S_{\text{universe}} \geq 0.$$

   $$\Delta S_{\text{isolated system}} \geq 0.$$

   Since a system whose energy, volume, and mass remain constant is in effect an isolated system,

   $$\Delta S_{E,V,M} \geq 0.$$

2. $\Delta E_{\text{total}} = 0.$
   $\Delta S_{\text{total}} > 0.$
   Irreversible.
3. Free expansion of a gas into a vacuum.
   Mixing of two gases or two liquids.

* Sometimes paraphrased: If you think things are mixed up now, just wait.

Solution of sugar or salt in pure water.
Spilled water evaporating from a hot stove.
A waterfall.
Neutralization of H_3O^+ by OH^- in a beaker.
Oxidation of I^- by MnO_4^- in a beaker.
Wood burning in a fireplace.
An explosion.
Production of helium from hydrogen in stars.
The flash of an electronic flash gun.
Radioactive decay.
Unwinding of a clock.
Making whipped cream.
Talking.
Frying an egg.
Processes previously mentioned: warm ice melting, cold water freezing, explosion of cold hydrogen and oxygen, dissociation of very hot steam, heat flow from hot to cold, and the complete conversion of the potential energy of a weight into heat.
Any non-friction-free process.

Humpty Dumpty's great fall. The words of the nursery rhyme, perhaps better than any figures, vividly express how irreversible this event was.

> All the king's horses, and all the king's men,
> Couldn't put Humpty together again.

Sir Arthur Eddington adds this explanatory note. When Humpty Dumpty fell,

> Something happened which could not be undone. The fall could have been undone. It is not necessary to invoke the king's horses and the king's men; if there had been a perfectly elastic mat underneath, that would have sufficed. At the end of his fall Humpty Dumpty had kinetic energy which, properly directed, was just sufficient to bounce him back on to the wall again. But, the elastic mat being absent, an irrevocable event happened at the end of the fall—namely, the introduction of a *random element* into Humpty Dumpty.*

Suppose the king's horses and the king's men could have put Humpty Dumpty together again. Would the chief source of the *random element* in Humpty Dumpty have been thereby removed?

4. Melting or freezing of a substance at its normal melting point.
Precipitation of an infinitesimal amount of salt from a saturated solution.

* *Nature of the Physical World*, Ann Arbor, 1958.

Evaporation or condensation of a liquid at its normal vapor pressure.
Controlled expansion of a gas against a (very nearly) balancing external pressure.
Controlled compression of a gas by a (very slightly) overbalancing external pressure.
Reaction of H_3O^+ with OH^- in an electrochemical cell to give the maximum possible cell voltage.
Flow of heat between two objects at the same temperature.
A heat engine, heat pump, or refrigerator working at maximum efficiency.
Any friction-free mechanical process.

5. The First Law is silent regarding the direction of natural processes.

   So far as the First Law alone is concerned, a perfect vacuum could be produced by the spontaneous self-compression of a gas, miscible liquids could unmix, fresh water could be obtained from salt water without the expenditure of any work, water vapor could condense on a hot stove, water could ionize spontaneously to form large concentrations of H_3O^+ and OH^-, I_2 and Mn^{++} could react spontaneously in acidic solutions to form large concentrations of I^- and MnO_4^-, the chimney gases and radiant energy from a fire could recombine to form wood and oxygen, the energy of the flash from an electronic flash gun could revert to its source and recharge the gun's condensors, warm water could freeze, cold ice could melt, liquid water could dissociate spontaneously to form high concentrations of molecular hydrogen and oxygen, and heat could flow spontaneously from a cold body to a warmer one or be converted entirely to mechanical energy.

6. The Second Law is silent about the rate of natural processes. A bouncing ball should eventually come to rest, it says. Supercooled water should eventually freeze. A cold mixture of hydrogen and oxygen should eventually produce water. But how long is eventually? Seconds? Or centuries? The Second Law does not say. Its NO is emphatic; its YES is only permissive.

7. The length of the rope.
   The distance between the horizontal supports.
   The gravitational attraction between the earth and the rope, which gives rise to a term called the potential energy of the rope.

   The Second Law, which states that the most stable configuration for the rope subject to the given constraints is that configuration which maximizes the entropy of the universe (the rope plus the surrounding air). Like the case of the falling weight, this configuration is the configuration in which the potential energy of the mechanical object (the rope) is a minimum and the temperature (and entropy) of the rope and its surroundings is a maximum.

8. Use thermal energy taken from a cold body to increase the altitude of a weight resting initially on a warmer body. When this has been accomplished, let the weight fall back to its original position on the warmer body. Continue hammering away in this fashion until the desired quantity of thermal energy has been transferred from the colder body to the warmer one.
9. No, no more than the electrolytic dissociation of water into hydrogen and oxygen contradicts the fact that the explosion of hydrogen and oxygen to form water is a spontaneous, irreversible process (cf. Chapter 2, problem 8).
10. No. Only part of the energy removed from the boiler appears as work. The remainder (neglecting friction) appears in a condenser. A steam engine cannot convert heat completely to work. This matter is discussed more fully in Chapter 9. The efficiency of a heat engine in converting heat to work is limited by the fact that the boiler cannot be made infinitely hot, or the condenser infinitely cold.

Chapter 5

1. Boron is very hard, although not as hard as diamond. Estimated entropy: 2–3 cal/deg-mole. Experimental entropy: 1.6 cal/deg-mole. Barium in hardness and atomic mass is similar to lead. Estimated entropy: 15.5 cal/deg-mole. Experimental entropy: 16.0 cal/deg-mole. Ammonia is isoelectronic with water; at 25°C and 1 atm. it is a gas. Estimated entropy: 45 cal/deg-mole. Experimental entropy: 46 cal/deg-mole.
2. Taking one-tenth the difference between the entropy of $Na_2SO_4 \cdot 10H_2O$ and Na_2SO_4 as a measure of the contribution to the entropies of crystalline hydrates of one mole of water, one obtains for $CaSO_4 \cdot 2H_2O$ an estimated entropy of 46.7 cal/deg-mole. The experimental entropy is 46.4 cal/deg-mole.
3. Monoclinic sulfur.
4. Calcite.
5. At the normal boiling point the entropy of the saturated vapor is for many substances about 20 to 23 cal/deg-mole larger than the entropy of the liquid. At the critical point the two entropies are the same.
6. The three types of motion are called, respectively, translational, rotational, and vibrational motion.
7. "Dry air" is air in which the concentration of water vapor is small; that is to say, air for which the ratio n_{H_2O}/V is small. Since $n_{H_2O}/V =$

P_{H_2O}/RT, "dry air" may also be described as air for which at a given temperature the partial pressure of the water vapor (P_{H_2O}) is small.

8. For an ideal gas $(V_2/V_1) = (n_2RT_2/P_2)/(n_1RT_1/P_1)$. If $n_2 = n_1$ and $T_2 = T_1$, $(V_2/V_1) = (P_1/P_2)$.
9. Melting and evaporation are endothermic processes; that is to say, during these processes the thermal energy and entropy of the surroundings decrease. According to the Second Law, however, the total entropy change in the universe (the melting or evaporating chemical plus its thermal surroundings) cannot be negative.

Chapter 6

1. 34.6 cal/deg-mole. This is the translational entropy of water vapor at 25°C and 1 atm. Actually, since a neon atom is slightly heavier than a water molecule, the translational entropy of neon, which is its total entropy, should be slightly more than 34.6 cal/deg-mole. The experimental value for the entropy of neon at 25°C and 1 atm. is 35.0 cal/deg-mole.
2. Yes. The entropy of sodium chloride per mole of independently vibrating nuclei ($17.3 \div 2 = 8.6$) is about the same as the entropy of a simple elemental substance of comparable hardness, e.g. copper (8.0). The fact that sodium bromide has a slightly larger entropy than sodium chloride is consistent with the rule that other things being equal the larger the mass of a particle the more closely spaced are the quantum-mechanically allowed energy levels for the particle (or for the system that contains the particle), the easier is thermal excitation of the particle's characteristic motion (in NaCl and NaBr lattice vibrations of the cations and anions), and the larger the entropy.
3. $[19.7 + (20.0 - 17.3)]$ cal/deg-mole $= 22.4$ cal/deg-mole. Experimental S: 23.0 cal/deg-mole.
4. $[32.6 + 2(20.0 - 17.3)]$ cal/deg-mole $= 38.0$ cal/deg-mole. Experimental S: 38.6 cal/deg-mole.
5. Yes. The entropy of Na_2SO_4 if each atom executed periodic displacements from its mean lattice position almost independently of the other atoms would be 7 times 8.5 (the approximate contribution of a mole of atoms or elementary ions of medium atomic weight to the entropy of a lattice of average hardness), or about 60 cal/deg-mole. On the other hand, if the five atoms of each sulfate group were to move about in the lattice as a rigid body, which on chemical grounds seems likely, the entropy of Na_2SO_4 would be more like 17 cal/deg-mole for the sodium ions plus about 12 cal/deg-mole for lattice vibrations of the sulfate

group as a whole, plus 5 to 7 cal/deg-mole for restricted rotations of the complex sulfate group. The fact that the sum of these figures is approximately equal to the observed entropy suggests that not many of the internal degrees of freedom of the sulfate group are excited at room temperature.

6. At 25°C and 1 atm., the entropy of oxygen is constituted as follows: translation 36.3, rotation 12.7, vibration 0, total 49.0. The corresponding values for xenon are 40.5, 0, 0, 40.5.
7. This difference is too large to be accounted for solely on the grounds that the molecule NO_2 is heavier than the N_2O molecule. Nor can a difference of this magnitude be due to differences in vibrational entropy, for the vibrational entropy of NO_2 like that of N_2O is less than 5 cal/deg-mole at 25°C. The difference stems mainly from the fact that the rotational entropy of NO_2 is greater than that of N_2O. N_2O is a linear molecule and has only two degrees of rotational freedom. NO_2 is a bent molecule and has three rotational degrees of freedom.
8. HF is perforce a linear molecule. H_2O is not. Therefore the rotational entropy of H_2O is larger than that of HF. At 25°C and 1 atm. the entropy of HF is constituted as follows: translation 35.0,* rotation 6.5, vibration 0, total 41.5.
9. Not because of the translational entropy certainly—this is larger for HCl than for oxygen—nor because of the vibrational entropy, which is zero for both gases at 25°C. It must be because HCl has the smaller rotational entropy. The reason for this is that in molecular rotation the nuclei rotate around the center of mass of the molecule. In oxygen the center of mass lies midway between the two nuclei; in HCl the center of mass lies close to the massive chlorine nucleus. Thus in the rotation of HCl, as in the rotation of the moon and the earth, or of the earth and the sun, the lighter partner (hydrogen, moon, earth, respectively) executes most of the motion. Therefore the quantum-mechanically allowed rotational energy levels of the HCl molecule are more widely spaced and less easily populated thermally than those of oxygen.
10. Because in a dilute gas individual molecular rotations and vibrations are not affected by isothermal changes in volume and pressure.
11. Translation: Large decrease. The free volume decreases by a factor of $\sim 10^4$ [$R \ln (1/10^4) = -18$ cal/deg-mole].

    Rotation: Moderate decrease. Free rotation in the gas becomes usually hindered rotation in condensed phases.

    Vibration: Little change for small amplitude vibrations in the absence of strong intermolecular interactions.

* At 25°C and 1 atm. the translational entropy of a gas is equal in calories per degree per mole to $26 + 6.9 \log_{10}$ (MW).

12. For the 400°K body: (a) $+1200$ calories/400°K = $+3$ cal/deg. (b) -1200 calories/400°K = -3 cal/deg.
   For the 300°K body: (a) $+1200$ calories/300°K = $+4$ cal/deg. (b) -1200 calories/300°K = -4 cal/deg.
13. -3 cal/deg + 4 cal/deg = $+1$ cal/deg. The entropy increase of the cold body is greater than the entropy decrease of the warm body.
14. Yes. Break up the total energy added into small increments ΔE_i—from the standpoint of accuracy the smaller each ΔE_i the better—divide each ΔE_i by the average temperature T_i of the substance at the time that increment of energy was added, and add together all $\Delta E_i/T_i$ to find the total change in S. In the limit of small ΔE_i this procedure corresponds to finding the area beneath the curve formed by plotting the reciprocal of the absolute temperature during warm-up against the energy uptake.
   (In integral calculus it is shown that

$$\lim_{\Delta E_i \to 0} \sum_{T=T_1}^{T=T_2} \left(\frac{1}{T_i}\right) \Delta E_i = \int_{T=T_1}^{T=T_2} \frac{dE}{T}.$$

This is equal to the value of $[S(T_2) - S(T_1)]$. In terms of the heat capacity

$$C \equiv \lim_{\Delta T \to 0} \frac{\Delta E}{\Delta T} = \frac{dE}{dT},$$

$$dE = CdT,$$

and

$$S(T_2) - S(T_1) = \int_{T=T_1}^{T=T_2} \frac{CdT}{T}.$$

Thus the entropy change can also be determined by plotting the value of (C/T) during warm-up against T, or C against $\ln T$—since $dT/T = d\ln T$. If C is essentially constant over the temperature interval T_1 to T_2,

$$S(T_2) - S(T_1) = C \ln (T_2/T_1).$$

Over the temperature interval 0° – 25°C, for example, the heat capacity of liquid water is about 18 cal/deg-mole; for this substance, therefore,

$$S(25°) - S(0°) \approx (18 \text{ cal/deg-mole})\ 2.303 \log_{10} \frac{298}{273}$$
$$= 1.5 \text{ cal/deg-mole.})$$

15. It is a quantitative measure of disorder.
   It is zero at absolute zero for perfectly crystalline substances.
   It always increases with increasing temperature.
   It always increases in a spontaneous change.

It is an extensive property.
It has the same units as heat capacity: cal/deg (or cal/deg-mole).
It changes on the addition of thermal energy ΔE by the amount $\Delta E/T$.
Its absolute value can be determined from the Third Law and the relation $\Delta S = \Delta E/T$.
Its value for any substance always increases during melting and evaporation.
It can sometimes be expressed as a sum of contributions from translational, rotational, and vibrational motions.
It always increases as the mass involved in the random thermal motion increases. For this reason:

Translational entropy increases with increasing molecular weight.*
Rotational entropy increases with increasing moment of inertia.
And electronic entropy is generally not important at room temperature.

It always increases as the volume involved in the random thermal motion increases. For this reason:

Hard substances have low entropies.
Vibrational entropy is often not important at room temperature.
Nuclear entropy is seldom important at any ordinary temperature.
Rotational entropy is usually more important than vibrational entropy and less important than translational entropy.
And the translational entropy of a gas always increases as the pressure decreases.

Chapter 7

1. (i) The irreversible character of the conversion of a weight's potential energy into thermal energy. (ii) The irreversible character of the flow of thermal energy from hot objects to colder ones.
2. See paragraphs 2 and 3.
3. From the definition of τ, $\Delta S = \Delta E/\tau$.
   (a) Here $\Delta E = Q$. Therefore $\Delta S = Q/\tau$.
   (b) From the additive property of entropy and the definition of the operator Δ, it follows that

$$\Delta(S_{\text{total}}) = \Delta(S_A + S_B)$$
$$= \Delta S_A + \Delta S_B .$$

* The factor is $(3/2)R \log_{10}(\text{MW}) = 6.9 \log_{10}(\text{MW})$.

Also, from the definition of τ, we have that $\Delta S_A = \Delta E_A/\tau_A$ and $\Delta S_B = \Delta E_B/\tau_B$, where in this problem, $\Delta E_A = -Q$ and $\Delta E_B = +Q$. Therefore

$$\Delta(S_{\text{total}}) = -\frac{Q}{\tau_A} + \frac{Q}{\tau_B}$$

$$= Q\,\frac{(\tau_A - \tau_B)}{\tau_A\tau_B}.$$

This last expression is positive if $\tau_A > \tau_B$.

(c) From part (b) we know that for the flow of Q calories from A to B,

$$\Delta S_{\text{total}} = Q\left(\frac{\tau_A - \tau_B}{\tau_A\tau_B}\right).$$

Q and the product $\tau_A\tau_B$ are positive. If it is also known that ΔS_{total} is positive, it follows that τ_A must be greater than τ_B.

(d) Thermal energy can flow spontaneously between two objects A and B in thermal contact with each other in the direction A to B if (part b) and only if (part c) $\tau_A > \tau_B$. Or, for the spontaneous flow of thermal energy from A to B, the condition $\tau_A > \tau_B$ is necessary (part c) and sufficient (part b).

4. "The change in S per unit change in E times the change in E." This product gives the total change in S. Note that it is numerically equivalent to the quotient Q/τ.
5. When a good thermometer is placed in thermal contact with an object whose temperature is to be determined, the thermometer quickly adopts the object's temperature without sensibly altering that temperature. When a warm-blooded person is placed in contact with an inanimate object, however, it frequently happens that the object adopts the person's temperature rather than the person adopting the object's temperature; should the latter occur, it might prove fatal to the "thermometer." A warm-blooded person is often a better thermo*stat* than thermo*meter*.

Chapter 8

1. Two bodies that are in thermal equilibrium with a third body are in thermal equilibrium with each other.
2. Both have large values for hot objects and small values for cold objects. (In other words, energy can flow spontaneously from object A to B if and only if T or $\Delta E/\Delta S$ for A is greater than T or $\Delta E/\Delta S$ for B. Two objects A and B are in thermal equilibrium if and only if T or $\Delta E/\Delta S$ for A is equal to T or $\Delta E/\Delta S$ for B.)

Both increase as thermal energy is added to a body.
Both approach the value zero as the thermal energy approaches zero.
Both are intensive properties.

3. Yes, for converting from energy to entropy units, and vice versa.
4. To calculate T given ΔE and ΔS.
   To calculate ΔE given T and ΔS.
   To calculate ΔS given T and ΔE.
5. Cold. 0.1°K.
6. 8600 to 9900 cal/mole. (Many substances have an entropy of vaporization at the normal boiling point of about 20 to 23 cal/deg-mole. This is called Trouton's Rule.)
7. Zero. Yes.
8. -1440 cal/273°K $= -5.3$ cal/deg. $+5.3$ cal/deg. $+5.5$ cal/deg.
9. $+68{,}021$ cal/298°K $= +228$ cal/deg.
10. Exceedingly hot. A substance with a negative absolute temperature is a substance whose entropy increases as its thermal energy decreases. It could lose energy spontaneously to any substance whose temperature is positive, for in the process the entropies of *both* substances would increase and, *a fortiori*, the entropy of the universe would increase.

Chapter 9

1. 50-50 (i.e. 1 to 1).
2. 50-50 (for a reversible process $\Delta S_{\text{total}} = 0$).
3. It needs to be shown that

$$\Delta S_{\text{total}} = -\frac{\Delta(\text{P.E.})}{T}.$$

Let θ represent the surroundings. By the First Law and the additivity of E

$$\Delta E_{\text{total}} = \Delta E_\theta + \Delta E_{\text{Wt.}} = 0,$$

where $$\Delta E_{\text{Wt.}} = \Delta(\text{P.E.}).$$

Therefore $$\Delta E_\theta = -\Delta(\text{P.E.}).$$

By the additivity of S

$$\Delta S_{\text{total}} = \Delta S_\theta + \Delta S_{\text{Wt.}},$$

where $$\Delta S_{\text{Wt.}} = 0.$$

Therefore $$\Delta S_{\text{total}} = \Delta S_\theta.$$

But by the definition of T

$$\Delta S_\theta = \frac{\Delta E_\theta}{T} = -\frac{\Delta(\text{P.E.})}{T}.$$

Therefore
$$\Delta S_{\text{total}} = -\frac{\Delta(\text{P.E.})}{T}.$$

4. $\Delta(\text{P.E.}) = mg\Delta h$

$$= (1\text{ gm})\left(980\,\frac{\text{cm}}{\text{sec}^2}\right)(1\text{ cm})$$

$$= 980\text{ ergs} \times \frac{1\text{ cal}}{4.18 \times 10^7\text{ ergs}}$$

$$= 2.3 \times 10^{-5}\text{ cal.}$$

The chance, therefore, is $10^{-10^{16}}$ to 1, or 1 in $10^{10^{16}}$.

Regarding the improbability of unlikely events, Eddington remarked:

> The law that entropy always increases—the second law of thermodynamics—holds, I think, the supreme position among the laws of Nature. If someone points out to you that your pet theory of the universe is in disagreement with Maxwell's equations—then so much the worse for Maxwell's equations. If it is found to be contradicted by observations—well, these experimentalists do bungle things sometimes. But if your theory is found to be against the second law of thermodynamics I can give you no hope; there is nothing for it but to collapse in deepest humiliation. This exaltation of the second law is not unreasonable. There are other laws which we have strong reason to believe in, and we feel that a hypothesis which violates them is highly improbable; but the improbability is vague and does not confront us as a paralysing array of figures, whereas the chance against a breach of the second law (i.e. against a decrease of the random element) can be stated in figures which are overwhelming.

On the improbability of the unmixing of gases, Boltzmann wrote:

> Not until after a time enormously long compared to $10^{10^{10}}$ years will there be any noticeable unmixing of the gases. One may recognize that this is practically equivalent to *never*, if one recalls that in this length of time, according to the laws of probability, there will have been many years in which every inhabitant of a large country committed suicide, purely by accident, on the same day, or every building burned down at the same time—yet the insurance companies get along quite well by ignoring the possibility of such events. If a much smaller probability than this is not practically equivalent to impossibility, then no one can be sure that today will be followed by a night and then a day.*

* Ludwig Boltzmann, *Lectures on Gas Theory*, Berkeley, Calif.: University of California Press, 1964.

5. $$\Delta S_{\text{total}} \equiv (S_{\text{total}})_{\substack{\text{final}\\ \text{state}}} - (S_{\text{total}})_{\substack{\text{initial}\\ \text{state}}}$$
$$= (R/N_0) \ln [(\Omega_{\text{total}})_{\substack{\text{final}\\ \text{state}}}/(\Omega_{\text{total}})_{\substack{\text{initial}\\ \text{state}}}].$$

Therefore

$$\frac{(\Omega_{\text{total}})_{\substack{\text{final}\\ \text{state}}}}{(\Omega_{\text{total}})_{\substack{\text{initial}\\ \text{state}}}} = e^{\frac{\Delta S_{\text{total}}}{R} \cdot N_0} = 10^{\frac{\Delta S_{\text{total}}}{2.303R} \cdot N_0}.$$

6. (a) From A to B the entropy of the gas increases. From C to D it decreases.

(b) Nothing—if the steps are reversible. These so-called adiabatic steps involve only two objects: the gas and the weight. Thus, for these adiabatic steps

$$\Delta S_{\text{total}} = \Delta S_{\text{gas}} + \Delta S_{\text{Wt.}}\,.$$

But $$\Delta S_{\text{Wt.}} = 0.$$

$$\therefore\ \Delta S_{\text{gas}} = 0$$

if $$\Delta S_{\text{total}} = 0.$$

During the reversible adiabatic expansion of a gas the positive increment in the entropy of the gas engendered by the increase in volume is balanced exactly by a negative increment engendered by a decrease in temperature.

(c)

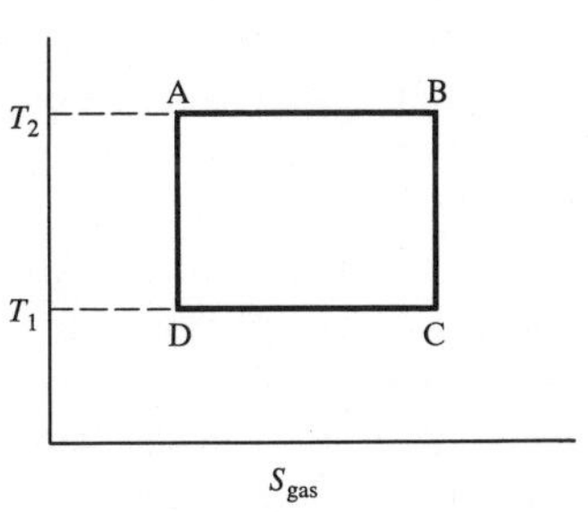

An interesting interpretation may be placed upon the area $ABCDA$. It represents the work done by the gas in one reversible passage around the cycle $ABCDA$. For if the cycle as a whole is reversible, ΔS_{total} must be zero for each individual step of the cycle, since a positive value for ΔS_{total} in one step of the cycle could never be balanced off by a negative value for ΔS_{total} in some other step of the cycle. The adiabatic steps BC and DA must therefore be represented on the T-S diagram of the gas (the only part of the universe whose entropy might change during an adiabatic step) by lines along which the entropy of the gas does not change, i.e. by vertical lines. Consequently the enclosed area $ABCDA$ is equal to the area beneath the line segment AB less the area beneath the line segment CD, i.e. to $T_2\Delta S - T_1\Delta S$, where ΔS represents the horizontal distance $AB = CD$. This horizontal distance is the change in entropy of the gas during the isothermal steps AB and CD. Since these steps are assumed to be reversible ($\Delta S_{\text{total}} = 0$), the change in entropy of the gas in these steps is exactly balanced by a change in entropy of the thermal surroundings. Thus, for steps AB and CD the horizontal distance ΔS is in magnitude equal to the entropy change of the hot and cold thermal reservoirs, respectively. Multiplied by the temperature of the thermal reservoir, this entropy change gives the thermal energy gained or lost by the reservoir in these steps. The term $T_2\Delta S$ represents the thermal energy lost by the hot thermal reservoir in the isothermal expansion AB; the term $T_2\Delta S$ represents the thermal energy gained by the colder thermal reservoir in the isothermal compression CD. The difference between these two terms, by the First Law, is the work done by the gas in one complete cycle $ABCDA$.

Actually, of course, the gas in traveling around a full cycle which restores the gas to its initial condition is quite incapable of delivering by itself any useful work at all. It merely serves, so to speak, as a catalyst—necessary to the operation, but in the end itself unchanged—in the transformation of thermal energy removed from the hot reservoir—and not later delivered to the colder reservoir—into work. The energy that appears eventually as work comes from the hot reservoir, not from the gas.

The energy that appears as work, $T_2\Delta S - T_1\Delta S$, stands to the energy withdrawn from the hot reservoir, $T_2\Delta S$, as the temperature difference $(T_2 - T_1)$ stands to T_2.

7. Consider one complete cycle taken in the sense $ABCDA$.

Let Q_2 represent the thermal energy removed *from* the hot reservoir whose temperature is T_2. In terms of Q_2 and T_2 the hot reservoir's change in entropy per cycle is

$$\frac{-Q_2}{T_2}.$$

Similarly, let Q_1 represent the thermal energy delivered *to* the cold reservoir whose temperature is T_1. In terms of Q_1 and T_1 the cold reservoir's change in entropy per cycle is

$$\frac{+Q_1}{T_1}.$$

If the cycle is reversible ($\Delta S_{\text{total}} = 0$)

$$\frac{-Q_2}{T_2} + \frac{Q_1}{T_1} = 0.$$

This implies that

$$Q_1 = Q_2 \times \frac{T_1}{T_2}.$$

The colder the cold reservoir, or the hotter the hot one, the smaller the amount of energy that needs to be diverted to the cold reservoir to compensate for the decrease in entropy of the hot reservoir.

Now let W be the work done by the system. By the First Law W is the difference between the thermal energy lost by the hot reservoir and the thermal energy gained by the cold reservoir.

$$W = Q_2 - Q_1.$$

Substituting for Q_1 from the previous expression and dividing by Q_2, one obtains the following expression for the efficiency of a reversible heat engine.

$$\frac{W}{Q_2} = 1 - \frac{T_1}{T_2} = \frac{T_2 - T_1}{T_2}.$$

If the temperatures T_1 and T_2 are laid out on a vertical scale that starts at absolute zero, a reversible heat engine's efficiency, which, in fact, is the maximum efficiency of any heat engine operating between the same two temperatures, may be expressed as the ratio of the length of the line segment T_1T_2 to the length of the line segment $0T_2$.

T_2

T_1

0

8. Make the boiler hotter and the condenser colder.
9. Let Q_1 = thermal energy removed *from* the lake and let Q_2 = thermal energy delivered *to* the house interior. The problem is to find values for Q_1 and Q_2 such that, first, the change in energy of the universe (lake bottom, house interior, and source of mechanical or electrical work) is zero, i.e. the problem is to find values for Q_1 and Q_2 such that the

energy delivered to the house interior is the sum of the energies lost by the lake bottom and the source of the mechanical or electrical energy; i.e. Q_1 and Q_2 must be such that

$$Q_2 = Q_1 + 100 \text{ cal.}$$

And, second, the problem is to find values for Q_1 and Q_2 such that, simultaneously, the change in entropy of the universe is equal to or greater than zero, i.e. the problem is to find for Q_1 and Q_2 values such that

$$\frac{Q_2}{300°\text{K}} - \frac{Q_1}{277°\text{K}} \geq 0,$$

from which it is found that

$$Q_1 \leq Q_2 \times \frac{277}{300}.$$

Substituting this expression for Q_1 into the expression above for Q_2, one discovers that

$$Q_2 \leq Q_2 \times \frac{277}{300} + 100 \text{ cal}$$

$$Q_2 \leq 100 \text{ cal} \times \frac{300}{300 - 277} = 1304 \text{ cal.}$$

This last equation says, in effect, that as much as 1304 calories of thermal energy could be delivered to a 27°C house interior by using 100 calories of mechanical or electrical energy to pump 1304 − 100 = 1204 calories of thermal energy from a lake bottom at 4°C. The change in entropy of the house interior is 1304 cal/300°K = 4.35 cal/deg and that of the lake bottom −1204 cal/277°K = −4.35 cal/deg. Neither the First nor the Second Law is violated by the process, which comes warmly recommended by thermodynamicists, who enthusiastically point out that pumping heat in this fashion liberates to the house many times more thermal energy than is the case when the consumed mechanical or electrical energy is degraded directly, through friction or with an electric heater, to an equivalent amount (100 cal.) of thermal energy. Still, today few houses are heated by heat pumps. Why is this?

10. Suppose there exist two thermal reservoirs, one at 400°K and the other at 300°K. By means of a heat engine, remove from the 400°K reservoir 1200 calories of energy, use 300 of this to do useful mechanical work, and dump 900 in the 300°K reservoir. Then induce 900 calories to flow spontaneously from the 300°K to the 400°K reservoir and repeat the

cycle until the desired quantity of thermal energy has been converted completely to work.

It has been remarked in jest that hell must be isothermal, for otherwise the resident engineers and physical chemists (of which there must be some) could set up a heat engine to run a refrigerator to cool off a portion of their surroundings to any desired temperature.

11. Consider this universe: a system σ, its thermal surroundings θ, and a weight. By the First Law any changes that occur within this self-contained universe must be such that

$$\Delta E_\sigma + \Delta E_\theta + \Delta E_{\text{Wt.}} = 0.$$

By the definitions of Q and W

$$\Delta E_\theta = -Q$$

$$\Delta E_{\text{Wt.}} = +W.$$

Call ΔE_σ simply ΔE. The First Law may then be stated as follows:

$$\Delta E = Q - W.$$

The increase in the internal energy of a system is equal to the decrease in thermal energy of its thermal surroundings less the increase in mechanical energy of its mechanical surroundings.

Chapter 10

1. When the change solid → liquid is reversible

$$\Delta S_{\text{total}} = \Delta S_\sigma + \Delta S_\theta = \Delta S_\sigma + \frac{\Delta E_\theta}{T} = \Delta S_\sigma - \frac{\Delta E_\sigma}{T} = 0.$$

Therefore at the normal melting point

$$T = \frac{\Delta E_\sigma}{\Delta S_\sigma}.$$

Melting is favored by a small energy of fusion and a large entropy of fusion.

2. A high melting point indicates a large energy of fusion or a small entropy of fusion, or both. The similarities in the boiling points of the four isomers suggest that there is not a large variation in intermolecular forces from one isomer to the next and that the remarkably high melting point of the last isomer stems more than anything else from a small entropy of fusion. A large body of independent data supports this

inference. The almost spherical shape of 2,2,3,3-tetramethylbutane (hexamethylethane) permits molecular rotation or libration in the solid and thereby greatly diminishes the driving force for melting, which lies solely in the greater entropy of the liquid over that of the solid. The molecules are in a sense able to pre-melt without leaving their lattice sites. Listed below are some additional melting-point comparisons of this type.

Substance	*Melting Point* (°C)	*Liquid Range* (°C)
Methylcyclohexane	−126	227
*Cyclohexane	+ 6	75
Tetrachloroethane	− 44	190
Pentachloroethane	− 29	191
*Hexachloroethane	—	Sublimes at +187
Pentane	−130	166
*Tetramethylmethane	− 17	26

In general with free, or nearly free, rotation in the solid goes a small entropy of fusion, a high melting point, a small liquid range, and (see later) a low solubility.

3. At the normal melting point $\Delta S_\sigma = \Delta E_\sigma/T = (5400 \text{ cal/mole})/329°\text{K} = 16.4 \text{ cal/deg-mole}$.
4. 10.9 cal/deg-mole (compare with ethylpentachlorobenzene).
5. Because the entropy of the universe is thereby increased.

   At low temperatures the highly exothermic reaction $2H = H_2$ gives rise in the surroundings to an immense increase in entropy that is sufficient to offset the fact that for this reaction the product has a smaller entropy than the starting material (compare with the freezing of cold water and the explosion of cold hydrogen and oxygen). At high temperatures, however, the entropy change in the surroundings is numerically much smaller; under these conditions the endothermic reaction $H_2 = 2H$ gives rise to an increase in entropy of the chemical system that is sufficient to offset the diminished entropy of the surroundings (compare with the melting of hot ice and the dissociation of very hot water vapor).
6. $\Delta S_\sigma = S_{\text{products}} - S_{\text{reactants}}$. In both reactions $S_{\text{products}} > S_{\text{reactants}}$. The reason for this is that both reactions produce simpler chemical species from more complicated ones. Hence on the average the atoms in the products are less constrained in their random thermal motions by chemical bonds than are those in the reactants. In other words, the volume involved in the motion of the individual atoms in the products

* Almost spherical molecules.

is on the average larger than that of the same atoms in the reactants—and the larger the volume involved in the thermal motions of the constituent particles, the larger tends to be the entropy.

Alternatively, one may say that since generally

$$S_{\text{trans.}} > S_{\text{rot.}} \gg S_{\text{vib.}} \approx 0,$$

the greater the number of particles with their full complement of translational entropy the greater the total entropy.

7. In the reaction

$$2\text{B(solid)} + 3\text{H}_2\text{(gas)} = 2\text{BH}_3\text{(gas)}$$

three moles of gas are consumed and only two moles of gas are produced; the sign of ΔS_σ for the reaction is, therefore, probably negative. At high temperatures (T large, $\Delta S_\theta = \Delta E_\theta/T$ small) ΔS_σ dominates the term ΔS_θ and determines the sign of ΔS_{total}.

8. For the reaction $H_2\text{(gas)} \rightarrow 2H\text{(gas)}$, $\Delta S_\sigma = S_{\text{products}} - S_{\text{reactants}} = 2S_H - S_{H_2}$. The entropies of atomic hydrogen and molecular hydrogen depend, through the translational entropies of these gases, upon the values of the partial pressures P_H and P_{H_2}, respectively. As these decrease, S_H and S_{H_2} increase. Owing to the factor 2 before S_H in the expression for ΔS_σ, a sudden decrease in pressure increases momentarily the value of ΔS_σ and, also, ΔS_{total}, if it is assumed, probably quite legitimately, that the heat of reaction is little affected by isothermal changes in pressure—for then the other part of ΔS_{total}, ΔS_θ, which is equal to the heat of reaction divided by the absolute temperature (assumed constant), will not be affected. Thus, if initially $\Delta S_{\text{total}} = 0$ (system at equilibrium), it follows that immediately after the pressure drop $\Delta S_{\text{total}} > 0$ for the change $H_2 \rightarrow 2H$, and the degree of dissociation of H_2 should increase. During dissociation P_H increases and P_{H_2} decreases; this causes S_H to decrease, S_{H_2} to increase and ΔS_σ to decrease, until eventually ΔS_σ returns to its former value, or very nearly this value, and ΔS_{total} is once again zero. At this new equilibrium point the partial pressures P_H and P_{H_2} have smaller values than previously; however the value of the ratio $P_H{}^2/P_{H_2}$ is unchanged.

9. No.

$$\begin{aligned}\Delta S_{\text{total}} &= \Delta S_\sigma + \Delta S_\theta \\ &= (45.1 - 16.7)\text{ cal/deg-mole} - \frac{10{,}500\text{ cal/mole}}{298^\circ\text{K}} \\ &= (28.4 - 35.3)\text{ cal/deg-mole} \\ &= -6.9\text{ cal/deg-mole.}\end{aligned}$$

10. Yes.

$$S^{\text{gas}}_{H_2O}\,(P = 0.01 \text{ atm.}) = S^{\text{gas}}_{H_2O}\,(P = 1 \text{ atm.}) - R \ln (0.01/1)$$
$$= 45.1 \text{ cal/deg-mole} + 9.2 \text{ cal/deg-mole.}$$

$$\therefore\ \Delta S_{\text{total}} = (37.6 - 35.3) \text{ cal/deg-mole}$$
$$= +2.3 \text{ cal/deg-mole.}$$

11. If $\Delta S_{\text{total}} = 0$, $\Delta S_\sigma + \Delta S_\theta = 0$. For the evaporation of pure water at 25°C (all entropy values in cal/deg-mole),

$$\Delta S_\theta = -\frac{10{,}520}{298.15} = -35.28 \qquad \text{and} \qquad \Delta S_\sigma = S^{\text{gas}}_{H_2O} - S^{\text{liq}}_{H_2O}\,,$$

where $S^{\text{liq}}_{H_2O} = 16.716$ and $S^{\text{gas}}_{H_2O} = 45.106 - R \ln (P^{\text{gas}}_{H_2O}$, in atm.). Evidently $P^{\text{gas}}_{H_2O}$ must be such that

$$[(45.106 - R \ln P^{\text{gas}}_{H_2O}) - 16.716] + [-35.28] = 0.$$

Therefore

$$\log P^{\text{gas}}_{H_2O} = \frac{(45.106 - 16.716) - 35.28}{4.5757} \qquad (P^{\text{gas}}_{H_2O} \text{ in atm.})$$
$$= -1.506,$$

and
$$P^{\text{gas}}_{H_2O} = 3.119 \times 10^{-2} \text{ atm.}$$
$$= 23.7 \text{ mm Hg.}$$

12. 23.7 mm Hg.
13. When the partial pressure of the vapor is equal to the so-called "vapor pressure," ΔS_{total} vanishes for the change liquid-to-vapor or vapor-to-liquid.
14. The normal boiling point is the temperature at which bubbles just begin to form in the liquid when the external pressure is 1 atmosphere. At this point the vapor pressure of the substance (which is the pressure of the vapor in the bubbles) is 1 atmosphere.
15. If at the normal boiling point ($P_{\text{vapor}} = 1$ atmosphere) the entropy of vaporization is approximately 28.39 cal/deg-mole (this is the entropy of vaporization at 25°C when the partial pressure of the vapor is 1 atmosphere) and if at the normal boiling point the heat of vaporization is approximately 10.52 kcal/mole, then the temperature T_{bp} at the boiling point must be such that

$$28.39 \text{ cal/deg-mole} - \frac{10{,}520 \text{ cal/mole}}{T_{bp}} \approx 0.$$

Therefore

$$T_{bp} \approx \frac{10{,}520 \text{ cal/mole}}{28.39 \text{ cal/deg-mole}} = 371°\text{K}.$$

Chapter 11

1. T_θ = temperature of the thermal surroundings.
   T_σ = temperature of the ideal gas.
   $(T_\theta)_1$ = initial temperature of the surroundings.
   $(T_\theta)_2$ = final temperature of the surroundings.
2. It has been assumed implicitly that in both initial and final states the gas and its thermal surroundings are in thermal equilibrium, i.e. it has been assumed that for both initial and final states $T_\theta = T_\sigma$.
3. $(T_\theta)_1 = (T_\theta)_2$.
4. $\Delta T_\theta \equiv (T_\theta)_2 - (T_\theta)_1$.
   $\Delta V_\sigma \equiv (V_\sigma)_2 - (V_\sigma)_1$.
5. $\Delta T_\theta = 0$; therefore (see 2) $\Delta T_\sigma = 0$. Also, $\Delta V_\theta = 0$. $\Delta V_\sigma > 0$.
6. Unless the rate of change of E with T and/or V is infinite (can you think of such a case?*), it can be concluded that $\Delta E = 0$. This conclusion applies to θ.
7. $E_{\text{total}} = E_\theta + E_\sigma$. This relation applies to all states of the universe.
8. $\Delta(E_{\text{total}}) \equiv (E_{\text{total}})_2 - (E_{\text{total}})_1$.
9. $\Delta(E_{\text{total}}) = 0$.
10. $\Delta E_\sigma = -\Delta E_\theta$.
11. $\Delta E_\sigma = 0$ (see 6).
12. It can be concluded (see statement of problem 6) that the rate of change of E with V as V (but not T) changes is zero. This conclusion applies to σ.

    To paraphrase Boltzmann, when the volume of an ideal gas is increased at constant temperature the kinetic energy of translational motion remains unchanged. The molecules just move farther apart.
13. Yes. (An adiabatic process is a process for which $\Delta E_\theta = 0$.)
14. Yes.
15. No.
16. Yes, merely immerse the expanding gas in a thermostat.
17. (a) Yes. Heat it. (b) Yes, if its temperature doesn't change (see 16). (c) Yes, if the gas is allowed to cool at the right rate as it expands (see 18). (e) What?!
18. The entropy of the gas would not change. For a reversible process, $\Delta S_{\text{total}} = 0$. For an adiabatic process, $\Delta S_\theta = 0$. For any process, $\Delta S_{\text{total}} = \Delta S_\theta + \Delta S_\sigma$. Hence, for a reversible, adiabatic process, $\Delta S_\sigma = 0$.
19. The spontaneous expansion of the gas.
20. No. During the original event, the free expansion, the only part of the universe affected (if the gas expanding is an ideal gas) is the gas itself.

* The rate of change of E when T (but not V) varies would be infinite at the melting of a substance whose solid had the same density as its liquid.

Compression of a gas, on the other hand, consumes mechanical energy; also, it leaves either the gas or the gas and its surroundings with thermal energy not present initially. Similarly, in condensing the gas with liquid nitrogen the energy that the gas liberates on condensation will cause some of the liquid nitrogen to evaporate. In both cases, part of the universe—the gas—is restored to its initial condition, but other parts of the universe—the surroundings of the gas—are not. This is to be expected. Once the entropy of the universe has increased, as it does in the spontaneous, free expansion of a gas, there is no way on earth to restore *all* parts of the universe to their initial states. To do so would be to say that the entropy of the universe could decrease.*

Chapter 12

1. (a) $\Delta(S_{\text{gas}}) = nR \ln (V_2/V_1)$

$= (1 \text{ mole})(1.987 \text{ cal/deg-mole}) [2.303 \log_{10} (10/1)]$

$= 4.6 \text{ cal/deg.}$

(b) 4.6 cal/deg. (c) 46 cal/deg. (d) 4.6 cal/deg.

2. $P_1 = 1$ (atm.), $P_2 = P$, $S_1 = S^0$, $S_2 = S$.
3. $3S^0 - R \ln P^3$.
4. The equation can be applied to any ideal gas whose temperature does not change as the volume changes from V_1 to V_2. It is correct to say that for any event the change in the entropy, or the energy, or the temperature, or the pressure, or the volume, or any other thermodynamic property of a specified chemical system, ideal gas or otherwise, depends only upon the system's initial and final states and not on what occurs concurrently in other parts of the universe. The change in pressure of an ideal gas during an isothermal expansion, for example, depends only upon the number of moles of gas, its temperature, and its initial and final volumes, and is the same if these are the same regardless of whether the gas expands reversibly (Fig. 12.1) or irreversibly (Fig. 11.2). A similar statement holds for the entropy. The net change in the microscopic disorder of a system depends on the over-all change in the system and not on the detailed manner in which this change occurs. To say or find that this is not true is tantamount to saying or discovering that the over-all change has not yet been fully characterized. Some additional

* The pressures of modern civilization, so it is said, have produced a peculiar form of neurosis, manokleptia. A person so afflicted runs backward into stores and leaves things. Is the behavior of a poor manokleptiac (so he becomes if not soon cured) the reverse of that of a kleptomaniac?

parameter needs to be specified. The energy and entropy changes that take place during the solution of iron wire in hydrochloric acid, for example, are not completely determined by specifying the mass and the chemical composition of the wire and the temperature and pressure of the system. Different heats and entropies of solution can be obtained for wires of equal mass and identical chemical composition. To predict correctly the heat and entropy of solution of iron wire in hydrochloric acid it is necessary to know something about the wire's thermal and mechanical history. Has the wire been mechanically strained or work-hardened?—tempered by quenching or carefully annealed?—and so forth.

The reader may wish to register at this point a mild protest. The openly pragmatic manner in which initial and final states are specified in thermodynamics seems to render redundant the statement that changes in the thermodynamic properties of a system depend solely upon the initial and final states of the system. Such a statement can hardly help but be true. Were it not true in a particular case, there would immediately ensue a careful search for some additional factor—the mechanical or thermal history of the iron wire, for example—which, upon being specified, would set the statement right. Put another way, the factors that need to be specified to characterize fully an event can be determined only by experiment. What then is the special contribution of classical thermodynamics to the study of chemical systems? For one thing, it sets forth a useful bookkeeping procedure. For another, this: thermodynamics reveals relationships, usually otherwise unsuspected, between the thermal properties of a system. It reveals, for example (cf. Chapter 37) that the rate at which the spring constant of a wire changes with temperature is related directly to the rate at which the entropy of the wire changes with length.

5. (1 mole) $R \ln 2 = 1.38$ cal/deg.
6. The free expansion of an ideal gas from V_1 to $2V_1$ (problem 5). For this event, $\Delta S_\sigma = 1.38$ cal/deg and $\Delta E_\theta/T = 0$.
7. In general, $\Delta S_{\text{total}} = \Delta S_\sigma + \Delta S_\theta$, where $\Delta S_\theta = \Delta E_\theta/T$. For irreversible events, $\Delta S_{\text{total}} > 0$. For such events, therefore,

$$\Delta S_\sigma > -\frac{\Delta E_\theta}{T}. \qquad \text{(cf. problem 6)}$$

When hydrogen and oxygen at 1 atm. partial pressure react spontaneously (and irreversibly) to produce a mole of liquid water at room temperature, for example, approximately 68,000 calories are liberated to the surroundings. For this event $\Delta S_\sigma \approx -39$ cal/deg and $-\Delta E_\theta/T \approx -227$ cal/deg.

8. The two processes produce identical changes in the thermodynamic

properties of the gas but different changes in the thermodynamic properties of the surroundings. In both processes the volume and entropy of the gas increase, its temperature and energy remain the same, and its pressure decreases. In the free expansion the surroundings are unaffected. In the reversible isothermal expansion the thermal energy of a thermostat decreases by an amount equal to the increase in potential energy of a weight and the entropy of the thermostat decreases by an amount equal to the increase in entropy of the gas.

9. As the volume increased, the temperature of the gas would drop, its energy would drop (by an amount equal to the increase in the weight's potential energy), and—if the process is truly reversible—the entropy of the gas (being the only entropy that could change) would not change. The process could be called an isentropic expansion; also, since it is an adiabatic process ($\Delta E_\theta = 0$), it could be called a reversible adiabatic expansion.

10. $\Delta S_\sigma \overset{\Delta S_{\text{total}}=0}{=} -\dfrac{\Delta E_\theta}{T} \left(\text{often written } \Delta S = \dfrac{q_{\text{rev.}}}{T}, \text{ where } q \equiv -\Delta E_\theta\right).$

11. All seven equations apply to the reversible isothermal expansion of an ideal gas.

    Those equations in which it has been assumed that $\Delta E_\sigma = 0$, namely $\Delta E_\theta = -\Delta E_{\text{Wt.}}$ and $\Delta S_\sigma = \Delta E_{\text{Wt.}}/T$, do not apply to all gases. This leaves the first equation, the third equation, and the equations $\Delta S_\sigma = -\Delta S_\theta$, $\Delta S_\theta = \Delta E_\theta/T$, and $\Delta S_\sigma = -\Delta E_\theta/T$. The last four equations apply to all reversible processes ($\Delta S_{\text{total}} = 0$). Only the first equation and the equation $\Delta S_\theta = \Delta E_\theta/T$ apply to all processes. (The last remark is in a way redundant. The label θ is used only for parts of the universe that behave as the passive thermal surroundings of chemical systems. This is tantamount to saying that the condition $\Delta S_\theta = \Delta E_\theta/T$ is always satisfied.)

12. The net effect is the lowering of a weight (or something mechanically equivalent to this) and the production in some object (a thermostat, for example) of an equivalent quantity of thermal energy. From the thermodynamic point of view this two-step event is equivalent to dropping a weight directly onto the floor. Note that both events are irreversible.

13. The net effect of the isothermal expansion of an ideal gas is not simply the extraction of thermal energy from an object and the raising of a weight. The net effect is the extraction of thermal energy from an object, the raising of a weight, and *the expansion of a gas*. The last effect is the driving force behind the other two.

14. No. See problem 8.

15. Yes. Once the initial and final states of the several parts of a universe

(chemical system, thermostat, and weight) have been specified, the changes that occur in these parts are determined.

16. (a) $\Delta V/V_1$
   (b) x
   (c) V_1 (or V_2)
   (d) $\Delta V/V_1$
   (e) $nRT(\Delta V/V_1)$
   (f) nRT/V_1
   (g) $P_1 \times \Delta V$
   (h) $A \times \Delta h$
   (i) mg/A
   (j) $mg(\Delta h)$
   (k) potential energy
   (l) mass m (or the "Wt.")

Taken in reverse these steps show that for small changes in V the change in the potential energy of the restraining weight on an ideal gas is equal to $nRT(\Delta V/V)$. In calculus it is shown that the sum of terms $\Delta V/V$ as V increases from V_1 to V_2 approaches in the limit as each ΔV becomes smaller and smaller the value of the natural logarithm of the ratio V_2/V_1; i.e.

$$\begin{pmatrix}\text{Change in P.E. of Wt. when gas}\\ \text{volume changes from } V_1 \text{ to } V_2\end{pmatrix} = \lim_{\Delta V \to 0} \sum_{V=V_1}^{V=V_2} nRT\left(\frac{\Delta V}{V}\right)$$

$$= nRT\left[\lim_{\Delta V \to 0} \sum_{V=V_1}^{V=V_2} \frac{\Delta V}{V}\right]$$

$$= nRT \ln\left(\frac{V_2}{V_1}\right).$$

Chapter 13

1. (a) two
   (b) chemical system
   (c) θ
   (d) thermal surroundings
   (e) chemical system
   (f) $\Delta S_\sigma + \Delta S_\theta$
   (g) products
   (h) reactants
   (i) $\Delta S_\sigma = S_{\text{products}} - S_{\text{reactants}}$

(j) $\Delta E_\theta/T$
(k) absolute temperature
(l) $\Delta E_\sigma + \Delta E_\theta = 0$; or, $\Delta E_\theta = -\Delta E_\sigma$
(m) $-\Delta E_\sigma/T$
(n) the chemical system σ
(o) $-\Delta E_\sigma/T + \Delta S_\sigma$
(p) ΔE_σ (or ΔE_θ)
(q) exothermic
(r) endothermic
(s) $-\Delta E_\sigma/T$ (or ΔS_θ)
(t) ΔS_σ
(u) products
(v) greater
(w) entropy
(x) reactants
(y) low energy
(z) high entropy
(a′) decrease
(b′) partial pressure (or concentration)

2. $P_H = (K \cdot P_{H_2})^{1/2} = (K)^{1/2} =$ (a) $10^{-35.5}$ atm.
(b) 1 atm. (approximately)
(c) $10^{1.2}$ atm. = 16 atm. (approximately)

3. K depends on T but not on P_{H_2}, P_H, P_{total}, or V.

4. (a) endothermic
(b) negative
(c) positive
(d) negative
(e) positive
(f) endothermic
(g) increases
(h) Le Chatelier
(i) Yes

5. $R \ln K = -(\Delta E_\sigma)_V/T + (\Delta S^0{}_\sigma)_V$. Therefore

$$K = e^{\frac{(\Delta S^0{}_\sigma)_V}{R}} \cdot e^{-\frac{(\Delta E_\sigma)_V}{RT}}.$$

This suggests that

$$A = e^{\frac{(\Delta S^0{}_\sigma)_V}{R}}$$

$$B = \frac{(\Delta E_\sigma)_V}{R}.$$

For the reaction $H_2 = 2H$, $A = 10^{4.7}$ and $B = 52{,}000$.

6. The equation of a straight line in slope-intercept form is

$$y = mx + b.$$

m is the slope of the line and b is its intercept on the y (or $x = 0$) axis. The equation

$$\log_{10} K = [-(\Delta E_\sigma)_V/2.3R](1/T) + [(\Delta S^0{}_\sigma)_V/2.3R]$$

is of this form with

$$y = \log_{10} K$$
$$x = 1/T$$
$$m = -(\Delta E_\sigma)_V/2.3R$$
$$b = (\Delta S^0{}_\sigma)_V/2.3R.$$

7. Let K_1 and K_2 be the known equilibrium constants at T_1 and T_2, respectively. Assume that over the temperature interval T_1 to T_2 $(\Delta E_\sigma)_V$ and $(\Delta S^0{}_\sigma)_V$ are constant. Then

$$R \ln K_1 = -(\Delta E_\sigma)_V/T_1 + (\Delta S^0{}_\sigma)_V$$
$$R \ln K_2 = -(\Delta E_\sigma)_V/T_2 + (\Delta S^0{}_\sigma)_V.$$

Subtracting the first equation from the second and solving for $(\Delta E_\sigma)_V$, one finds that

$$(\Delta E_\sigma)_V = \frac{R \ln \left(\frac{K_2}{K_1}\right)}{\left(\frac{1}{T_1} - \frac{1}{T_2}\right)} = \frac{RT_1T_2 \ln \left(\frac{K_2}{K_1}\right)}{(T_2 - T_1)}.$$

If $K_2 > K_1$ when $T_2 > T_1$, $(\Delta E_\sigma)_V > 0$. $(\Delta S^0{}_\sigma)_V$ can be found by substituting this value of $(\Delta E_\sigma)_V$ back into either the first or the second equation.

Chapter 14

1. (a) For this equation to be valid, the quantities ΔS and ΔE should refer to (i) a chemical system that is (ii) at equilibrium with respect to the reaction to which the symbol Δ refers; furthermore, this chemical system should be (iii) surrounded by a thermostat with which it is in thermal equilibrium and which maintains its temperature at the constant value T, and it should be (iv) confined to a

container that keeps its volume constant. Or, the quantities ΔS and ΔE could refer to the thermostat in (iii).

(b) Same as in (a), except that for "volume constant" in (iv) read "pressure constant," and add (v) that the system should be in hydrostatic equilibrium with its surroundings.

2. In equation 1(b) the second term

$$\left(-\frac{\Delta E + P\Delta V}{T}\right)$$

is equal to ΔS_θ, the entropy change of the thermal surroundings. This is always equal to $\Delta E_\theta/T$. When the volume of σ does not change, $\Delta E_\theta = -\Delta E_\sigma$ (part (a)). If the volume of σ does change at constant pressure, then (by the definition of P_σ and ΔV_σ),

$$\Delta E_\theta = -(\Delta E_\sigma + P_\sigma \Delta V_\sigma).$$

3. The column of air above a reaction mixture in a test tube, beaker, or any other reaction vessel that is open to the atmosphere. At sea level the mass of this air column times g divided by A is normally about 14.7 lb/in^2 = 1,013,250 dynes/cm^2 = 1 "atm."

4. It appears as thermal energy in the surroundings of the chemical system. Often the thermal surroundings of a chemical system is the atmosphere itself.

5. $Q = -\Delta E_\sigma$ when $\Delta V_\sigma = 0$ (or when $P_\sigma = 0$). $Q \neq -\Delta E_\sigma$ when $\Delta V_\sigma \neq 0$ (if $P_\sigma \neq 0$). If ΔV_σ is negative ($V_{\text{products}} < V_{\text{reactants}}$), Q is greater (more positive) than $-\Delta E_\sigma$. Conversely, if ΔV_σ is positive ($V_{\text{products}} > V_{\text{reactants}}$), Q is less (more negative) than $-\Delta E_\sigma$. (Throughout this discussion it has been assumed that during the reaction to which the symbol Δ refers no "useful work" has been produced or consumed. This matter is discussed more fully in Part IV.)

6. (a) In general, $H - E = PV$.

(b) For an ideal gas, $PV = nRT$ (parts (i) and (ii)). At 300°K (part (iii)), $PV/n = 600$ cal/mole.

(c) When $P = 0$, $PV = 0$; hence (part (i)), $H - E = 0$; i.e. when $P = 0$, $H = E$. When $P = 1$ atm. (and $T = 300$°K), the volume of liquid water is about 18 cc/mole; thus $H - E \approx$ (1 atm.) (18 cc/mole) = 0.44 cal/mole ($R = 1.987$ cal/deg-mole = 0.0821 liter-atm./deg-mole = 82.1 cc-atm./mole).

These differences are very small compared to ordinary heats of reaction.

7. (a) $Q_P = (\Delta E_\theta)_P = -(\Delta E_\sigma + P\Delta V_\sigma)_P = -(\Delta H_\sigma)_P = -(H_{\text{products}} - H_{\text{reactants}}) = -(H_{H_2} - 2H_H) = 2H_H - H_{H_2} = 104.178$ kcal.

(b) $H_{\text{H}} = \dfrac{104.178}{2}$ kcal/mole $= 52.089$ kcal/mole.

(c) $P\Delta V = P(V_{\text{products}} - V_{\text{reactants}}) = P\left(\dfrac{RT}{P} - \dfrac{2RT}{P}\right)$

$= -RT = -593$ cal/mole $= -0.593$ kcal/mole.

(d) $E_{\text{H}} = H_{\text{H}} - RT$ (per mole).* Similarly, $E_{\text{H}_2} = H_{\text{H}_2} - RT$. Therefore

$$2E_{\text{H}} - E_{\text{H}_2} = (2H_{\text{H}} - H_{\text{H}_2}) - RT$$
$$= 104.178 - 0.593$$
$= 103.585$ kcal. (In Chapter 13 this number was rounded off to 104 kcal.)

(e) $Q_V = (\Delta E_\theta)_V = -(\Delta E_\sigma)_V = -(E_{\text{products}} - E_{\text{reactants}})$
$= -(E_{\text{H}_2} - 2E_{\text{H}}) = 103.585$ kcal.

Because the potential energy of the atmosphere diminishes slightly when the reaction $2\text{H} = \text{H}_2$ occurs at constant pressure, owing to the fact that only one mole of gas is produced for every two moles of gas consumed, slightly more thermal energy is liberated to the surroundings of the reaction mixture when the reaction occurs at constant pressure than when it occurs at constant volume. The opposite is true for reactions for which ΔV_σ, measured at constant pressure, is greater than zero, rather than, as here, less than zero.

8. The units on K are atm.$^{[(c+d)-(a+b)]}$

$$P\Delta V_\sigma = [(c + d) - (a + d)]RT.$$

9. The reaction is thermodynamically possible when $\Delta S_{\text{total}} > 0$. In terms of ΔS^0, ΔE_σ, $P\Delta V_\sigma$, and the partial pressures of A, B, C, D,

$$\Delta S_{\text{total}} = \Delta S^0 - RT \ln \frac{P_C{}^c P_D{}^d}{P_A{}^a P_B{}^b} - \frac{\Delta E_\sigma + P\Delta V_\sigma}{T}.$$

Let K be defined by the equation

$$\Delta S^0 - RT \ln K - \frac{\Delta E_\sigma + P\Delta V_\sigma}{T} = 0.$$

* The symbol E, H, V, S, G, with a subscript H_2, H_2O, etc., means "the contribution that the chemical H_2, H_2O, etc., makes per mole to the value of E, H, V, S, or G of the phase in question." (This contribution is sometimes called a "partial molal quantity." To stress its derivative character, authors frequently place a bar over the symbol E, H, V, S, or G.)

K is the value of $P_C{}^cP_D{}^d/P_A{}^aP_B{}^b$ that makes $\Delta S_{\text{total}} = 0$. $\Delta S_{\text{total}} > 0$ whenever, owing to a small numerator or a large denominator,

$$\frac{P_C{}^cP_D{}^d}{P_A{}^aP_B{}^b} < K.$$

A small numerator (small P_C and/or small P_D) implies that S_{products} is large. A large denominator (large P_A and P_B) implies that $S_{\text{reactants}}$ is small. Large S_{products} and small $S_{\text{reactants}}$ tends to make $(S_{\text{products}} - S_{\text{reactants}}) = \Delta S_\sigma$ large. Large positive ΔS_σ tends to make $\Delta S_{\text{total}} = \Delta S_\sigma + \Delta S_\theta > 0$. And $\Delta S_{\text{total}} > 0$ is the thermodynamic condition for spontaneity.

10. $\Delta S_{\text{total}} = \Delta S_\sigma + \Delta S_\theta = \Delta S_\sigma + (\Delta E_\theta/T) = \Delta S_\sigma + (Q/T)$. For a reversible reaction ($\Delta S_{\text{total}} = 0$),

$$\Delta S_\sigma = -\frac{Q}{T}.$$

For an irreversible reaction ($\Delta S_{\text{total}} > 0$),

$$\Delta S_\sigma > -\frac{Q}{T}.$$

Often authors use the symbol "q" to represent the thermal energy absorbed *from* θ *by* σ. Call this quantity q_σ. $q_\sigma = -\Delta E_\theta \equiv -Q$. Then

$$\Delta S_\sigma = \frac{(q_\sigma)_{\text{rev.}}}{T} \qquad \Delta S_\sigma > \frac{(q_\sigma)_{\text{irrev.}}}{T}.$$

For the irreversible reaction

$$2H_2\ (1\text{ atm.}) + O_2\ (1\text{ atm.}) = 2H_2O(\text{liq}) + 136.6\text{ kcal},$$

what are the values at 25°C of ΔS_σ, q_σ, and q_σ/T?

Chapter 17

1. The thermodynamic implications of the statements in this problem may be expressed in terms of equalities or inequalities between or among free energies or linear combinations of free energies.

   (a) At 1800°C (and 1 atm.), $G_{\text{Ti}}^{\text{solid}} = G_{\text{Ti}}^{\text{liq}}$.

   (b) At 120°C and 0.025 mm Hg, $G_{S_8}^{\text{solid}} = G_{S_8}^{\text{liq}} = G_{S_8}^{\text{gas}}$.

   (c) At the temperature at which $G_{\text{Ag}}^{\text{solid}} = G_{\text{Ag}}^{\text{liq}}$, $G_{\text{Au}}^{\text{solid}} < G_{\text{Au}}^{\text{liq}}$.

Fusible substances can serve as useful temperature indicators. Pieces of silver and gold placed on an object to be heat treated may later reveal how hot the object got. If the silver pieces emerge as rounded buttons while the gold pieces emerge unchanged, it may be stated with reasonable assurance that the temperature reached 960.5° but not 1063°C. Phase changes form, in fact, the basis of international scales of temperature and pressure. In *A Tramp Abroad*, Mark Twain relates that during an attempt to ascend the Riffleberg a four-legged individual in the party ate a can of nitroglycerine and blew himself sky high. Soon thereafter those below were pelted by frozen mule meat. "This shows," Twain says, "better than any estimate in figures, how high the experimenter went."

(d) At room temperature and atmospheric pressure

$$G_{S_8}^{\text{rhombic}} < G_{S_8}^{\text{monoclinic}},\ G_{S_8}^{\text{liq}},\ \text{and}\ G_{S_8}^{\text{gas}}\ .$$

(e) Under ordinary conditions $G_{H_2}^{\text{gas}} + (1/2)G_{O_2}^{\text{gas}} > G_{H_2O}^{\text{liq or gas}}$.

(f) Under ordinary conditions

$$G_{Zn}^{\text{solid}} + 2G_{H^+}^{\text{aq HCl}} > G_{Zn^{++}}^{\text{aq HCl}} + G_{H_2}^{\text{gas}} \qquad \text{while}$$

$$G_{Cu}^{\text{solid}} + 2G_{H^+}^{\text{aq HCl}} < G_{Cu^{++}}^{\text{aq HCl}} + G_{H_2}^{\text{gas}}\ .$$

Adding the latter inequality, turned around, to the former, one finds that under ordinary conditions

$$G_{Zn}^{\text{solid}} + G_{Cu^{++}}^{\text{aq}} > G_{Zn^{++}}^{\text{aq}} + G_{Cu}^{\text{solid}}\ ;$$

that is, under ordinary conditions zinc should displace copper from solution.

(g) At 1120 atm. and -10°C, $G_{H_2O}^{\text{solid I}} = G_{H_2O}^{\text{liq}}$.

(h) At 22,400 atm. and 81.6°C, $G_{H_2O}^{\text{solid VII}} = G_{H_2O}^{\text{liq}}$.

(i) At 25°C the escaping tendency of the pure liquid solvent is equal to the escaping tendency of the solvent from solution when the pressure on the solution is 7.6 atm. greater than the pressure on the pure solvent. Symbolically, with the latter pressure taken as 1 atm.,

$$G_1^{\text{liq}}(N_1 = 1, P = 1\ \text{atm.}) = G_1^{\text{liq}}(N_1 < 1, P = 8.6\ \text{atm.}).$$

(j) At 246°C and 1 atm. pure solid lead *and* pure solid antimony can exist in equilibrium with a liquid phase that is 80 mole per cent lead, 20 mole per cent antimony. A liquid of this composition is saturated with respect to both lead and antimony. Symbolically, at 246°C and 1 atm.,

$$G_{Pb}^{\text{solid}}(N_{Pb} = 1) = G_{Pb}^{\text{liq}}(N_{Pb} = 0.8, N_{Sb} = 0.2)$$

$$G_{Sb}^{\text{solid}}(N_{Sb} = 1) = G_{Sb}^{\text{liq}}(N_{Pb} = 0.8, N_{Sb} = 0.2).$$

(k) $G^{\text{gas}}_{H_2O} > G^{\text{liq and/or solid}}_{H_2O}$.

(l) At 10°C, $G^{\text{gas}}_{H_2O} = G^{\text{liq}}_{H_2O}$ ($N_{H_2O} = 1$).

2. The equation is dimensionally homogeneous. The dimensions of E are the same as the dimensions of PV; the dimensions of S are the same as the dimensions of PV/T.
3. For the transition H_2O(solid) = H_2O(liq) at 0°C and 1 atm.

$$\Delta V = (1.0001 - 1.0908)\,\frac{\text{cc}}{\text{gm}} = -0.0907\,\frac{\text{cc}}{\text{gm}}$$

$$\Delta E \approx \Delta H = 79.7\,\frac{\text{cal}}{\text{gm}}$$

$$\Delta S = \frac{79.7\,\dfrac{\text{cal}}{\text{gm}}}{273^\circ\text{K}} = 0.292\,\frac{\text{cal}}{^\circ\text{K-gm}}\,.$$

Therefore

$$P \approx \frac{(0.292\ \text{cal}/^\circ\text{K-gm})}{-(0.0907\ \text{cc/gm})}\,T - \frac{(79.7\ \text{cal/gm})}{-(0.0907\ \text{cc/gm})}$$

$$= -\left(3.22\,\frac{\text{cal}}{^\circ\text{K-cc}}\right)T + 878\,\frac{\text{cal}}{\text{cc}}$$

$$= -133T + 36{,}200 \qquad T \text{ in } ^\circ\text{K}, \quad P \text{ in atm.}$$

$$\overset{T=263^\circ\text{K}}{=} \quad 1200 \text{ atm.} \qquad \text{(cf. problem 1(g))}$$

4. The standard free energies of dissociation of calcium carbonate at 1115.4°K and 1210.1°K could be calculated and the average values of ΔH^0 and ΔS^0 over this temperature interval determined. These results might be compared with values estimated from tabulated data at 298°K.

Thus, at any temperature,

$$\Delta G^0(T) = -RT \ln K_{eq}(T).$$

For the reaction

$$CaCO_3(\text{solid}) = CaO(\text{solid}) + CO_2(\text{gas}),$$

$$K_{eq} = \frac{a^{\text{solid}}_{CaO}\, a^{\text{gas}}_{CO_2}}{a^{\text{solid}}_{CaCO_3}} = P_{CO_2}\,. \qquad (P \text{ in atm.})$$

* See Chapter 28.

Therefore,

$$\Delta G^0\,(1115°\text{K}) = -R\,(1115°\text{K})\ln(0.4513) = +1764\,\frac{\text{cal}}{\text{mole}}$$

$$\Delta G^0\,(1210°\text{K}) = -R\,(1210°\text{K})\ln(1.770) = -1373\,\frac{\text{cal}}{\text{mole}}\,.$$

Average values for ΔH^0 and ΔS^0 over the interval 1115° to 1210°K may be determined in the following manner. Suppose that ΔH^0 and ΔS^0 are constant over this temperature interval. Then ($T_1 = 1115°$K, $T_2 = 1210°$K),

$$\Delta G^0(T_1) = \Delta H^0 - T_1\Delta S^0$$
$$\Delta G^0(T_2) = \Delta H^0 - T_2\Delta S^0\,.$$

Solving these two equations for ΔS^0 and ΔH^0 one finds that

$$\Delta S^0 = -\frac{\Delta G^0(T_2) - \Delta G^0(T_1)}{T_2 - T_1}$$
$$= +33.0 \text{ cal/deg-mole.}$$
$$\Delta H^0 = \Delta G^0(T_i) + T_i\Delta S^0 \qquad (i = 1, 2)$$
$$= +38{,}600 \text{ cal/mole.}$$

The corresponding values at 298°K (Appendix I) are 38.4 cal/deg-mole and 42,500 cal/mole.

Chapter 18

(a) solid
(b) liquid
(c) gas
(d) lowest
(e) $E + PV - TS$
(f) E
(g) molar
(h) energy
(i) solid
(j) left
(k) T
(l) E
(m) temperature
(n) large
(o) molar entropy
(p) large
(q) large
(r) pressure (or partial pressure)
(s) low
(t) right
(u) bottom
(v) liquid
(w) stable
(x) coexist
(y) $\Delta S/\Delta V$
(z) entropy
(a′) transition
(b′) $\Delta S = \Delta H/T$
(c′) heat
(d′) transition
(e′) vapor pressure
(f′) $PV = nRT$
(g′) vapor pressure
(h′) heats
(i′) vaporization

Chapter 19

1. $A = -20.2°C$ (the temperature at which the vapor pressure of ice is 0.001 atm.). $B = -1.86°C$. $C = 0°C$. $D = 100°C$. $E = 100.52°C$.
2. (a)

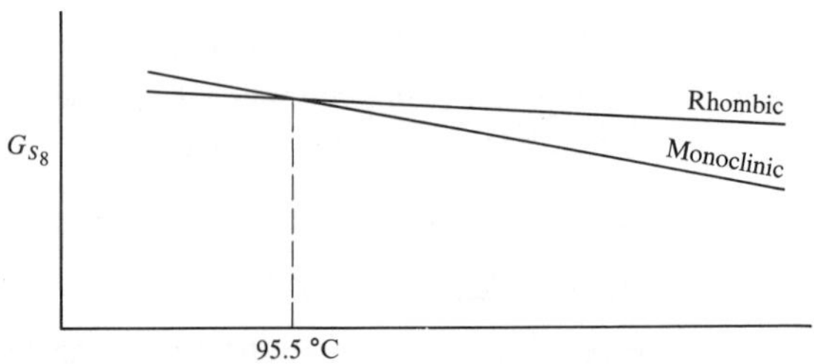

Below 95.5°C rhombic sulfur is more stable than monoclinic sulfur. Above 95.5°C monoclinic sulfur is more stable than rhombic sulfur.

(b)

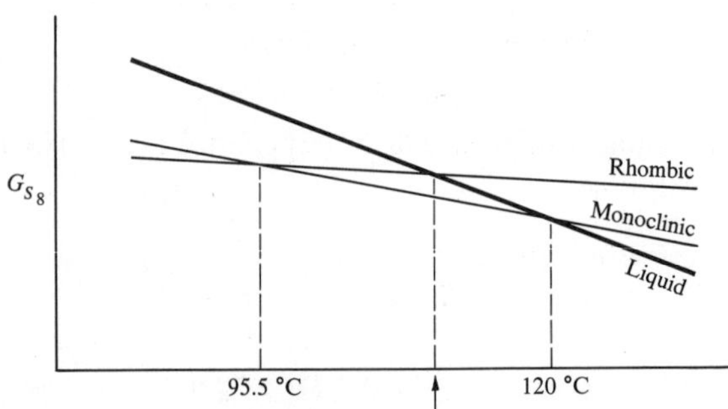

(c) Monoclinic sulfur has the higher melting point (see the diagram for part (b); the arrow indicates the melting point of rhombic sulfur).
(d) Monoclinic sulfur has the larger molar entropy—7.78 cal/deg-mole of atoms at 25°C, compared to 7.62 for rhombic sulfur. Monoclinic sulfur also has a slightly larger molar volume, 16.3 cm³/mole of atoms, compared to 15.5 for rhombic sulfur.

In any phase transition the higher temperature form must have the larger molar entropy; often, but not always, it also has the larger molar volume. Steam, for example, has a larger molar volume than liquid water; on the other hand, at 0°C, liquid water is more, not less dense than ice. The well-known grey tin - white tin transition exhibits similar characteristics: in going from the low-temperature, low-entropy grey tin form to the higher-temperature, higher-entropy white tin form the density increases, from 5.75 to 7.28 gm/cm³. The structural reasons for this behavior are interesting; they are similar to the reasons for th anomalous behavior of water at 0°C. Grey tin (like ice) has the diamond structure. Each atom in grey tin has four relatively tightly bound neighbors at 2.80 Å. The distance to the next-nearest neighbors of a tin atom in grey tin is more than 4.5 Å. In white tin each tin atom has four nearest-neighbors at 3.016 Å. Evidently the tin-tin bonding in white tin is somewhat weaker than in grey tin; presumably this is the reason for white tin's greater entropy. In addition, however, each tin atom in white tin has two more neighbors at 3.175 Å; this is the reason for white tin's greater density. These structural differences are reflected in the chemical properties of tin. Sn(II) is formed when white tin dissolves in concentrated hydrochloric acid; under similar conditions grey tin gives Sn(IV).

(e) Monoclinic sulfur, the form with the larger molar free energy at 25°C, should show the larger solubility. At 25°C its solubility in diethyl ether, ethyl bromide, and ethyl alcohol is, in fact, 30 per cent greater than that of rhombic sulfur.

If rhombic sulfur were more soluble than monoclinic sulfur at 25°C it would be possible to effect spontaneously the transition from rhombic to monoclinic sulfur at 25°C by saturating a solvent with respect to rhombic sulfur and placing in this rhombic-saturated solvent a seed crystal of monoclinic sulfur.

(f) The melting point of rhombic sulfur lies above the rhombic-monoclinic transition temperature (see the diagram in (b)).

3. The free energy curve for the liquid phase can cross the free energy curves for the solid phases either to the right or to the left of the cross-over point of the two solid phase free energy curves; there are no other possibilities (see the diagram for problem 2(b)). The melting points of the two solids must lie on the same side of the solid-solid transition temperature: either both above it or both below it.

4. Like sulfur, benzophenone can exist in two crystalline forms whose free energies are disposed with respect to the free energy of the liquid phase as shown below.

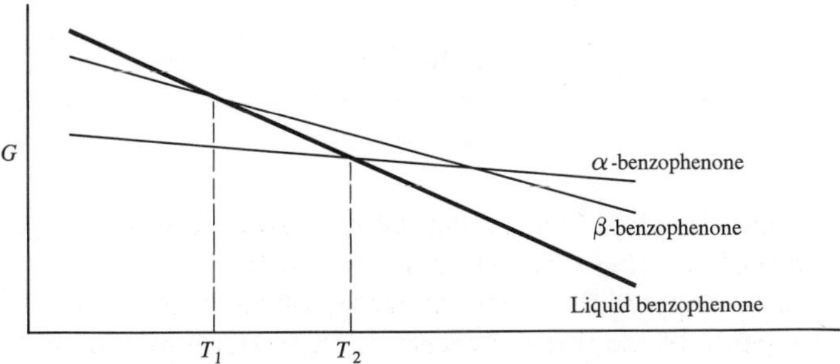

Below temperature T_2 α-benzophenone is the most stable form. Above T_2 liquid benzophenone is the most stable form. β-benzophenone is metastable at all temperatures. On being heated β-benzophenone should melt at T_1 (if reversion to more stable α-benzophenone does not occur first). At T_1, however, liquid benzophenone is less stable than α-benzophenone, which, given time, should crystallize from the melt. Later, at T_2, α-benzophenone should melt.

5. Solid-solution formation between solvent and solute lowers the free energy curve for the solvent in the solid phase and diminishes the solute's freezing-point-lowering effect.

Chapter 20

1. $\Omega = e^{S/k} = (10^{1/2.303})^{S/k} = 10^{S/2.303k}$

$$= 10^{\frac{10 \text{ cal/deg}}{7.60 \times 10^{-24} \text{ cal/deg}}}$$

$$= 10^{1.32 \times 10^{24}}.$$

Ω is the number 21 followed by a septillion zeros.

2. $\Omega_1 = e^{S_1/k}$ and $\Omega_2 = e^{S_2/k}$. Therefore

$$\frac{\Omega_2}{\Omega_1} = e^{\frac{S_2 - S_1}{k}} = e^{\frac{\Delta S_{\text{total}}}{k}}.$$

When ΔS_{total} is equal to 10, 1, and 10^{-6} cal/deg., Ω_2/Ω_1 is equal to (problem 1) ten raised to the power 1.32 times 10^{24}, 10^{23}, and 10^{17}, respectively. When $\Delta S_{\text{total}} = 0$, $\Omega_2/\Omega_1 = 1$.

3. Conversion of 300 microcalories of thermal energy at 27°C entirely to work would alter the entropy of the universe by -10^{-6} cal/deg. Therefore (problem 2),

$$\frac{\Omega_{\text{final}}}{\Omega_{\text{initial}}} = e^{\frac{\Delta S_{\text{total}}}{k}}$$

$$= 10^{-1.32 \times 10^{17}}$$

$$= \frac{1}{10^{1.32 \times 10^{17}}}.$$

If all the individual micromolecular configurations are equally probable, the chance of effecting this conversion is 1 in $10^{1.32 \times 10^{17}}$.

4. Were this to happen, the potential energy of the weight would increase (at the expense of the thermal energy of the system) by the amount

$$mg \cdot \Delta h = (100 \text{ gm}) \underbrace{\left(32 \frac{\text{ft}}{\text{sec}^2} \times \frac{12 \text{ in}}{1 \text{ ft}} \times \frac{2.54 \text{ cm}}{1 \text{ in}}\right)}_{980 \frac{\text{cm}}{\text{sec}^2}} (10^{-1} \text{ cm}) \times \frac{1 \text{ cal}}{4.18 \times 10^7 \frac{\text{gm cm}^2}{\text{sec}^2}}$$

$$\approx 300 \text{ microcalories.}$$

Therefore (problem 3) the chance is about 1 in $10^{1.32 \times 10^{17}}$.

5. The spectroscopic entropy might be in error for one or more of the following reasons: the equations are incorrect; the calculations have been improperly carried out; the moment of inertia has been improperly determined; the vibrational frequency has been improperly determined; the gas is not an ideal gas.

The Third Law entropy might be in error if during warm-up there were to occur in the substance irreversible, entropy-producing processes; for contributions to the entropy from such sources would not be included in the $(1/T) \times$ (energy absorbed) computation. This would tend to make the Third Law entropy less than the actual entropy.

The interpretation of the warm-up data might also be in error. For in arriving at the figure 46.2 cal/deg-mole, it is assumed that the entropy of the carbon monoxide crystals used in the experiment would have vanished had it been possible to cool them all the way down to 0°K. But if each carbon monoxide molecule during the initial cooling stage were to freeze out at random in either one of two possible orientations, CO or OC, and if this orientational disorder were to persist to the lowest temperatures reached in the experiment, one should add to the summation of terms of the form $(1/T) \times$ (energy absorbed) 1.38 cal/deg-mole to allow for the "configurational" entropy already possessed by the crystals before warm-up.

(That the discrepancy calculated is 1.38 cal/deg-mole rather than the observed value, 1.1 cal/deg-mole, might be accounted for by assuming that on crystallization there occurred partial ordering of the carbon monoxide molecules. This effect might be homogeneous—the same degree of ordering in each crystal of the polycrystalline sample—or heterogeneous, owing to differences in the thermal histories of different layers of the sample in the calorimeter. For in filling the calorimeter, which for this purpose was kept at liquid-hydrogen temperatures, those crystals that formed first were formed from molecules condensing directly, or almost directly, on the cold calorimeter walls; later other crystals probably formed on top of these. Owing to the energy liberated on condensation, and to the relatively poor thermal conductivity of polycrystalline carbon monoxide, crystals overlying other crystals would have been formed at effectively higher temperatures than those beneath them. This might be important. Both the rate at which films of nitric oxide are deposited on cold surfaces and subsequent annealing of such films has been shown to have a marked effect upon the spectroscopic properties of such films.)

The determination of the entropy of carbon monoxide by both spectroscopic and calorimetric methods was undertaken with a view to checking the validity of the Third Law. If the last view stated above is correct, the validity of the Third Law has been checked within fairly close limits. In addition, the usefulness of Boltzmann's relation has been illustrated. Furthermore, an unsuspected fact about the structure of crystalline carbon monoxide has been revealed. Also, the validity of the spectroscopic assignments for carbon monoxide, the validity of a number of theoretical equations, the correctness of the arithmetical computations, and the skill of the investigators, have been confirmed.*

To attribute to a single experiment verification of such numerous and diverse conjectures may seem overly generous. In fact, a single experiment out of context never makes anything certain—any more than one sentence makes a book, or one brush-stroke a painting. Nonetheless, most scientists felt at the time that these results greatly increased the plausibility of the Third Law and that, also, they lent considerable support to the then young science of statistical thermodynamics.

The entropy of ethane has been the subject of a similar study. For this molecule the value of the entropy at 184.1°K calculated (with the aid of spectroscopic data) for translation and over-all rotation (48.634 cal/deg-mole) and for 17 of the $3 \times 8 - 6 = 18$ degrees of internal freedom (0.058 cal/deg-mole) is 0.85 cal/deg-mole less than the calorimetric value. The difference is attributed to internal rotation of the methyl groups about the carbon-carbon bond. From the figure 0.85 cal/deg-mole the magnitude of the potential barrier that hinders this rotation (and whose exact origin is still unknown) has been calculated to be 2830 cal/mole.

6. For an ideal gas doubling V keeping T, M, and n constant increases Ω by the factor 2^{nN_0}. In an isothermal expansion of an ideal gas from V_1 to V_2 (T, M, and n constant)

$$\Delta S = S_2 - S_1 = k \ln \frac{\Omega_2}{\Omega_1} = k \ln \left(\frac{V_2^{nN_0}}{V_1^{nN_0}}\right)$$

$$= nR \ln \left(\frac{V_2}{V_1}\right).$$

* One of the investigators, Professor William F. Giauque, received a Nobel prize for his work in this field.

7. For a small flow of thermal energy from A to B let the changes that occur in Ω_A and Ω_B be designated as $-\delta\Omega_A$ and $+\delta\Omega_B$, respectively ($\delta\Omega_A$ and $\delta\Omega_B$ are both positive quantities; by assumption, they are much smaller than Ω_A and Ω_B). In this notation,

$$(\Omega_{\text{total}})_{\text{initial}} = (\Omega_A)_{\text{initial}} \cdot (\Omega_B)_{\text{initial}}$$
$$= \Omega_A \cdot \Omega_B$$

$$(\Omega_{\text{total}})_{\text{final}} = (\Omega_A)_{\text{final}} \cdot (\Omega_B)_{\text{final}}$$
$$= (\Omega_A - \delta\Omega_A) \cdot (\Omega_B + \delta\Omega_B)$$
$$= \Omega_A\Omega_B + \Omega_A \cdot \delta\Omega_B - \Omega_B \cdot \delta\Omega_A - \delta\Omega_A \cdot \delta\Omega_B\,.$$

Dropping the small, second-order quantity $\delta\Omega_A \cdot \delta\Omega_B$ and imposing the Second Law condition that

$$\frac{(\Omega_{\text{total}})_{\text{final}}}{(\Omega_{\text{total}})_{\text{initial}}} > 1,$$

one finds after rearrangement that

$$\frac{\delta\Omega_B}{\Omega_B} > \frac{\delta\Omega_A}{\Omega_A}\,.$$

This may be written (Chapter 15) as

$$\delta(\ln \Omega_B) > -\delta(\ln \Omega_A)$$

or

$$\delta(k \ln \Omega_B) > -\delta(k \ln \Omega_A)\,.$$

The quantity on the left represents the increase in entropy of B; that on the right the decrease in entropy of A. Only if the latter is less than the former can energy flow from A to B.

8. Summarized below are some pertinent properties of $O(x)$, e, and $\ln x$.

(1) $O(xy) = O(x) + O(y)$

(2) $nO(x) = O(x^n)$

(3) $e = \lim\limits_{x\to 0} (1 + x)^{1/x}$

(4) $\dfrac{d(\ln x)}{dx} = \dfrac{1}{x}$

(5) $O(1) = \ln (1) = 0.$

From these relations and the substitution $y = 1 + (\delta x/x)$, it follows that

$$O(x + \delta x) - O(x) = O\left(1 + \frac{\delta x}{x}\right) \qquad \text{Property (1), (2)} \quad [n = -1]$$

$$\frac{O(x + \delta x) - O(x)}{\delta x} = O\left(1 + \frac{\delta x}{x}\right)^{1/\delta x} \qquad \text{Property (2)}$$

$$= \frac{1}{x} \cdot O\left(1 + \frac{\delta x}{x}\right)^{\frac{1}{\delta x/x}} . \qquad \text{Property (2)}$$

$$\therefore \quad \frac{d[O(x)]}{dx} = \frac{d(\ln x)}{dx} \cdot O(e), \qquad \begin{array}{l}\text{Definition (3)}\\ \text{Property (4)}\end{array}$$

$$\text{or} \quad d[O(x)] = O(e)d(\ln x).$$

Adding up the increments on the left and on the right from $x = 1$ to $x = x$, one obtains

$$O(x) - O(1) = O(e)[\ln (x) - \ln (1)].$$

Hence,

$$O(x) = O(e) \cdot \ln x \qquad \text{Property (5)}$$
$$= \ln x^{O(e)} .$$

That is to say

$$x^{O(e)} = e^{O(x)} \qquad \text{Definition of a logarithm.}$$

Therefore,

$$x = [e^{O(x)}]^{\frac{1}{O(e)}} = \left[e^{\frac{1}{O(e)}}\right]^{O(x)} .$$

In other words, $O(x)$ is the logarithm of x to the base $e^{1/O(e)}$.

$$O(x) = \log_{e^{1/O(e)}} x$$

There is essentially no difference between $O(x)$ and $\ln (x)$. One is proportional to the other for all values of x.

Chapter 21

1. (a) The number of different distributions is 5.

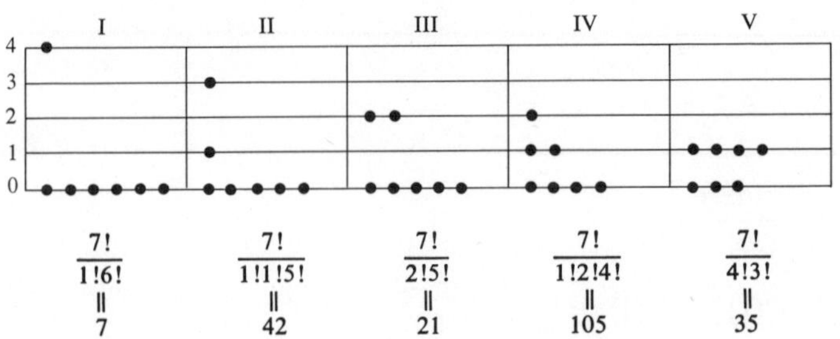

(b) $W = 105$

(c) $\Omega = 7 + 42 + 21 + 105 + 35 = 210$

(d) $\dfrac{\log \Omega - \log W}{\log \Omega} \times 100 = \dfrac{2.322 - 2.021}{2.322} \times 100 = 13$

(e) (i) $\left(\dfrac{n_1}{n_0}\right)_{\text{ave.}} = \dfrac{7 \times 0 + 42 \times 1 + 21 \times 0 + 105 \times 2 + 35 \times 4}{7 \times 6 + 42 \times 5 + 21 \times 5 + 105 \times 4 + 35 \times 3} =$

$$\frac{392}{882} = 0.44$$

(ii) $\left(\dfrac{n_1}{n_0}\right)_{\text{IV}} = \dfrac{2}{4} = 0.50$

2. (a) The number of different distributions is 5.

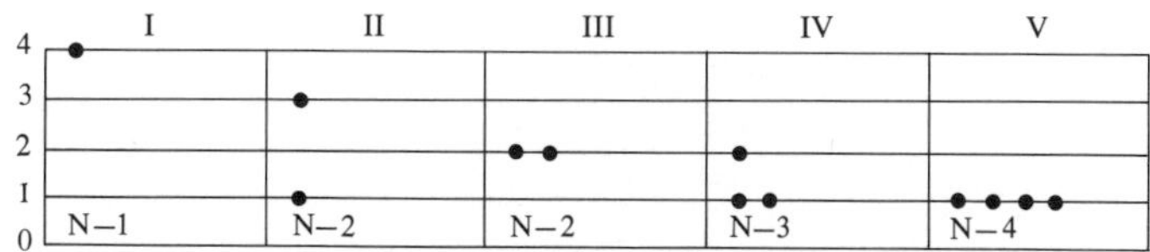

$N \quad N(N-1) \quad \dfrac{N(N-1)}{2!} \quad \dfrac{N(N-1)(N-2)}{2!} \quad \dfrac{N(N-1)(N-2)(N-3)}{4!}$

(b) For large N the last distribution is the most probable.

$$W = \frac{(N)(N-1)(N-2)(N-3)}{4!} \approx \frac{N^4}{4!} \qquad (N \gg 3)$$

$$\Omega = \frac{N^4}{4!} + \frac{N^3}{2!} + N^2\left(\frac{1}{2!} + 1\right) + N \qquad (N \gg 3)$$

(c) For large N,

$$\frac{N^4}{4!} \gg \frac{N^3}{2!} \gg N^2\left(\frac{3}{2}\right) \gg N;$$

that is, for large N,

$$\Omega \approx \frac{N^4}{4!}.$$

Thus, in this case,

$$\lim_{N \to \infty} \frac{W}{\Omega} = 1.$$

(d) For large N

$$\left(\frac{n_1}{n_0}\right)_{\text{ave.}} \approx \left(\frac{n_1}{n_0}\right)_{\substack{\text{distribution}\\ V}} = \frac{4}{N-4} \approx \frac{4}{N}. \qquad (N \gg 4)$$

3. (a) $\dfrac{n_1}{n_0} = 0.50$ for the most probable distribution.

$= 0.44$ when all distributions are counted.

$$\frac{\Omega_{n=4}}{\Omega_{n=5}} = \frac{\left[\dfrac{(7+4-1)!}{(7-1)!4!}\right]}{\left[\dfrac{(7+5-1)!}{(7-1)!5!}\right]} = \frac{5}{11} = 0.45$$

(b) $\dfrac{n_1}{n_0} = \dfrac{4}{N} = \dfrac{n}{N} \qquad (N \gg 1)$

$$\frac{\Omega_{n=4}}{\Omega_{n=5}} \overset{N\gg 1}{=} \frac{\left(\dfrac{N^4}{4!}\right)}{\left(\dfrac{N^5}{5!}\right)} = \frac{5}{N} = \frac{n+1}{N}$$

It would appear that when N and n are both large

$$\left(\frac{n_1}{n_0}\right)_{\text{ave.}} = \left(\frac{n_1}{n_0}\right)_{\substack{\text{most probable}\\ \text{distribution}}} = \frac{\Omega_n}{\Omega_{n+1}}.$$

4. $T = \delta E/\delta S$.

Take $\delta E = \varepsilon$.

Then $\delta S = k \ln \Omega_{n+1} - k \ln \Omega_n = k \ln \left(\dfrac{\Omega_{n+1}}{\Omega_n}\right)$.

Therefore,

$$T = \frac{\varepsilon}{k \ln \left(\dfrac{\Omega_{n+1}}{\Omega_n}\right)} = -\frac{h\nu}{k \ln \left(\dfrac{\Omega_n}{\Omega_{n+1}}\right)}.$$

5. Combining the conclusions of problems 3 and 4, one finds that

$$\frac{n_1}{n_0} = e^{-h\nu/kT}.$$

6. $W = \dfrac{(n_0 + n_1 + n_2 + n_3 + \cdots)!}{n_0!\, n_1!\, n_2!\, n_3! \cdots}$

7. Promotion of an oscillator from its ground level to its first excited state decreases n_0 by 1, increases n_1 by 1, and leaves unchanged the values of $n_2, n_3, \ldots$ and the value of the sum $(n_0 + n_1 + n_2 + n_3 + \cdots)$. Therefore,

$$\frac{W_n}{W_{n+1}} = \frac{\left[\dfrac{1}{n_0!n_1!}\right]}{\left[\dfrac{1}{(n_0 - 1)!(n_1 + 1)!}\right]} = \frac{n_1 + 1}{n_0} \overset{n_1 \gg 1}{=} \frac{n_1}{n_0}.$$

8. The new distribution of energy is described by the numbers $n_0 + 1$, $n_1 - 2$, $n_2 + 1$, n_3, n_4, ... This distribution has the same energy as the distribution from which it was derived. The number of associated microstates is

$$\frac{(n_0 + n_1 + n_2 + n_3 + n_4 + \cdots)!}{(n_0 + 1)!(n_1 - 2)!(n_2 + 1)!\, n_3!\, n_4! \cdots}.$$

By definition, this number cannot be greater than W. It will be nearly equal to W, however, if the change described is a relatively small one, i.e. if the numbers n_0, n_1, n_2 are large.

9. If the number of microstates associated with the distribution $n_0 + 1$, $n_1 - 2$, $n_2 + 1$, n_3, n_4 ... (problem 8) is almost equal to the number of microstates associated with the most probable distribution n_0, n_1, n_2, n_3, n_4 (problem 6),

$$(n_0 + 1)!(n_1 - 2)!(n_2 + 1)! \approx n_0!n_1!n_2!.$$

Canceling common factors, one finds that

$$(n_0 + 1)(n_2 + 1) \approx (n_1)(n_1 - 1).$$

When $n_0, n_1, n_2 \gg 1$, this can be written

$$\frac{n_1}{n_0} \approx \frac{n_2}{n_1}.$$

It would appear that for an ensemble of harmonic oscillators

$$\frac{n_{i+1}}{n_i} = \frac{n_1}{n_0} \overset{\text{(Problem 5)}}{=} e^{-h\nu/kT} \quad i = 0, 1, 2, 3, \ldots.^*$$

For a more detailed discussion of this result and for the origin of several techniques and [id]eas employed in the present chapter the reader is referred to Ronald Gurney's *Introduction* [to] *Statistical Mechanics*, New York: McGraw-Hill Book Co., 1949. The heuristic method [o]f the present chapter, particularly, is patterned after George Polya's *Mathematics and* [Pl]*ausible Reasoning. Vol. I. Induction and Analogy in Mathematics. Vol. II. Patterns of* [Pl]*ausible Inference*, Princeton: Princeton University Press, 1954.

10. $$T = \left(\frac{\varepsilon}{2.303k}\right) \cdot \left(\log \frac{\Omega_{n+1}}{\Omega_n}\right)^{-1} = \left(\frac{\varepsilon}{2.303k}\right) \tau,$$

where $$\tau \equiv \left(\log \frac{\Omega_{n+1}}{\Omega_n}\right)^{-1}.$$

Tabulated below are values for Ω, τ, and $\Delta\tau$ for $N = 2$, 3, 4 and $n = 0, \ldots, 6$.

	$N = 2$			$N = 3$			$N = 4$		
n	Ω	τ	$\Delta\tau$	Ω	τ	$\Delta\tau$	Ω	τ	$\Delta\tau$
0	1			1			1		
		3.32			2.095			1.66	
1	2		2.36	3		1.23	4		0.84
		5.68			3.32			2.50	
2	3		2.33	6		1.19	10		0.82
		8.01			4.51			3.32	
3	4		2.31	10		1.17	20		0.80
		10.32			5.68			4.12	
4	5		2.31	15		1.17	35		0.78
		12.63			6.85			4.90	
5	6		2.31	21		1.16	56		0.78
		14.94			8.01			5.68	
6	7			28			84		

These calculations show that (1) T increases as E or $n\,(E = n\varepsilon)$ increases, N constant; (2) T decreases as N increases, E constant (neither of these conclusions is very surprising); and that (3) the increase in T per unit increase in E decreases as E increases and appears to approach, rather quickly, a constant value. The implication of this observation regarding the heat capacities of these systems is considered in the next problem.

11. Let $$C = \frac{E_{n+1} - E_n}{T_{n+1} - T_n} \qquad (*)$$

$$= \frac{(n+1)\varepsilon - n\varepsilon}{\left(\dfrac{\varepsilon}{2.303\,k}\right)(\tau_{n+1} - \tau_n)}$$

$$= \frac{2.303\,k}{\Delta\tau}.$$

* C is the average heat capacity at constant volume over the temperature interval $T_n - T_{n+}$

$$C \text{ (per mole)} = \frac{2.303}{N \cdot \Delta_T} \cdot R \qquad (N_0 k = R)$$

$$= 0.499\ R \qquad N = 2$$

$$= 0.662\ R \qquad N = 3$$

$$= 0.738\ R \qquad N = 4$$

These values are approximately equal to 1/2, 2/3, and 3/4 R, respectively. It would appear that at high temperatures the heat capacity per mole is equal to

$$\frac{N-1}{N} R \approx R. \qquad (N \gg 1)$$

12. It is very hot. It can lose energy without suffering any change whatsoever in its entropy—which, regardless of its energy content, is identically zero ($S = k \ln \Omega = k \ln 1 = 0$). By the usual definition of T,

$$T = \frac{\delta E}{\delta S},$$

its temperature is infinite.

13.

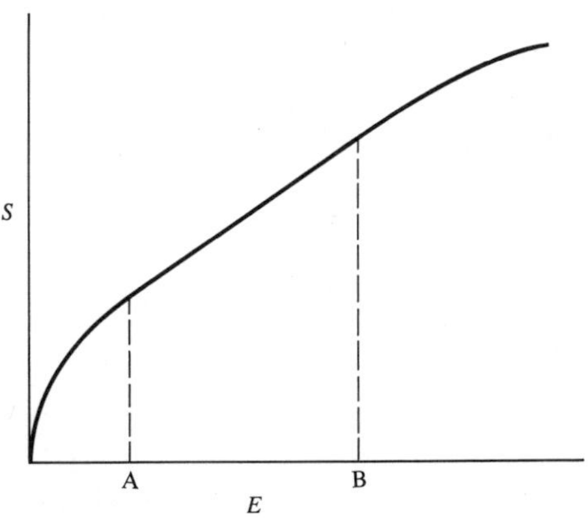

The three regions, left-of-A, A-to-B, and right-of-B represent, respectively, warm-up of the solid, melting, and warm-up of the liquid. The rate at which S increases as E increases is rapid at first (large change in S for given change in E), constant during melting, and small when S is large. At any point the inclination of the line tangent to the curve is a measure of the reciprocal of the absolute temperature. (The tangent line's slope, $\delta S/\delta E$, is $1/T$.)

14. The maximum energy the system can have is $N\varepsilon$. At this point $n_1 = N$, $n_0 = 0$, and

$$S = k \ln \Omega = k \ln \frac{(n_0 + n_1)!}{n_0!n_1!} = k \ln \frac{N!}{0!N!} = 0.$$

It can be shown that Ω has its maximum value when $n_0 = n_1 = N/2$. A plot of S against E must look something like this.

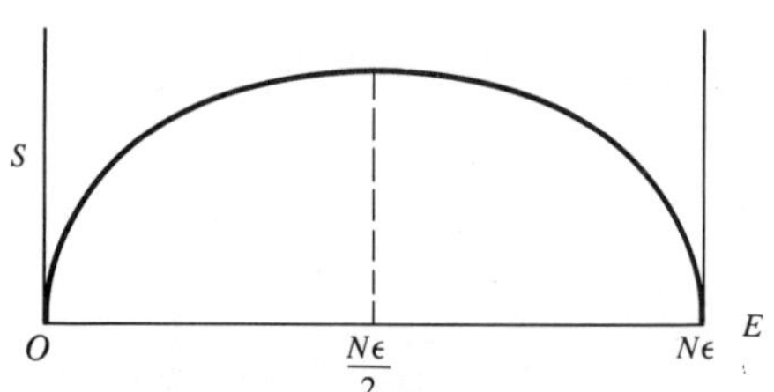

To the left of $E = N\varepsilon/2$, S increases as E increases; in this region $T\ (= \delta E/\delta S)$ is positive. At $N\varepsilon/2$, S does not change as E changes; at this point (where $n_0 = n_1$) T is infinite. To the right of $N\varepsilon/2$, S decreases as E increases; in this region T is negative.

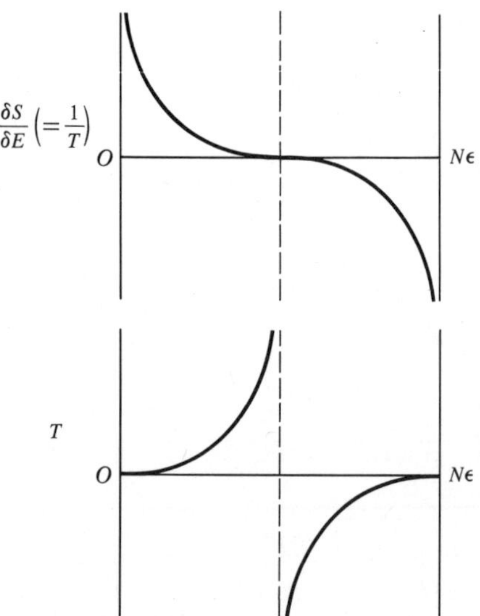

Chapter 22

1. According to the law of Dulong and Petit, the molar heat capacity of simple solids at room temperature is about $3R$. This value was derived

in the present chapter on the assumption that $kT \gg h\nu$. The law is obeyed reasonably well for non-abrasive substances; at room temperature the heat capacities of copper, zinc, lead, calcium, strontium, and barium, for example, are 5.85, 5.99, 6.4, 6.28, 6.0, 6.30 cal/deg-mole, respectively. The law is usually poorly obeyed by hard substances; diamond's heat capacity at room temperature, for example, is only 1.45 cal/deg-mole; at 1150°K, however, the heat capacity of diamond is 5.2 cal/deg-mole. One might expect the law not to be valid when $h\nu > kT$; i.e. when $h\nu/kT > 1$.

2. No. When $\nu = 3 \times 10^{13}$ cps and $T = 300°K$,

$$\frac{h\nu}{kT} = \frac{(6.62 \times 10^{-27}\ \text{gm cm}^2\ \text{sec}^{-1})(3 \times 10^{13}\ \text{sec}^{-1})}{(1.38 \times 10^{-16}\ \text{gm cm}^2\ \text{sec}^{-2}\ \text{deg}^{-1})(300°\text{K})}$$
$$= 4.79 > 1.$$

3. When $h\nu/kT = 4$, $e^{h\nu/kT} = 54.6$ and ($T = 300°K$) $3N_0h\nu = 3N_0(4kT) = 12RT = 7200$ cal/mole. Therefore,

$$E = \frac{7200\ \text{cal/mole}}{54.6 - 1} = 134\ \text{cal/mole}$$

$$S = -3R \ln\left(1 - \frac{1}{54.6}\right) + \frac{134}{300}\frac{\text{cal}}{\text{deg-mole}}$$
$$= [-13.82 \log (0.9817) + 0.447]\frac{\text{cal}}{\text{deg-mole}}$$
$$= 0.548\frac{\text{cal}}{\text{deg-mole}}.$$

4. When $e^{h\nu/kT} \gg 1$, $e^{h\nu/kT} - 1 \approx e^{h\nu/kT}$. Therefore,

$$E = \frac{3N_0h\nu}{e^{h\nu/kT} - 1} \approx \frac{3N_0(kT)\dfrac{h\nu}{kT}}{e^{h\nu/kT}}$$
$$= 3RT\left(\frac{h\nu}{kT}\right)e^{-h\nu/kT}.$$

Also, when $e^{h\nu/kT} \gg 1$, $e^{-h\nu/kT} \ll 1$. Therefore, since for small x,

$$\ln (1 + x) \approx x$$

$$S = -3R \ln (1 - e^{-h\nu/kT}) + \frac{E}{T}$$
$$= -3R(-e^{-h\nu/kT}) + 3R\left(\frac{h\nu}{kT}\right)e^{-h\nu/kT}$$
$$= 3R\left(1 + \frac{h\nu}{kT}\right)e^{-h\nu/kT}.$$

For $h\nu/kT = 3, 4, 5$, these formulas yield for E and S the values 270, 126,* 60.7 cal/mole and 1.20, 0.549,* 0.243 cal/deg-mole, respectively.

5. The atoms in diamond are bound to each other by single bonds. The natural frequency of vibration of these bonds is

   $$(1/\text{wave length of absorbed radiation})(3 \times 10^{10}\ \text{cm/sec}) \approx 3 \times 10^{13}\ \text{sec}^{-1}.$$

   (This is the frequency of a typical bond stretching vibration; bond bending vibrations occur at somewhat lower frequencies.) If each atom in diamond were able to execute in three directions vibrations of this frequency, one would estimate from the calculations in problems 2, 3, and 4 that diamond would have an entropy between 0.24 and 0.55 cal/deg-mole. (Allowance for the lower frequencies of the bending vibrations would increase slightly this estimate.) The measured Third Law entropy of diamond is 0.5829 cal/deg-mole at 298.15°K.

6. $h\nu/kT$ for diamond at 300°K is about 4.8 (problems 2 and 5). To bring diamond well into the region of the law of Dulong and Petit, a temperature of at least 3000°K would probably be required (see problem 14).

7. (a) 3(1) − 3(for translation) = 0.

   (b) 3(2) − 3(for translation) − 2(for rotation) = 1.

   (c) 3(3) − 3(for translation) − 2(for rotation) = 4.

   The vibrations corresponding to these four degrees of freedom are pictured below.†

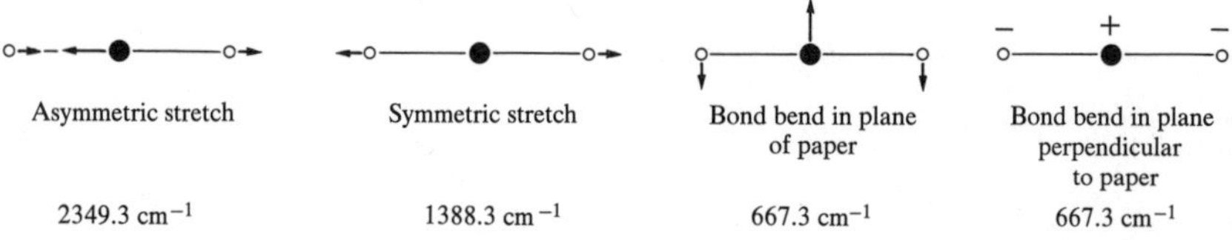

Asymmetric stretch	Symmetric stretch	Bond bend in plane of paper	Bond bend in plane perpendicular to paper
2349.3 cm^{-1}	1388.3 cm^{-1}	667.3 cm^{-1}	667.3 cm^{-1}

   (d) 3(3) − 3(for translation) − 3(for rotation) = 3 (see Fig. 5.1).

8. No. vib. deg. freedom = $3N - 3 = 0$ monatomic molecule ($N = 1$)

   $3N - 5$ linear molecule ($N > 1$)

   $3N - 6$ non-linear molecule ($N > 2$)

9. The discussion of thermal excitation of quantized motion in Chapter 6 suggests that at room temperature translational and rotational, but not

* Cf. problem 3.

† The earth radiates energy to outer space, chiefly in the infrared at frequencies near the bond-bending vibration of CO_2, whose presence in the atmosphere therefore influences the earth's temperature through the "greenhouse effect." It has been suggested that increasing combustion of fossil fuels may increase the concentration of CO_2 in the atmosphere sufficiently to raise the average temperature of the earth several degrees, enough to melt the polar ice caps and flood large land masses, including most of Texas—unless the CO_2 is absorbed by the oceans.

vibrational, degrees of freedom make their full, classical contribution to the heat capacities of these gases. Listed below in the second column are the gases' experimental heat capacities at constant volume. In the third column is the sum of the expected classical contributions from translational and rotational degrees of freedom. The differences between the calculated values and the experimental values are listed in the last column.

Molecule	$C_V = C_P - R$	$C_{\text{trans.}} + C_{\text{rot.}}$	$C_V - (C_{\text{trans.}} + C_{\text{rot.}})$
Ar	2.981	$\left(\frac{3}{2} + \frac{0}{2}\right) R = 2.981$	0.000
N_2	4.973	$\left(\frac{3}{2} + \frac{2}{2}\right) R = 4.968$	0.005
CO_2	6.88	$\left(\frac{3}{2} + \frac{2}{2}\right) R = 4.97$	+1.91
H_2O	6.03	$\left(\frac{3}{2} + \frac{3}{2}\right) R = 5.96$	+0.07

Good agreement is obtained for argon and nitrogen. For carbon dioxide—and, to a lesser extent, for water—vibrational degrees of freedom appear to contribute to C_V at room temperature. A calculation of this contribution is made in the "Concluding Comments" for problem 14 of this chapter.

10.

N	Vib. deg. freedom (per molecule)	$(C_{\text{trans.}} + C_{\text{rot.}})$*	$C_{\text{vib.}}$*	C_{total}†
1	0	$\frac{3}{2} R$	0	$\frac{1}{1}\left(\frac{3}{2} R\right) = 1.50R$
2	1	$\frac{5}{2} R$	R	$\frac{1}{2}\left(\frac{7}{2} R\right) = 1.75R$
3	3	$3R$	$3R$	$\frac{1}{3}(6R) = 2.00R$
4	6	$3R$	$6R$	$\frac{1}{4}(9R) = 2.25R$
5	9	$3R$	$9R$	$\frac{1}{5}(12R) = 2.40R$
12	30	$3R$	$30R$	$\frac{1}{12}(33R) = 2.75R$
N	$3N$–6	$3R$	$(3N\text{–}6)R$	$\frac{1}{N}(3N\text{–}3)R = 3\left(\frac{N\text{–}1}{N}\right)R$‡

* Per mole of molecules.
† Per mole of atoms.
‡ Cf. problem 11, Chapter 21 and problem 14(b) of this Chapter.

11. N = 100 1000 10^6 6×10^{23}

C_{total} = 2.97R 2.997R 2.999997R 3R

In problems 10 and 11, the number of molecules has been adjusted so that there is in the system in each instance the same number of atoms —6×10^{23}—and (hence) the same number of degrees of freedom— $3(6 \times 10^{23})$. With the monatomic gas, all $3(6 \times 10^{23})$ degrees of freedom are translational degrees of freedom and the heat capacity of the system is $(3/2)R$. As these atoms form dimers, trimers, and higher polymers, vibrational degrees of freedom are substituted for translational and rotational degrees of freedom and the heat capacity, in the classical limit, steadily increases. When polymerization is complete, one has a single, gigantic, 6×10^{23}-atomic molecule, perhaps better called a crystal, with 3 translational degrees of freedom, 3 rotational degrees of freedom, $3(6 \times 10^{23}) - 6$ vibrational degrees of freedom, and a heat capacity (in agreement with the law of Dulong and Petit) of $3R$.

The interpretation of the heat capacities of substances in terms of their internal structures was a problem that greatly troubled Willard Gibbs, the chief architect of statistical mechanics. In his introduction to *Elementary Principles in Statistical Mechanics*, Gibbs writes, "... we do not escape difficulties in as simple a matter as the number of degrees of freedom of a diatomic gas. It is well known that while theory would assign to the gas six degrees of freedom per molecule, in our experiments on specific heat we cannot account for more than five."* These difficulties were further compounded by J. J. Thomson's discovery of the electron as a constituent of all matter. Suddenly the surfeit of degrees of freedom became very large. The complications introduced by this fact were such as to cause Gibbs, then late in his life, to suggest that perhaps it was time he passed on.

Meanwhile, Planck was finding that, to explain why the degrees of freedom associated with high frequency oscillations of atoms in solids do not contribute as classical theory says they should to the radiant energy emitted by hot objects, it seemed necessary to suppose that the energy of an oscillator was proportional to its frequency. Einstein showed how this hypothesis could be used to account for departures from the law of Dulong and Petit, particularly at low temperatures.

* This is true for hydrogen because its small mass, and for oxygen and nitrogen because their strong bonds—double and triple bonds, respectively—render the vibrational frequencies of these molecules so great as to make thermal excitation of their vibrational degrees of freedom at room temperature insignificant. On the other hand, all six molecular degrees of freedom are classically excited at room temperature for the heavier, less-strongly bonded diatomic molecules Cl_2, Br_2, and I_2. The full six degrees of freedom of hydrogen, oxygen, and nitrogen are also classically excited at very high temperatures.

Einstein's equation provides also—as will be illustrated in the comments to problem 14—a quantitative account of the contributions of molecular vibrations to the heat capacities of gases.

12. When N and n are large numbers,

$$S = k \ln \frac{(N+n)!}{N!n!}$$
$$\approx k[(N+n)\ln(N+n) - (N+n) - N\ln N + N - n\ln n + n]$$
$$= k\left[N\ln\frac{N+n}{N} + n\ln\frac{N+n}{n}\right].$$

13. In the text, it was found that

$$\frac{N+n}{N} = (1 - e^{-\varepsilon/kT})^{-1}$$

and that
$$\frac{N+n}{n} = e^{\varepsilon/kT}.$$

Substituting from these equations into the entropy formula obtained in the previous problem, and setting $N = N_0$ (this makes n equal to the number of units of thermal energy per mole), one finds that

$$S(\text{per mole of 1-D osc.}) = -R\ln(1 - e^{-\varepsilon/kT}) + kn\left(\frac{\varepsilon}{kT}\right).$$

The last term is equal to

$$\frac{1}{T} \times E\,(\text{per mole of 1-D osc.}).$$

Multiplication by 3 yields the result derived in the text.

14. From the result

$$T_n = \frac{\varepsilon/k}{\ln\dfrac{N+n}{n+1}},$$

one finds that

$$T_{n+1} - T_n = \frac{\varepsilon}{k}\left[\frac{1}{\ln\dfrac{N+n+1}{n+2}} - \frac{1}{\ln\dfrac{N+n}{n+1}}\right]$$

$$= \frac{\varepsilon}{k}\left[\frac{\ln\dfrac{N+n}{n+1} - \ln\dfrac{N+n+1}{n+2}}{\left(\ln\dfrac{N+n+1}{n+2}\right)\left(\ln\dfrac{N+n}{n+1}\right)}\right].$$

The change in energy that corresponds to this change in temperature is

$$E_{n+1} - E_n = (n + 1)\varepsilon - n\varepsilon = \varepsilon.$$

The heat capacity of the N-oscillator system is

$$\frac{E_{n+1} - E_n}{T_{n+1} - T_n}.$$

This times $1/N$ is the heat capacity per oscillator; the later times N_0 is the heat capacity on a molar basis.* Thus,

$$C(\text{per mole}) = \frac{(N_0k)\left(\ln \dfrac{N+n+1}{n+2}\right)\left(\ln \dfrac{N+n}{n+1}\right)}{N\left(\ln \dfrac{N+n}{n+1} - \ln \dfrac{N+n+1}{n+2}\right)}.$$

(a) When $n = 0$,

$$C = \frac{R\left(\ln \dfrac{N}{2}\right)(\ln N)}{N\left(\ln N - \ln \dfrac{N}{2}\right)} = \frac{R\left(\ln \dfrac{N}{2}\right)(\ln N)}{N \ln 2}.$$

Now the logarithm of N—or the square of the logarithm of N—increases much less rapidly than does N itself when N is large.† Thus, as $N \to \infty$ ($n = 0$), $C \to 0$. (When $N = 10^{23}$, for example, $(\log_{10} N)^2/N = (23)^2 \times 10^{-23}$.)

In other words, as the thermal energy of a macroscopic ensemble of oscillators approaches zero, the ensemble's heat capacity approaches zero.

(b) As $n \to \infty$, the ratios $\dfrac{N+n+1}{n+2}$ and $\dfrac{N+n}{n+1}$ approach 1. Because of the form of the denominator in the expression for the heat capacity, the exact rate at which these two ratios approach 1 is important. For this reason, we begin by expressing each ratio in the form $1 + x$; when n is large, x is small. By direct solution of the equations

$$\frac{N+n+1}{n+2} = 1 + x \qquad \text{and} \qquad \frac{N+n}{n+1} = 1 + x,$$

* More accurately, the heat capacity of N_0/N N-oscillator systems.

† $\dfrac{dN}{dN} = 1.\quad \dfrac{d(\ln N)}{dN} = \dfrac{1}{N} \ll 1$ when $N \gg 1$.

one finds that

$$\frac{N + n + 1}{n + 2} = 1 + \frac{N - 1}{n + 2}$$

$$\frac{N + n}{n + 1} = 1 + \frac{N - 1}{n + 1} \cdot$$

Making use of the fact that when x is small, $\ln (1 + x) \approx x$, one finds that when $n \gg N$,

$$C \text{ (per mole)} = \frac{R\left(\dfrac{N - 1}{n + 2}\right)\left(\dfrac{N - 1}{n + 1}\right)}{N\left(\dfrac{N - 1}{n + 1} - \dfrac{N - 1}{n + 2}\right)}$$

$$= R\,\frac{N - 1}{N} \cdot$$

(cf. problem 11, Chapter 21 and problem 10, this Chapter).

When $N \gg 1$, C (per mole) $= R$. At high temperatures the heat capacity of an ensemble of one-dimensional oscillators is 1.987 cal/deg-mole.

Concluding comments (Derivation and applications of Einstein's heat capacity expression).

When the ratio n/N is neither very small (part (a)) nor very large (part (b)), an ensemble of oscillators is neither very cold nor very hot and its heat capacity is neither 0 nor R, but some intermediate value, which can be found as follows.

In the general expression for the heat capacity (problem 14), the term in parentheses in the denominator can be written

$$\ln \frac{(N + n)(n + 2)}{(N + n + 1)(n + 1)} \cdot$$

For large N and n, the argument of the logarithm is approximately 1. Setting

$$\frac{(N + n)(n + 2)}{(N + n + 1)(n + 1)} = 1 + x,$$

one finds that

$$x = \frac{(N - 1)}{(N + n + 1)(n + 1)}$$

$$\approx \frac{N}{n(N + n)} \qquad (N, n \gg 1).$$

Thus, for large N and n,

$$C \text{ (per mole)} = \frac{R\left(\ln \dfrac{N+n}{n}\right)^2}{\dfrac{N^2}{n(N+n)}}$$

$$= R\left(\ln \frac{N+n}{n}\right)^2 \left(\frac{n}{N+n}\right) \left(\frac{N+n}{N}\right)^2$$

$$= R\,(\ln e^{\varepsilon/kT})^2\,(e^{-\varepsilon/kT})(1 - e^{-\varepsilon/kT})^{-2}$$

$$= Rx^2\, e^{-x}(1 - e^{-x})^{-2}\,,$$

where $$x = \frac{\varepsilon}{kT} = \frac{h\nu}{kT}\cdot$$

When $kT \gg h\nu$, x is small, $e^{-x} \approx 1 - x$, and C (per mole) $\approx R$. When $x = 0.48$—its value, approximately, for diamond at 3000°K (problem 6)—C (per mole of 1-D osc.) $= 0.98\, R$.

The vibrational heat capacities of CO_2, H_2O, and N_2 at 298°K are calculated below from their vibrational frequencies and the formula $C_{\text{vib.}} = Rx^2\, e^{-x}(1 - e^{-x})^{-2}$.

$$\begin{aligned} CO_2.\quad \nu &= 2349.3,\ 1388.3,\ 667.3,\ 667.3\ \text{cm}^{-1} \\ C_{\text{vib.}} &= 0.003 + 0.111 + 0.895 + 0.895 \\ &= 1.91\ \text{cal/deg-mole.} \end{aligned}$$

$$\begin{aligned} H_2O.\quad \nu &= 3755.8,\ 3651.7,\ 1595.0\ \text{cm}^{-1} \\ C_{\text{vib.}} &= 0.000 + 0.000 + 0.053 \\ &= 0.053\ \text{cal/deg-mole} \end{aligned}$$

$$\begin{aligned} N_2.\quad \nu &= 2359.6\ \text{cm}^{-1} \\ C_{\text{vib.}} &= 0.003\ \text{cal/deg-mole.} \end{aligned}$$

These numbers should be compared with those in the last column of the table in the discussion of problem 9.

Summary of thermodynamic functions for one-dimensional oscillators.

N = number of loosely coupled one-dimensional oscillators

n = number of units of thermal energy

ν = oscillator frequency

ε = spacing between allowed energy levels $= h\nu$

$$x = \frac{h\nu}{kT} = 1.4385\,\frac{\nu}{T} \qquad (\nu \text{ in cm}^{-1})$$

$$\Omega = \frac{(N + n - 1)!}{(N - 1)!n!}$$

$$\frac{N + n}{N} = (1 - e^{-x})^{-1} \qquad \frac{N + n}{n} = e^{x} \qquad n = N(e^{x} - 1)^{-1}$$

In the equations below the values for E, S, and C_V are per mole of one-dimensional oscillators.

$$E = n\varepsilon = N_0 h\nu(e^{x} - 1)^{-1} = RTx(e^{x} - 1)^{-1}$$

$$= RT \qquad \text{Large } T \text{ (small } x)$$

$$= RT\,xe^{-x} \qquad \text{Small } T \text{ (large } x)$$

$$S = -R\ln(1 - e^{-x}) + R\,x(e^{x} - 1)^{-1}$$

$$= -R\ln x + R \qquad \text{Large } T \text{ (small } x)$$

$$= R(1 + x)e^{-x} \qquad \text{Small } T \text{ (large } x)$$

$$C_V = \frac{R\left(\ln\dfrac{N + n + 1}{n + 2}\right)\left(\ln\dfrac{N + n}{n + 1}\right)}{N\ln\dfrac{(N + n)(n + 2)}{(N + n + 1)(n + 1)}}$$

$$= 0 \qquad n = 0,\ N \text{ large}$$

$$= R\,\frac{N - 1}{N} \qquad n \gg N$$

$$= R\,x^2e^{-x}(1 - e^{-x})^{-2} \qquad n \text{ and } N \text{ large.}$$

Chapter 23

1. As T approaches zero n_i/n_j approaches 0. As T approaches infinity n_i/n_j approaches 1.

2. $n_i = n_0 e^{-\frac{(\varepsilon_i - \varepsilon_0)}{kT}}$

$$n_0 = \frac{N}{\sum_{i=0}^{\infty} e^{-\frac{(\varepsilon_i - \varepsilon_0)}{kT}}}$$

As used in this chapter—particularly in the expressions for n_i (in terms of n_0), n_0 and Q—ε_i stands for the energy of the ith energy level with respect to the energy of the ground level.

3. 1.
4. As T approaches zero Q approaches 1. As T approaches infinity Q approaches ∞.
5. As T approaches zero and infinity, n_0 approaches N and 0, respectively.
6. As T approaches zero or infinity, n_i approaches 0 ($i > 0$).
7. Diminishing the spacing between energy levels, T constant, increases Q and decreases n_0.
8. Doubling N, T constant, doubles n_i, $i = 0, 1, 2, 3, \ldots$. Q is not affected.

9. $\Delta S_\sigma = k \ln \dfrac{n_1}{n_2}$ $\qquad (n_2 \gg 1)$

$$\Delta S_\theta = -\frac{(\varepsilon_2 - \varepsilon_1)}{T_\theta}$$

$\Delta S_\sigma > 0$ (if $n_1 > n_2$). $\Delta S_\theta < 0$. If $T_\theta = T_\sigma = T$, $\Delta S_{\text{univ.}} \geq 0$ if and only if

$$k \ln \frac{n_1}{n_2} - \frac{(\varepsilon_2 - \varepsilon_1)}{T} \geq 0.$$

In other words, thermal excitation of particles from the first to the second excited states can occur if and only if

$$\frac{n_2}{n_1} \leq e^{-\frac{(\varepsilon_2 - \varepsilon_1)}{kT}}.$$

10. $S = k \ln \dfrac{N!}{n_0! n_1! n_2! \ldots}$

$$= k[\ln N! - \ln n_0! - \ln n_1! - \ln n_2! - \cdots]$$

$$= k[(N \ln N - N) - (n_0 \ln n_0 - n_0) - (n_1 \ln n_1 - n_1) - (n_2 \ln n_2 - n_2) - \cdots].$$

The entropy formula cited in the problem follows from this expression and the relation $N = n_0 + n_1 + n_2 + \cdots$.

11. Substituting from the equations

$$\frac{n_0}{N} = Q^{-1}$$

$$\frac{n_1}{N} = Q^{-1}e^{-\varepsilon_1/kT}\,; \qquad \ln\frac{n_1}{N} = -\frac{\varepsilon_1}{kT} - \ln Q$$

$$\frac{n_2}{N} = Q^{-1}e^{-\varepsilon_2/kT}\,; \qquad \ln\frac{n_2}{N} = -\frac{\varepsilon_2}{kT} - \ln Q$$

$$\cdot$$
$$\cdot$$
$$\cdot$$

into the expression given in problem 10, one finds that

$$S = -k[-n_0 \ln Q - \frac{n_1\varepsilon_1}{kT} - n_1 \ln Q - \frac{n_2\varepsilon_2}{kT} - n_2 \ln Q - \cdots]$$

$$= k(n_0 + n_1 + n_2 + \cdots)\ln Q + \frac{(n_1\varepsilon_1 + n_2\varepsilon_2 + \cdots)}{T}$$

$$= R \ln Q + \frac{E\,(\text{per mole})}{T} \qquad \left(\text{when} \sum_{i=0}^{\infty} n_i = N_0\right).$$

12. Let $x = h\nu/kT$.

$$Q = 1 + e^{-x} + e^{-2x} + e^{-3x} + \cdots$$
$$e^{-x}Q = \qquad e^{-x} + e^{-2x} + e^{-3x} + \cdots$$

Subtracting and solving for Q, one finds that

$$Q = \frac{1}{1 - e^{-x}} = \frac{1}{1 - e^{-h\nu/kT}}.$$

When $T = 0$, $x = \infty$, and $Q = 1$. When $kT \gg h\nu$, $x \ll 1$, and $Q = 1/x = kT/h\nu$.

13. With $\varepsilon_0 = 0$, and $x = h\nu/kT$,

$$E = n_1\varepsilon_1 + n_2\varepsilon_2 + n_3\varepsilon_3 + \cdots$$

$$= \frac{N}{Q}e^{-x}h\nu + \frac{N}{Q}e^{-2x}2h\nu + \frac{N}{Q}e^{-3x}3h\nu + \cdots$$

$$= \frac{N}{Q}e^{-x}h\nu(1 + 2e^{-x} + 3e^{-2x} + \cdots)$$

$$e^{-x}E = \frac{N}{Q}\,e^{-x}h\nu(e^{-x} + 2e^{-2x} + \cdots)$$

$$E(1 - e^{-x}) = \frac{Nh\nu e^{-x}}{Q}(1 + e^{-x} + e^{-2x} + \cdots)$$

$$= \frac{Nh\nu e^{-x}}{Q}\cdot Q$$

$$E = \frac{Nh\nu e^{-x}}{1 - e^{-x}} = \frac{Nh\nu}{e^{x} - 1} = \frac{Nh\nu}{e^{h\nu/kT} - 1}\,.$$

14. $n_1/N = e^{-\varepsilon_1}/Q = (1 - e^{-h\nu/kT})\,e^{-h\nu/kT}$. At 300°K, $\dfrac{h\nu}{kT} = 1.438 \times \dfrac{667.3}{300}$ $= 3.20$, $e^{-h\nu/kT} = 0.041$, and $(n_1/N)100 = 3.9\,\%$. At 1000°K, $(n_1/N)100$ $= 23.6\,\%$.

15. If $\dfrac{n_1}{n_0} = e^{-h\nu/kT} = \dfrac{1}{e}$, $\dfrac{h\nu}{kT} = \dfrac{1.438\nu\ (\text{cm}^{-1})}{T\ (°\text{K})} = -\ln\dfrac{1}{e} = 1$, and $T(°\text{K}) =$ $1.438\ (667.3) = 958$.

16. An inverted population, i.e. one in which $n_1 > n_0$, $n_2 > n_1$, etc.

Chapter 24

1. $$v_e = \frac{h}{m_e\lambda_e} = \left(7.27\,\frac{\text{cm}^2}{\text{sec}}\right)\frac{1}{\lambda_e}$$

(a) $$= \left(7.27\,\frac{\text{cm}^2}{\text{sec}}\right)\frac{1}{2\times10^{-5}\ \text{cm}} = 3.63\times10^{5}\,\frac{\text{cm}}{\text{sec}}$$

(b) $$= \left(7.27\,\frac{\text{cm}^2}{\text{sec}}\right)\frac{1}{2\pi(0.529\times10^{-8}\ \text{cm})} = 2.19\times10^{8}\,\frac{\text{cm}}{\text{sec}}$$

(c) $$(\text{K.E.})_{\substack{\text{H 1}s\\ \text{electron}}} = \frac{1}{2}(9.11\times10^{-28}\ \text{gm})\left(2.19\times10^{8}\,\frac{\text{cm}}{\text{sec}}\right)^2$$

$$= 2.18\times10^{-11}\,\frac{\text{gm cm}^2}{\text{sec}^2} = 13.6\ \text{e.v.} = 314\,\frac{\text{kcal}}{\text{mole}}\,.$$

2. $\lambda\cdot\dfrac{h\nu}{c} = \dfrac{h}{c}(\lambda\nu) = \dfrac{h}{c}(c) = h.$

An amusing account of Heisenberg's uncertainty principle is given by Banesh Hoffmann in *The Strange Story of the Quantum*, New York: Dover Publications, Inc., 1959.

3. The length of the "orbit" is $2l$. Therefore, $n\lambda = 2l$, $\lambda = \frac{2l}{n}$, and

$$\text{K.E.} = \frac{1}{2}mv^2 = \frac{1}{2}m\left(\frac{h}{m\lambda}\right)^2 = \frac{n^2h^2}{8ml^2}.$$

In the lowest state, $n = 1$ ($n = 0$ corresponds to no particle at all). When $m = 1$ amu (1.66×10^{-24} gm),

$$(\text{K.E.})_{n=1} = \frac{(1)^2(6.62 \times 10^{-27})^2}{8(1.66 \times 10^{-24})l^2_{\text{cm}}} \frac{\text{gm cm}^2}{\text{sec}^2} \Big/ \text{particle}$$

$$= \frac{3.30 \times 10^{-30}}{l_{\text{cm}}{}^2} \frac{\text{ergs}}{\text{particle}} = \frac{4.75 \times 10^{-14}}{l_{\text{cm}}{}^2} \frac{\text{cal}}{\text{mole}}$$

$$= 4.75 \frac{\text{cal}}{\text{mole}} \times \text{(a) } 10^{-16}, \quad \text{(b) } 10^{0}, \quad \text{(c) } 10^{10}.$$

$(3/2)RT$, the average translational kinetic energy per mole of gas molecules, is equal to these values when T is (a) 1.6×10^{-16} °K, (b) 1.6°K, (c) 1.6×10^{10}°K (cf. Table 24.1).

For a particle in a conventional three-dimensional box,

$$\text{K.E.} = \frac{n_x{}^2h^2}{8ml_x{}^2} + \frac{n_y{}^2h^2}{8ml_y{}^2} + \frac{n_z{}^2h^2}{8ml_y{}^2}.$$

If $l_x = l_y = l_z = l$, where $l^3 = V$ (the volume of the box),

$$\text{K.E.} = \frac{h^2}{8mV^{2/3}}(n_x{}^2 + n_y{}^2 + n_z{}^2).$$

4. $n\lambda = 2\pi r$, $\lambda = 2\pi r/n$. Hence, K.E. $= n^2h^2/8\pi^2mr^2$. The product $m \cdot r^2$ is the system's moment of inertia, I. The newer quantum theory replaces n^2 by $n(n + 1)$. Thus, the energy levels of a rigid rotor may be written $\frac{n(n+1)h^2}{8\pi^2 I}$. From spectroscopic observations of molecular rotational energy levels, the several moments of inertia of a molecule (and the isotopically substituted molecule) can be determined and from these (and the nuclear masses), the geometry of the molecule—its bond lengths and bond angles—can often be calculated.

5. Let L = length of the orbit ($2l$ for a particle in a box of length l; $2\pi r$ for a rigid rotor). Then

$$\nu = \frac{v}{L} = \frac{v}{n\lambda} = \frac{v}{n(h/mv)} = \frac{mv^2}{nh} = \frac{2\text{ K.E.}}{nh};$$

hence,
$$\text{K.E.} = (1/2)nh\nu.$$

For particles in boxes and for rigid rotors, ν is not a constant of motion. As the quantum number n increases, the kinetic energy increases, v increases, and, since L remains constant, ν increases.

For an harmonic oscillator, K.E. = P.E. and E_{total} = K.E. + P.E. = 2 K.E. = $nh\nu$. In this case, L increases as n increases and ν (it turns out) is a constant of motion. The newer quantum theory replaces n in the harmonic oscillator expression by $(n + 1/2)$.

6. $$\lambda = \frac{c}{\nu} = \frac{\text{c}mL^2}{h}$$

$$= (7.5 \times 10^{12}) \cdot m_{\text{amu}} \cdot L_{\text{cm}}{}^2 \qquad (\lambda \text{ in cm})$$

Type of Motion	$\lambda = \frac{\text{c}mL^2}{h}$ (cm)	*Spectral Region*
Translation	3.75×10^{16}	Beyond ultra-long radio waves
Rotation	7.5×10^{-1}	Microwave (0.75 cm)
Vibration	6.7×10^{-4}	Infrared (1500 cm^{-1})
Electronic	4.1×10^{-5}	Visible (4100 Å)
Nuclear	7.5×10^{-12}	Gamma rays

7. Yes. This example illustrates the fact that dimensionally correct equations are not necessarily physically correct. Physically correct equations, however, are always dimensionally correct.
8. The dimensions of momentum are (mass)(length)(time)$^{-1}$; the dimensions of $h\nu/c$ are (mass)(length)^2(time)$^{-1}$ · (time)$^{-1}$/(length)(time)$^{-1}$ = dimensions of momentum. Energy has the dimensions (mass)(length)^2 (time)$^{-2}$ (in the cgs system, for example, energy is expressed in gm cm^2 sec^{-2}); mc^2 has the dimensions (mass)[(length)(time)$^{-1}$]^2 = dimensions of energy.

9. $[\nu] = T^{-1}$. $[v] = [c] = LT^{-1}$. $[p] = [m \cdot v] = MLT^{-1}$. $[a] = LT^{-2}$. $[A] = L^2$. $[V] = L^3$.

10.
$$[F] = [m \cdot a] = MLT^{-2}$$
$$[Q] = [(F \cdot d^2)^{1/2}] = M^{1/2}L^{3/2}T^{-1}$$
$$[E] = [\text{K.E.}] = \left[\frac{1}{2}mv^2\right] = ML^2T^{-2}$$
$$= [\text{P.E.}] = [mgh] = \text{,,}$$
$$[W] = [F \cdot d] = \text{,,}$$
$$[P] = [F/A] = ML^{-1}T^{-2}$$
$$[S] = [\delta S] = \left[\frac{\delta E}{T}\right] = ML^2T^{-2}\theta^{-1}$$
$$[R] = \left[\frac{PV}{nT}\right] = ML^2T^{-2}\theta^{-1}\,\text{mole}^{-1}$$

11. The only combination of the dimensional formula of h, ML^2T^{-1}, and of ν, T^{-1}, that yields the dimensional formula of ε, ML^2T^{-2}, is $ML^2T^{-1} \cdot T^{-1}$. Hence, $\varepsilon \propto h\nu$. Alternatively, this relation may be used to obtain the dimensional formula of h. The constant of proportionality is unity.

12. To introduce properly into the problem the units of time, which occurs in the dimensional formula of energy as T^{-2}, in the dimensional formula of h as T^{-1}, and in the dimensional formulas of m and l as T^0, Planck's constant must be squared; the dimensions of mass and length will then be correct if one divides h^2 by m and l^2:

$$\frac{[h^2]}{[m][l^2]} = \frac{M^2L^4T^{-2}}{ML^2} = ML^2T^{-2} = [\varepsilon].$$

Hence,
$$\varepsilon \propto \frac{h^2}{ml^2}.$$

The constant of proportionality is 1/8 (problem 3).

13. To introduce properly the units of time, which occurs in the dimensional formula of force as T^{-2}, in the dimensional formula of velocity as T^{-1}, and in the dimensional formulas of m and r as T^0, v must be squared; the dimensions of mass and length will then be correct if one multiplies v^2 by m and divides by r:

$$\frac{[v^2][m]}{[r]} = MLT^{-2} = [F].$$

Hence,
$$F \propto \frac{mv^2}{r}.$$

The constant of proportionality is unity.

14. (a) Assume that $r \propto e^x m_e^y h^z$. For dimensional homogeneity,

$$[r] = [e]^x [m_e]^y [h]^z$$

$$M^0 L T^0 = M^{(1/2)x+y+z} L^{(3/2)x+2z} T^{-x-z}$$

$$\left.\begin{aligned} M:&\quad 0 = \frac{1}{2}x + y + z \\ L:&\quad 1 = \frac{3}{2}x \qquad + 2z \\ T:&\quad 0 = -x \qquad - z \end{aligned}\right\} \Rightarrow \begin{aligned} x &= -2 \\ y &= -1 \\ z &= 2 \end{aligned}$$

$$\therefore \qquad r \propto \frac{h^2}{m_e e^2}.$$

(b) Assume that $E \propto e^x m_e^y h^z$. The following equations are then obtained.

$$\left.\begin{aligned} M:&\quad 1 = \frac{1}{2}x + y + z \\ L:&\quad 2 = \frac{3}{2}x \qquad + 2z \\ T:&\quad -2 = -x \qquad - z \end{aligned}\right\} \Rightarrow \begin{aligned} x &= 4 \\ y &= 1 \\ z &= -2 \end{aligned}$$

$$\therefore \qquad E \propto \frac{m_e e^4}{h^2}.$$

15. v decreases. When $r = \infty$, $v = 0$.

16. v increases.

17. (a) P.E. $\to 0$, (b) P.E. $\to -\infty$.

18. For a classically stable orbit, $\dfrac{mv^2}{r} = \dfrac{Ze^2}{r^2}$. Therefore,

$$\text{K.E.}\left(= \frac{1}{2}mv^2\right) = \frac{1}{2}\frac{Ze^2}{r} = \frac{1}{2}(-\text{P.E.}) = -\frac{1}{2}(\text{P.E.}).$$

19. $E = \text{K.E.} + \text{P.E.}$

$$= \text{K.E.} + (-2\,\text{K.E.}) = -\text{K.E.} \qquad \text{(a)}$$

$$= -\frac{1}{2}\text{P.E.} + \text{P.E.} \;= \frac{1}{2}\text{P.E.} \qquad \text{(b)}$$

20. $\text{P.E.} = -\dfrac{Ze^2}{r} = -\dfrac{(4.803 \times 10^{-10})^2}{(0.529 \times 10^{-8})} \dfrac{\text{gm cm}^2}{\text{sec}^2} \Big/ \text{molecule}$

$$= -4.36 \times 10^{-11} \frac{\text{ergs}}{\text{molecule}} = -628 \frac{\text{kcal}}{\text{mole}}.$$

$$E = \frac{1}{2}\text{P.E.} = -314 \frac{\text{kcal}}{\text{mole}}. \qquad \text{(cf. problem 1(c))}$$

21. r may have any value greater than zero provided the velocity is chosen to make $mv^2/r = Ze^2/r^2$.

22. From Bohr's condition,

$$r = \frac{nh}{2\pi m_e v}.$$

Substituting this value into the equation $m_e v^2 = Ze^2/r$, one finds that

$$m_e v^2 = \frac{2\pi Z e^2 m_e v}{nh}.$$

Hence,
$$v = \frac{2\pi Z e^2}{nh}.$$

Therefore,
$$r = \frac{n^2 h^2}{4\pi^2 Z m_e e^2}.$$

$$E = -\frac{mv^2}{2} \left(\text{or } -\frac{1}{2}\frac{Ze^2}{r}\right)$$

$$= -\frac{2\pi^2 Z^2 m_e e^4}{n^2 h^2}.$$

23. As Z increases (n constant), v increases, r decreases, and E decreases (becomes more negative). As n increases (Z constant), v decreases, r increases, and E increases (becomes less negative). $n = \infty$ represents an ionized atom.

24. $v_{\text{H } 1s} = \dfrac{2\pi e^2}{h} = \dfrac{2\pi(4.803 \times 10^{-10})^2}{(6.62 \times 10^{-27})} \dfrac{\text{cm}}{\text{sec}}$

$$= 2.19 \times 10^8 \frac{\text{cm}}{\text{sec}}. \qquad \text{(cf. problem 1(b))}$$

$$r_{\text{H } 1s} = \frac{h^2}{4\pi^2 m_e e^2} = 0.529 \text{ Å}.$$

25. A photon is emitted whose frequency is such that

$$h\nu = E_{n=2} - E_{n=1}.$$

26. $$\frac{1}{\lambda} = \frac{\nu}{c} = \frac{E_{n=2} - E_{n=1}}{hc}$$

$$= -\left(\frac{2\pi^2 Z^2 m_e e^4}{h^3 c}\right)\left(\frac{1}{2^2} - \frac{1}{1^2}\right)$$

$$= -(109{,}737 \text{ cm}^{-1})^* \left(-\frac{3}{4}\right) = 82{,}300 \text{ cm}^{-1}$$

$$\lambda = 1.215 \times 10^{-5} \text{ cm} = 1215 \text{ Å}.$$

SUMMARY OF FORMULAS

$\lambda = h/mv$	de Broglie's relation
$\Delta x \cdot \Delta p \geq h/4\pi$	Heisenberg's uncertainty relation
$n\lambda = L$ (L = orbit length)	Bohr's quantum condition

For photons

$$\lambda\nu = c = 3 \times 10^{10} \text{ cm/sec (in a physical vacuum)}$$

$$h\nu = \text{energy/photon}$$

$$h\nu/c = \text{momentum/photon}$$

Energy levels for particles

$n^2h^2/8ml^2$	Particle in a box
$n(n+1)h^2/8\pi^2 I$	Rigid rotor
$\left(n + \frac{1}{2}\right) h\nu$	Harmonic oscillator

Virial Theorem

$$\text{K.E.} = -\frac{1}{2}\text{P.E.}$$

$$E = -\text{K.E.} = \frac{1}{2}\text{P.E.}$$

* This is the value of the so-called "Rydberg constant" for a nucleus of infinite mass. The correct Rydberg constant for hydrogen is 109,677.58 cm^{-1}.

Bohr atom

$$v = 2\pi Ze^2/nh = 2.19 \times 10^8 \left(\frac{Z}{n}\right) \text{cm/sec}$$

$$r = n^2h^2/4\pi^2 Zm_e e^2 = 0.529 \left(\frac{n^2}{Z}\right) \text{Å} \qquad (*)$$

$$E = -2\pi^2 Z^2 m_e e^4/n^2h^2$$

$$\frac{1}{\lambda} = 109{,}678 \left(\frac{1}{n_1^2} - \frac{1}{n_2^2}\right) \text{cm}^{-1} \qquad (\text{for H}^1)$$

POSTSCRIPT TO CHAPTER 24

Broglie's relation is a simple and powerful way to introduce Planck's constant into small-particle problems when the potential energy over the particle's path is constant. This is the case for particles in boxes, rigid rotors, and circular Bohr orbits. Since the total energy—the potential energy plus the kinetic energy—is a constant of motion, a constant potential energy implies a constant kinetic energy and, therefore, a constant momentum; hence, by de Broglie's relation, λ is constant. Under these conditions fitting wave lengths into the available space is a relatively simple problem. The problem becomes more complex when the potential energy is variable—as it is in the problem of the harmonic oscillator, for example. For these problems Schrödinger's generalization of de Broglie's relation is useful. For the hydrogen atom, Schrödinger's equation gives a description that is more satisfactory than that given by Bohr's early theory, which improperly ascribes electronic orbital angular momentum to the ground state.

Chapter 26

(a) positive
(b) First
(c) negative
(d) $e^{\Delta S^0{}_\sigma/R} \cdot e^{-\Delta H_\sigma/RT}$
(e) positive
(f) decreases
(g) decreases
(h) Principle of Le Chatelier

The figure 0.529 Å is called the Bohr radius. It is the value of r when $n = Z = 1$.

Chapter 27

(a) greater
(b) positive
(c) greater
(d) positive
(e) $R \ln \dfrac{K_{eq}(T_2)}{K_{eq}(T_1)} \Big/ \left(\dfrac{1}{T_1} - \dfrac{1}{T_2}\right)$
(f) positive
(g) endothermic
(h) Le Chatelier's Principle

Chapter 28

1. $\mathsf{G} \equiv \mathsf{H} - T\mathsf{S}$. Hence, with $\mathsf{S} = \mathsf{S}^0 - R \ln a$, $\mathsf{G} = (\mathsf{H} - T\mathsf{S}^0) + RT \ln a$, where the quantity within the parentheses is a function of the temperature, but not the concentration.

   For further discussion of activities, activity coefficients, and standard states, the reader is referred to the later volumes on chemical thermodynamics in the Bibliography.
2. Substitute from the last expression into the one above it, noting that $x \ln y = \ln y^x$ and that $(\ln x) \pm (\ln y) = \ln xy^{\pm 1}$.

   For other problems similar to the examples discussed in this chapter, see problem 72 of the Additional Problems.

Chapters 29, 30, and 31

1. The reaction H_2(gas) + (1/2)O_2(gas) = H_2O(liq) decreases the entropy of the chemical system. To compensate for this, part of $-\Delta H_\sigma$ must be diverted to the thermal surroundings; in this case $W_{\max} < -\Delta H_\sigma$. No energy need be diverted to the thermal surroundings, however, if the reaction increases the entropy of the chemical system. When $\Delta S_\sigma > 0$, all of $-\Delta H_\sigma$ can be converted to work; indeed, without violating the Second Law $-\Delta H_\sigma$ could be augmented by an additional $T\Delta S_\sigma$ calories extracted from the thermal surroundings; i.e.

$$W_{\max} (= -\Delta G_\sigma = -\Delta H_\sigma + T\Delta S_\sigma) > -\Delta H_\sigma \qquad \text{when } \Delta S_\sigma > 0.$$

2. (a) H_2 - H^+ (0.1 molar) is the better electron donor. H_2 - H^+ (1.0 molar) is the better electron acceptor.

(b) The half-cell reactions are

$$1/2\ H_2(1\text{ atm.}) = H^+(0.1\text{ molar}) + e$$

$$H^+(1.0\text{ molar}) + e = 1/2\ H_2(1\text{ atm.}).$$

(c) The net cell reaction is

$$H^+(1.0\text{ molar}) = H^+(0.1\text{ molar}).$$

(d)
$$\mathscr{E}_{\text{max}} = \mathscr{E}^0 - \frac{0.0592}{n} \log \frac{a_{H^+(0.1\text{ molar})}}{a_{H^+(1.0\text{ molar})}}$$

where
$$\mathscr{E}^0 \equiv -\frac{\Delta G^0}{n\mathscr{F}}$$

$$\begin{aligned}\Delta G^0 &= G^0_{\text{product}} - G^0_{\text{reactant}}\\ &= G^0_{H^+(aq)} - G^0_{H^+(aq)}\\ &= 0.000\end{aligned}$$

and where
$$a_{H^+(0.1\text{ molar})} \approx 0.1$$

$$a_{H^+(1.0\text{ molar})} \approx 1.0$$

Therefore,
$$\begin{aligned}\mathscr{E}_{\text{max}} &\approx 0.000 - \frac{0.0592}{1} \log \frac{0.1}{1.0}\\ &= 0.0592\text{ volts.}\end{aligned}$$

(e) The driving force behind the reaction is the spontaneous diffusion of hydrogen ions from a high concentration (1.0 molar) to a lower concentration (0.1 molar). The configurational entropy of the chemical system is thereby increased by the amount $R \ln (1/0.1) =$ 4.6 cal/deg per mole of hydrogen ion transferred. The thermal entropy and enthalpy of the chemical system suffer little change. For this reaction, therefore,

$$\Delta G_\sigma \approx -T\Delta S_\sigma$$

where
$$\begin{aligned}\Delta S_\sigma &= S_{\text{product}} - S_{\text{reactant}}\\ &= S_{H^+(0.1\text{ molar})} - S_{H^+(1.0\text{ molar})}\\ &= 4.6\text{ cal/deg.}\end{aligned}$$

Thus,
$$\begin{aligned}\mathscr{E}_{\text{max}} \left(= -\frac{\Delta G_\sigma}{n\mathscr{F}} \approx \frac{T\Delta S_\sigma}{n\mathscr{F}}\right) &= \frac{(298°\text{K})(4.6\text{ cal/deg})}{1 \times 96{,}500\text{ coul.}}\\ &= 0.0592\text{ volts,}\end{aligned}$$

in agreement with the calculation in part (d).

(f) The source of the electrical energy is the thermal surroundings. By the First Law, $W + \Delta E_\theta + \Delta H_\sigma = 0$. If $\Delta H_\sigma = 0$, $W = -\Delta E_\theta$. The Second Law will not be violated if W is such that

$$\Delta S_{\text{total}} = -\frac{W}{T} + \Delta S_\sigma \geq 0.$$

Cells whose driving force is the spontaneous diffusion of a chemical from a high to a lower concentration are called concentration cells. They produce in the universe energy and entropy changes analogous to those produced in the universe during the reversible, isothermal expansion of an ideal gas. When a mole of hydrogen gas expands reversibly from 1.0 atm. to 0.1 atm., for example, $RT \ln (1.0/0.1)$ calories of mechanical work are produced at the expense of the thermal energy of the surroundings. This work could also be extracted electrochemically by means of the two half-cell reactions

$$H_2(1.0 \text{ atm.}) = 2H^+(1 \text{ molar}) + 2e$$

$$2H^+(1 \text{ molar}) + 2e = H_2(0.1 \text{ atm.})$$

for which the net cell reaction is

$$H_2(1.0 \text{ atm.}) = H_2(0.1 \text{ atm.})$$

and for which $W_{\max} (= -\Delta G_\sigma) = RT \ln (1.0/0.1)$ and

$$\mathscr{E}_{\max} (= W_{\max}/n\mathscr{F}) = 0.0296 \text{ volts.}$$

3. $Q = e^{-\Delta G^0/RT} = K_{eq}$.

4. $\mathscr{E}_{\max}$ can be increased by increasing the activities of H_2 and Cu^{++} and by decreasing the activities of H^+ and Cu (the latter through alloy formation with silver and gold, for example). $\mathscr{E}_{\max} = 0$ whenever the activities a_{H_2}, $a_{Cu^{++}}$, a_{H^+}, and a_{Cu} are such that

$$\frac{a_{H^+}^2 a_{Cu}}{a_{H_2} a_{Cu^{++}}} = e^{-\Delta G^0{}_\sigma/RT} = 10^{+11.4} \qquad (T = 298°\text{K})$$

This condition can be satisfied by making the activity of the cupric ion small—by the addition, for example, of hydrogen sulfide or excess ammonia.

5. The half-cell reactions are

$$\begin{array}{rcll} Ag^+(aq) + e &=& Ag(c) & \\ \tfrac{1}{2}\, H_2(gas) &=& H^+(aq) + e & \\ \hline Ag^+(aq) + \tfrac{1}{2}\, H_2(gas) &=& Ag(c) + H^+(aq) & \text{Net cell rx.} \end{array}$$

$$\mathscr{E}^0 \left(= -\frac{\Delta G^0}{n\mathscr{F}}\right) = \frac{18{,}430 \text{ cal}}{1 \times 96{,}500 \text{ coul.}} = 0.799 \text{ volts.} \qquad (*)$$

Electrons flow spontaneously from the hydrogen electrode to the silver electrode. The standard H_2 - H^+ couple is a better electron donor, by 0.799 volts, than the standard Ag - Ag^+ couple.

6. The silver ion activity about the silver electrode will be reduced from its previous value of 1 (in the calculation of $\mathscr{E}^0$) to its value in a molar solution of HCl saturated with AgCl. The latter value can be calculated from the solubility product of silver chloride, which in turn can be calculated from the standard free energies of Ag^+, Cl^-, and AgCl. For the reaction

$$AgCl(c) = Ag^+(aq) + Cl^-(aq)$$

$$K_{eq}\,(= K_{sp}) = (Ag^+)(Cl^-) = 10^{\frac{-[(18{,}430-31{,}350)-(-26{,}224)]}{1{,}364}}$$
$$= 10^{-9.75}\,.$$

Thus, at the silver electrode

$$(Ag^+) = \frac{K_{sp}}{(Cl^-)} = 10^{-9.75}\,.$$

At this low silver ion concentrate the silver-silver ion couple is a relatively poor electron acceptor, as shown by the fact that $\mathscr{E}_{max}$ for $a_{H_2} = a_{Ag} = a_{H^+} = 1$ drops to the value

$$\mathscr{E}^0 - \frac{0.0592}{n} \log \frac{a_{Ag}a_{H^+}}{a_{Ag^+}a_{H_2}^{1/2}} = 0.799 - \frac{0.0592}{1} \log \frac{1 \cdot 1}{10^{-9.75} \cdot 1}$$
$$= 0.222 \text{ volts.}$$

Alternatively, in recognition of the fact that the ultimate source of the silver ions discharged at the silver electrode must over the long run be the $+1$ silver in the silver chloride, since there is little $+1$ silver in solution—only $10^{-9.75}$ moles/liter—one may write for the net half-cell reaction at the silver electrode the equation

$$AgCl(c) + e = Ag(c) + Cl^-(aq).$$

* If $\mathscr{E}^0$ calculates out to be a negative number, the half-cell reactions should be reversed.

Combined with the half-cell reaction at the hydrogen electrode, this yields for the net cell reaction the equation

$$AgCl(c) + \tfrac{1}{2}\,H_2(gas) = Ag(c) + Cl^-(aq) + H^+(aq)$$

for which when

$$a_{AgCl} = a_{H_2} = a_{Ag} = a_{Cl^-} = a_{H^+} = 1$$

$$\mathscr{E}_{max} = \mathscr{E}^0 = -\frac{\Delta G^0}{n\mathscr{F}} = 0.222 \text{ volts.}$$

Conversely, if the voltages 0.222 and 0.799 for the two cells Ag(c), AgCl(c) | HCl(1 molar) | H_2(gas, 1 atm.), (Pt) and Ag(c) | $AgNO_3$(1 molar) || KCl salt bridge || HCl(1 molar) | H_2(gas, 1 atm.), (Pt) are known, the concentration of Ag^+ in equilibrium with AgCl and 1 molar Cl^- can be calculated; i.e. from the voltages of these two cells can be calculated the solubility product of silver chloride.

7. (a) Cu - Cu^+ - Cu^{++} (b) Hg - Hg_2^{++} - Hg^{++} (c) Fe - Fe^{++} - Fe^{+++}

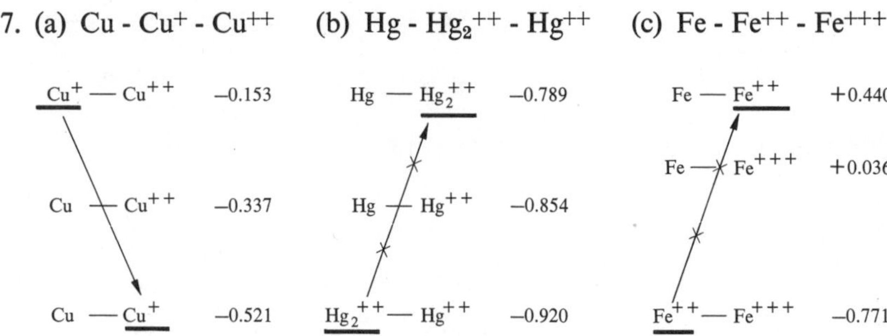

Because copper loses its second electron more easily than the first, Cu^+ can (and does) disproportionate; Hg_2^{++} and Fe^{++} do not.

Iron's three voltages (also those of copper and mercury) are interrelated in a simple manner. Since free energy changes are additive, the sum of the free energy changes when iron loses two electrons and then a third is, as indicated below, the free energy change for the loss of three electrons.

$$\begin{array}{ll} Fe^0 = Fe^{+2} + 2e & \Delta G_{02} \\ Fe^{+2} = Fe^{+3} + e & \Delta G_{23} \\ \hline Fe^0 = Fe^{+3} + 3e & \Delta G_{03} = \Delta G_{02} + \Delta G_{23} \end{array}$$

Introducing the relation $\Delta G = -n\mathscr{F}\mathscr{E}$, one finds that

$$-3 \cdot \mathscr{F}\mathscr{E}_{03} = -2 \cdot \mathscr{F}\mathscr{E}_{02} - 1 \cdot \mathscr{F}\mathscr{E}_{23}\,.$$

In other words,

$$\mathscr{E}_{03} = \frac{2 \cdot \mathscr{E}_{02} + 1 \cdot \mathscr{E}_{23}}{3}.$$

The potential of the Fe - Fe^{+++} couple is the weighted average of the potentials of the couples Fe - Fe^{++} and Fe^{++} - Fe^{+++}.

Very useful compilations and discussions of thermodynamic data and half-cell potentials are contained in the book *Oxidation Potentials* by Wendell M. Latimer (Prentice-Hall, Inc., 2nd ed., 1952).

8. From this chapter, $\mathscr{E} = -\Delta G_\sigma / n\mathscr{F}$. Hence,

$$\frac{d\mathscr{E}}{dT} = -\frac{1}{n\mathscr{F}} \frac{d(\Delta G_\sigma)}{dT}$$

$$= -\frac{1}{n\mathscr{F}} \left(\frac{dG_{\text{products}}}{dT} - \frac{dG_{\text{reactants}}}{dT} \right).$$

From Chapter 19 (and Chapter 33),

$$\frac{dG_x}{dT} = -S_x \qquad \text{(Pressure and composition constant).}$$

Therefore,

$$\frac{d\mathscr{E}}{dT} = -\frac{1}{n\mathscr{F}} [(-S_{\text{products}}) - (-S_{\text{reactants}})]$$

$$= \frac{\Delta S_\sigma}{n\mathscr{F}}.$$

9. From the data given in the problem

$$\Delta G^0 \, (= -n\mathscr{F}\mathscr{E}^0) = -26{,}210 \text{ cal}$$

$$\Delta S^0 \left(= n\mathscr{F} \frac{d\mathscr{E}^0}{dT} \right) = -13.72 \text{ cal/deg.}$$

Therefore,

$$\Delta H^0 \, (= \Delta G^0 + T\Delta S^0) = -26{,}210 + (298.15)(-13.72)$$

$$= -30{,}300 \text{ cal.}$$

From the data of Appendix I the values obtained for ΔG^0, ΔS^0, and ΔH^0 are, respectively, $-26{,}224$ cal, -13.88 cal/deg, and $-30{,}362$ cal.

Chapter 32

1. Crystalline, high melting, in-the-molten-state ionic conductors called "salts," listed in order of increasing base strength of the anion.*
2. "Hydrobromic acid" is an equi-molar aqueous mixture of H_3O^+ and Br^-. It contains only trace amounts of molecular HBr. (If as estimated the equilibrium constant for the reaction $HBr + H_2O = H_3O^+ + Br^-$ is 10^6 m/l, What is the concentration of HBr in 1 molar "hydrobromic acid"? In 1 molar sodium bromide? (Ans: 10^{-6} and 10^{-13} m/l).)
3. Donor species appear on the left (strong at top, weak at bottom), acceptor species on the right (strong at bottom, weak at top). Transfer species—electrons or protons—tend to fall spontaneously from upper occupied levels to lower vacant levels. The distance between two levels represents the standard free energy change for the transfer of electrons or protons between these levels.
4. An acid (proton donor) is to a base (proton acceptor) as a reducing agent (electron donor) is to an oxidizing agent (electron acceptor).
5. *Acid-Base Theory*

NEUTRALIZATION.—The degradation of protons. Mass migration of protons from upper occupied levels to lower vacant levels.

IONIZATION.—Promotion of protons to the H_3O^+ - H_2O level. A configurational entropy effect. The slight undoing of neutralization by bases of H_3O^+.

HYDROLYSIS.—Upward promotion of protons from the H_2O - OH^- level. A configurational entropy effect. The slight undoing of neutralization by acids of OH^-.

BUFFER.—A partially filled proton level. (Are hydrochloric acid and sodium hydroxide solutions buffers?†)

INDICATOR.—A proton donor whose light absorption visibly changes when deprotonated. A proton level that is one color when occupied and a different color when vacant.

Indicators are used in trace amounts to indicate where the proton population in solution is poised. If the majority of protons lie in levels above the indicator level, the indicator will be kept primarily in its occupied or acid form. If, on the other hand, a significant number of proton vacancies exist below the indicator level, the indicator will be kept primarily in its vacant or base form. In the protonated form an indicator is merely a weak acid which when neutralized exhibits a color

* For the addition of calcium metal to the bottom of this list, see "Tangent-Sphere Model of Molecules. IV. Estimation of Internuclear Distances, and a Note on the Electronic Structure of Metals," H. A. Bent, *Journal of Chemical Education*, in press (Spring 1965)

† Yes. Hydrochloric acid is the partially filled H_3O^+ - H_2O level. Sodium hydroxide is the partially filled H_2O - OH^- level.

change visible to the naked eye. The methyl orange (MO) and phenolphthalein (Ph) indicator levels are shown in the accompanying chart.

H_3O^+ —— H_2O

HMO —— MO^-
(*r*) (*y*)

HAc —— Ac^-

HPh —— Ph^-
(*c*) (*r*)

H_2O —— OH^-

c = colorless

r = red

y = yellow

If methyl orange in its base form (NaMO) is added to hydrochloric acid, protons fall spontaneously from the occupied H_3O^+ - H_2O level to the HMO - MO^- level, and the solution turns red. Addition of sodium hydroxide removes protons from the H_3O^+ - H_2O level first, then from the HMO - MO^- level, at which point the solution turns yellow. In this titration phenolphthalein would begin colorless and end up red.

If phenolphthalein (or methyl orange) in its acid form is added to sodium hydroxide, protons fall from the indicator level to the vacant H_2O - OH^- level, and the solution turns red (or yellow). Addition of hydrochloric acid fills up the H_2O - OH^- level first, then the HPh - Ph^- (or HMO - MO^-) level, at which point the solution turns colorless (or red).

Any indicator whose acid strength lies between (and preferably not too close) to that of H_3O^+ and H_2O can be used in the titration of H_3O^+ by OH^-, and vice versa. On the other hand, methyl orange is a poor indicator for the titration of acetic acid by hydroxide ion. It would turn yellow at the beginning of the titration and remain yellow right through the titration's end-point. Phenolphthalein would be more suitable. It would start colorless, remain that color until most of the acetic acid had been deprotonated, and then turn red.

Which indicator should be used in the titration of Ac^- by H_3O^+? H_3O^+ by Ac^-? OH^- by HAc?*

MO, MO, Ph.

Oxidation-Reduction Theory

REDUCTION.—Filling vacant levels. Electronation.
OXIDATION.—Emptying occupied levels. De-electronation.
STRONG OXIDIZING AGENT.—Low-lying vacant level.
STRONG REDUCING AGENT.—High-lying occupied level.
ELECTROLYSIS.—Electron promotion by external means.

6. For the reaction

$$NH_4^+ + H_2O = H_3O^+ + NH_3,$$

$$\Delta G^0 = [G^0_{H_3O^+} + G^0_{NH_3(aq)}] - [G^0_{NH_4^+} + G^0_{H_2O}]$$

$$= +12{,}640 \text{ cal.}$$

Therefore,

$$K_a \left(= \frac{(H_3O^+)\,(NH_3)}{(NH_4^+)} = 10^{-\Delta G^0/1364} \right)$$

$$= 10^{-9.27} = 5.9 \times 10^{-10}\,.$$

For the "hydrolysis of ammonia" ($NH_3 + H_2O = NH_4^+ + OH^-$),

$$K_{\text{``}h\text{''}} \left[= \frac{(NH_4^+)(OH^-)}{(NH_3)} \right] = \frac{K_w}{K_a} = 1.7 \times 10^{-5}\,.$$

These results are summarized in the accompanying figure.

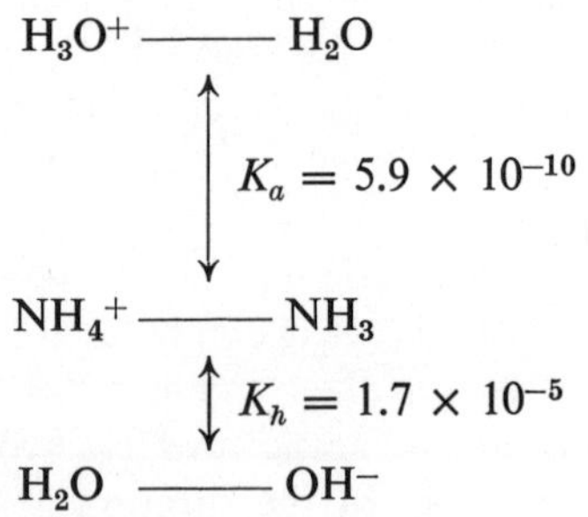

H_3O^+ ——— H_2O

↕ $K_a = 5.9 \times 10^{-10}$

NH_4^+ ——— NH_3

↕ $K_h = 1.7 \times 10^{-5}$

H_2O ——— OH^-

A STRATEGY FOR ESTIMATING THE HYDROGEN ION CONCENTRATION IN SOLUTIONS

(1) List all the potential proton donors in the system, together with their conjugate acceptors, in order of decreasing donor strength.

(2) To start with, assume that the protons are in the lowest possible levels. In pure water, for example, this assumption amounts to supposing that all the protons are in the H_2O - OH^- level, none in the H_3O^+ - H_2O level; i.e. that the only species present is H_2O. For aqueous acetic acid, the assumption amounts to supposing that the protons reside solely in the HAc - Ac^- level, which is full, and the H_2O - OH^- level, which is full; i.e. that the only species present are HAc and H_2O (no H_3O^+, hence no Ac^- or OH^-). This assumption is a generally good first approximation to the location of protons in solution. (At absolute zero it would be a perfect approximation.)

(3) Underline those species present in the hypothetical solution described in (2).

(4) Write down the proton-transfer reaction for the reaction of the strongest acid in the system (the top underlined species on the left) with the strongest base in the system (the bottom underlined species on the right).

(5) Set equal to each other the expressions for the proton escaping tendencies from the two levels considered in (4).

(6) From the stoichiometry of the reaction in (4) and the mathematical expression obtained in (5), calculate the proton-donor and proton-acceptor concentrations in the two levels under consideration. The ratio (*P*-donor)/(*P*-acceptor), multiplied by the level's escaping-tendency constant, is the value of (H_3O^+).

This strategy will be illustrated by its application to the questions in problem 7.

7. For parts (a) and (b) the strategy outlined above is applicable, but scarcely necessary. For part (a), a very strong acid in water, $(H_3O^+) = 0.5\ M$. For part (b), a very strong base in water, $(OH^-) = 1.0\ M$; hence $(H_3O^+) = 10^{-14}\ M$. For parts (c) through (h) the accompanying diagrams (p. 372) summarize steps (1) through (3).

If all protons in solutions (c) through (h) were in their lowest level(s), the only species present would be those underlined. This is a highly hypothetical situation, of course, for the escaping tendency of protons from a completely full level is infinite. Some protons should migrate immediately from fully occupied levels to higher vacant levels, as indicated by the solid arrows in diagrams (c) through (h). The shorter such an arrow, the greater the upward migration of protons required to make the escaping tendency of protons from the upper, initially vacant level equal to the proton escaping tendency from the occupied level immediately beneath it. Dashed arrows indicate less extensive proton transfer reactions; these generally can be ignored when computing the concentration of protons in the H_3O^+ - H_2O level.

	(c)	(d)	(e)
		HCl - Cl^-	
	H_3O^+ - H_2O	H_3O^+ - H_2O	H_3O^+ - H_2O
			HAc - Ac^-
	H_2O - OH^-	H_2O - OH^-	
	C_3H_8 - $C_3H_7^-$		H_2O - OH^-
Solute:	Propane	NaCl	HAc

	(f)	(g)	(h)
	H_3O^+ - H_2O	H_3O^+ - H_2O	H_3O^+ - H_2O
	HAc - Ac^-	HAc - Ac^-	H_2CO_3 - HCO_3^-
			HCO_3^- - CO_3^{--}
	H_2O - OH^-	H_2O - OH^-	H_2O - OH^-
Solute:	NaAc	HAc + NaAc	$NaHCO_3$

Under the starting conditions specified above—lower levels full, upper levels vacant—the proton transfer reactions represented by the solid arrows produce proton acceptor concentrations in the lower levels and proton donor concentrations in the upper levels that are equal to each other; i.e. (c) and (d), $(OH^-) = (H_3O^+)$; (e), $(Ac^-) = (H_3O^+)$; (f), $(OH^-) = (HAc)$; and (h), $(CO_3^{--}) = (H_2CO_3)$.

When equilibrium is attained with respect to the distribution of protons between upper and lower levels, the proton escaping tendency from the upper level is equal to the proton escaping tendency from the lower level; i.e. (c) and (d), $(H_3O^+) = K_w/(OH^-)$; (e), $(H_3O^+) = K_a(HAc)/(Ac^-)$; (f), $K_a(HAc)/(Ac^-) = K_w/(OH^-)$; and (h), $K_1(H_2CO_3)/(HCO_3^-) = K_2(HCO_3^-)/(CO_3^{--})$. Combining these two sets of results one finds that (c) and (d), $(H_3O^+) = K_w^{1/2}$; (e), $(H_3O^+) = (K_aC)^{1/2}$; (f), $(OH^-) = [(K_w/K_a)C]^{1/2}$—hence $(H_3O^+) = (K_aK_w/C)^{1/2}$—and (h), $(H_2CO_3) = (CO_3^{--}) = (K_2/K_1)^{1/2}C$—hence $(H_3O^+) = K_1(H_2CO_3)/(HCO_3^-) = K_2(HCO_3^-)/(CO_3^{--}) = (K_1K_2)^{1/2}$. In writing these last equations it has been assumed that the extent of upward proton migration in parts (e), (f), and (h) is small compared to the concentration of added solute C, here $0.5M$. In part (g), $(H_3O^+) = K_a(HAc)/(Ac^-) = K_a$.

These results may be generalized. Since the solid-arrow perturbation on the zeroth-order, all-protons-in-the-lowest-level(s) approximation renders the proton acceptor activity in the lower level equal to the proton donor activity in the upper level

$$a_{P\text{-acceptor}}^{\text{lower level}} = a_{P\text{-donor}}^{\text{upper level}}$$

and since equilibrium with respect to the distribution of protons between upper and lower levels requires that

$$K_{\text{lower level}} \frac{a_{P\text{-donor}}^{\text{lower level}}}{a_{P\text{-acceptor}}^{\text{lower level}}} = K_{\text{upper level}} \frac{a_{P\text{-donor}}^{\text{upper level}}}{a_{P\text{-acceptor}}^{\text{upper level}}} = a_{H_3O^+},$$

it follows that

$$a_{P\text{-acceptor}}^{\text{lower level}} = a_{P\text{-donor}}^{\text{upper level}} = \left[\frac{K_{\text{lower level}}}{K_{\text{upper level}}} a_{P\text{-donor}}^{\text{lower level}} a_{P\text{-acceptor}}^{\text{upper level}}\right]^{1/2}$$

and, hence, that

$$a_{H_3O^+} = (K_{\text{lower level}} K_{\text{upper level}} a_{P\text{-donor}}^{\text{lower level}}/a_{P\text{-acceptor}}^{\text{upper level}})^{1/2}.$$

The terms in this expression for the cases studied in problem 7 are given in the accompanying table.

Case	*Solute*	$K_{\text{lower level}}$	$K_{\text{upper level}}$	$a_{P\text{-donor}}^{\text{lower level}}$*	$a_{P\text{-acceptor}}^{\text{upper level}}$*	$a_{H_3O^+}$
(c)	Very weak acid	K_w	1	1	1	$(K_w)^{1/2}$
(d)	Very weak base	K_w	1	1	1	$(K_w)^{1/2}$
(e)	Weak acid	K_a	1	C	1	$(K_aC)^{1/2}$
(f)	Weak base	K_w	K_a	1	C	$(K_wK_a/C)^{1/2}$
(h)	HCO_3^-	K_2	K_1	C	C	$(K_2K_1)^{1/2}$

* $a_{H_2O} = 1$.

Postscript to Chapters 31 and 32

Electron- and proton-level diagrams display thermodynamic data in a form suitable for instant use. Transfer species other than elementary electrons and protons that have been treated in this fashion include oxygen atoms in studies of flame reactions and metallurgical processes, and various cations and anions in geochemical studies of ore differentiation. The practice of tabulating thermodynamic data in this manner is not new. Primitive tables of chemical affinity were widely circulated among chemists early in the eighteenth century. Not until the present century, however, was the quantitative relation between these tabulations and the Second Law of Thermodynamics fully appreciated. The thermodynamic theory of transfer reactions —with special reference to temperature and concentration effects—has been discussed by the author in an article, "Thermodynamic Data in Donor-Acceptor Form," in the *Journal of Physical Chemistry*, Vol. 61, p. 1419, 1957.

Chapters 33 and 34

1. (a) $G^{\text{solid}}_{\text{H}_2\text{O}}$
(b) increases
(c) configurational
(d) decrease
(e) insoluble
(f) solid
(g) solid
(h) greater
(i) $G^{\text{liq}}_{\text{H}_2\text{O}}$
(j) solid
(k) melt
(l) decreasing
(m) $G^{\text{solid}}_{\text{H}_2\text{O}}$
(n) $G^{\text{liq}}_{\text{H}_2\text{O}}$
(o) increase
(p) $G^{\text{liq}}_{\text{H}_2\text{O}}$
(q) $G^{\text{solid}}_{\text{H}_2\text{O}}$
(r) $S^{\text{liq}}_{\text{H}_2\text{O}}$
(s) greater
(t) $S^{\text{solid}}_{\text{H}_2\text{O}}$
(u) $G^{\text{liq}}_{\text{H}_2\text{O}} = G^{\text{solid}}_{\text{H}_2\text{O}}$
(v) increases
(w) temperature
(x) equilibrium
(y) pure solid
(z) decreases

2. At the triple point of water $G^{\text{solid}}_{\text{H}_2\text{O}} = G^{\text{liq}}_{\text{H}_2\text{O}} = G^{\text{gas}}_{\text{H}_2\text{O}}$. These equations define two conditions on the temperature and pressure of the system; e.g., $G^{\text{solid}}_{\text{H}_2\text{O}}(T, P) = G^{\text{liq}}_{\text{H}_2\text{O}}(T, P)$ and $G^{\text{liq}}_{\text{H}_2\text{O}}(T, P) = G^{\text{gas}}_{\text{H}_2\text{O}}(T, P)$. Consequently, in a system that contains pure solid, liquid, and gaseous water neither T nor P can be changed. There are no degrees of freedom. For this reason the region of coexistence of the three pure phases is referred to as "the triple *point*." The triple point of water occurs at +0.008°C and 4.6 mm Hg. (One often hears the region where pure water and pure ice coexist referred to as the freezing "point." More accurately

it should be referred to as the "freezing line." Usually what is meant, however, is the freezing point *at one atmosphere*. For the region where all three pure phases coexist this qualification is neither necessary nor correct. Water has no triple point at one atmosphere. When $P = 1$ atm. (and $N_1 = 1$), it is not possible to find a value for T that makes $G^{\text{solid}}_{\text{H}_2\text{O}} = G^{\text{liq}}_{\text{H}_2\text{O}} = G^{\text{gas}}_{\text{H}_2\text{O}}$.)

Postscript

"The word 'point' is an old and formerly respected term in thermodynamics," write Giauque, Brodale, Fisher, and Hornung.* They add, however, that "Recently there has been a rapidly growing downgrading of this accurate word in connection with various phenomena which are obviously gradual in nature"—such as the softening of a plastic, whose quasi-crystalline state on heating does not suddenly vanish but, like an old warrior, just gradually fades away.

Continuing, they say, "There is undoubtedly a real requirement for some brief way in which to refer to approximate temperatures in connection with gradual transitions. The Greeks probably did not have a name for this one, but there is a type of substandard English which seems sufficiently imprecise to cover such indefinite situations. We refer to the vernacular in which oil is 'erl' and pearl is 'poil.' From this dialect we have mined the word 'pernt,' which seems loose enough to cover the approximate temperature range of gradual effects.

"Scientists who are at least reputed to favor accuracy of thought and expression should welcome an elastic substitute for the word point, which would allow its return to its proper status. Thus, we suggest ferromagnetic or Curie pernt; lambda pernt; ferroelectric pernt; antiferromagnetic or Néel pernt, etc. . . . This is illiterally a Pernt of Order, and Disorder, for those who do not mind stretching a point. It is solidly based on Dialectical Magnetic Materialism and should be adopted.

"To lexicographers we offer: *Pernt*, a transitive noun; a meandering point; a mobile, shaped like a small period; a point that is both here and there; a point that is indeterminate due to uncertainty; a pernt is a pernt is a pernt.

"Only those scientists who believe that their use of the word point is above reproach, and whose public writings bear witness, are entitled to cast the first lodestone."

* W. F. Giauque, G. E. Brodale, R. A. Fisher, and E. W. Hornung, "Magnetothermodynamic Properties of $MnCl_2$ from 1.3° to 4.4°K at 90 kG. A Zero Entropy Reference. The Magnetomechanical Process at Absolute Zero," *Journal of Chemical Physics*, Vol. 42, pp. 1–20, 1965. Copyright 1965 by the American Institute of Physics. Reprinted by permission of the authors and the publisher.

3. To obtain the given equation, note that along an isobaric equilibrium curve $\delta T \approx (RT_{ntp}/\Delta S)N_2$, where $\Delta S = \Delta H/T_{ntp}$ and where $N_2 \equiv \dfrac{n_2}{n_2 + n_1} = m/(m + 1000/M_1) \approx m\, M_1/1000$ when $m \ll 1000/M_1$. δT has the same sign as ΔH. For the liquid-to-gas equilibrium, ΔH is positive; for the liquid-to-solid equilibrium, ΔH is negative.
4. $(\delta P/\delta T)_{N_2} = \Delta S/\Delta V = \Delta H/T\Delta V$. At a solid-gas or liquid-gas equilibrium point in the neighborhood of 0°C, $V^{\text{gas}}_{\text{H}_2\text{O}} \gg V^{\text{solid}}_{\text{H}_2\text{O}} \approx V^{\text{liq}}_{\text{H}_2\text{O}}$. Thus, for both cases $\Delta V \approx V^{\text{gas}}_{\text{H}_2\text{O}}$. Therefore, differences in ΔH are reflected directly in the rate of change of the vapor pressure with temperature.
5. (a) $P_1(100°\text{C}) = P_1{}^0(100°\text{C}) \cdot N_1 = P_1{}^0(100°\text{C}) \cdot (1 - N_2)$

$$= \left(760 - \frac{760}{56.5}\right) \text{ mm Hg}$$

$$= 746.6 \text{ mm Hg.}$$

(b) $P_1(101°\text{C}) = P_1{}^0(101°\text{C}) \cdot (1 - N_2)$

$$P_1{}^0(101°\text{C}) = P_1{}^0(100°\text{C}) + \left(\frac{\delta P_1{}^0}{\delta T}\right)_{\text{at } 100°\text{C}} \times \Delta T \qquad (\Delta T = 1°)$$

$$= 760 \text{ mm Hg} + 26.6 \frac{\text{mm Hg}}{\text{deg}} \times 1 \text{ deg} \quad \text{(Chapter 33)}$$

$$= 787 \text{ mm Hg.}$$

$$\therefore \; P_1(101°\text{C}) = \left(787 - \frac{787}{56.5}\right) \text{ mm Hg}$$

$$= 773.1 \text{ mm Hg.}$$

(c) An increase in temperature of 1°C increases the vapor pressure of the solution 773.1 − 746.6 = 26.5 mm Hg. Thus, to increase the vapor pressure of the solution 1 mm Hg, the temperature must be raised approximately 1/26.5 °C. Now, the solution's vapor pressure at 100°C is 760 − 746.6 = 13.4 mm Hg short of the normal boiling pressure. Therefore, the temperature must be raised

$$13.4 \times \frac{1}{26.5} = 0.51°\text{C}.$$

6. For a substance that is in internal equilibrium, $\delta E = T\delta S - P\delta V$. Setting $\delta E = 0$, and noting this fact by a right-hand subscript E on the variables δS and δV, one finds that $T\delta S_E - P\delta V_E = 0$, or

$$\frac{\delta S_E}{\delta V_E} = \left(\frac{\delta S}{\delta V}\right)_E = \frac{P}{T}.$$

For an ideal gas $P/T = nR/V$. Also, for an ideal gas E is a function of T only. Thus, to say that E is constant is the same thing as saying that T is constant; i.e. for an ideal gas $\delta(\ )_E = \delta(\ )_T$. Therefore, for an ideal gas

$$\delta S_T = \frac{nR}{V}\,\delta V_T = nR\delta(\ln V)_T\,.$$

7. (a) $G^{\text{liq}}_{\text{H}_2\text{O}}$
(b) $G^{\text{solid}}_{\text{H}_2\text{O}}$
(c) $G^{\text{liq}}_{\text{H}_2\text{O}}$
(d) $G^{\text{solid}}_{\text{H}_2\text{O}}$
(e) increase
(f) $G^{\text{solid}}_{\text{H}_2\text{O}}$
(g) $G^{\text{liq}}_{\text{H}_2\text{O}}$
(h) $V^{\text{solid}}_{\text{H}_2\text{O}}$
(i) greater
(j) $V^{\text{liq}}_{\text{H}_2\text{O}}$
(k) $G^{\text{solid}}_{\text{H}_2\text{O}}$
(l) greater
(m) $G^{\text{liq}}_{\text{H}_2\text{O}}$
(n) solid
(o) melt
(p) decreasing
(q) temperature
(r) $G^{\text{solid}}_{\text{H}_2\text{O}}$
(s) $G^{\text{liq}}_{\text{H}_2\text{O}}$
(t) increase
(u) $G^{\text{liq}}_{\text{H}_2\text{O}}$
(v) $G^{\text{solid}}_{\text{H}_2\text{O}}$
(w) $S^{\text{liq}}_{\text{H}_2\text{O}}$
(x) greater
(y) $S^{\text{solid}}_{\text{H}_2\text{O}}$
(z) $G^{\text{liq}}_{\text{H}_2\text{O}} = G^{\text{solid}}_{\text{H}_2\text{O}}$
(a′) increases
(b′) temperature
(c′) equilibrium
(d′) solid
(e′) decreases

8. $$\delta T = \frac{\Delta V}{\Delta S} \times \delta P = \frac{V^{\text{liq}}_{\text{H}_2\text{O}} - V^{\text{solid}}_{\text{H}_2\text{O}}}{S^{\text{liq}}_{\text{H}_2\text{O}} - S^{\text{solid}}_{\text{H}_2\text{O}}} \times (P_{\text{final}} - P_{\text{initial}})$$

$$= \frac{(18.01 - 19.68)\ \text{cc/mole}}{5.28\ \text{cal/deg-mole}} \times 10\ \text{atm.} \times \frac{1\ \text{cal}}{41.3\ \text{cc-atm.}}$$

$$= -0.077\ \text{deg.}$$

9. (a) G^{liq}_1
(b) G^{gas}_1
(c) increased
(d) $V^{\text{liq}}_1\delta P^{\text{liq}}$
(e) $G^{\text{liq}}_1 = G^{\text{gas}}_1$
(f) increased
(g) $\delta P^{\text{gas}} = (V^{\text{liq}}_1/V^{\text{gas}}_1)\delta P^{\text{liq}}$
(h) increases
(i) vapor pressure
(j) less

(k) molar volume

(l) gas

(m) P_1^{gas} (or Vapor pressure)

(n) less

(o) P_1^{liq}

(p)
$$\frac{18.0\ \text{cc/mole}}{\left[\dfrac{RT}{\text{VapPres at }25^\circ\text{C}}\right]} \times 10\ \text{atm.} = 0.17\ \text{mm Hg.}$$

10. (a) greater
   (b) zero
   (c) RN_2
   (d) less
   (e) pure solvent
   (f) RTN_2
   (g) solvent
   (h) the pure solvent
   (i) the solution
   (j) high
   (k) lower
   (l) increased
   (m) increase
   (n) solution
   (o) pressure
   (p) pure solvent
   (q) molar free energies of the solvent in the two phases
   (r) increase
   (s) $V_1^{\text{liq}}\pi$
   (t) RTN_2/V_1^{liq}
   (u) $n_2/(n_1 + n_2)$
   (v) n_2/n_1
   (w) $n_1V_1^{\text{liq}}$
   (x) volume
   (y) solution
   (z) n_2/V^{soln}
   (a′) moles per liter
   (b′) RTC_2
   (c′) intensive (or colligative)

(d′)
$$RTC_2 = \left(0.0820\,\frac{\text{liter-atm.}}{\text{deg-mole}}\right)(298\ \text{deg})\left(1\,\frac{\text{mole}}{\text{liter}}\right)$$
$$= 24.4\ \text{atm.}$$

"There is a world-wide interest in means for obtaining more fresh water," P. H. Abelson has written in a recent editorial in *Science* (Vol. 146, p. 1533, 1964). "President Johnson recently indicated his views, in part, by quoting a statement made by President Kennedy: 'There is no scientific breakthrough, including the trip to the moon, that will mean more to the country which first is able to bring fresh water from salt water at a competitive rate. . . .' "

One method under study is reverse osmosis. "Given a membrane permeable only by water, it is possible to obtain fresh water from sea water by exerting a differential pressure of about 24 atmospheres on salt water [most ocean water is about 0.5 molar in sodium chloride]. The present cellulose acetate membranes, however, are not perfect. A pressure of 100 atmospheres is required to produce water at the rate of 370 liters per square meter of membrane surface per day. The effluent, while potable, is not entirely free of salt. The membranes have only a few weeks of service life."

11. $$\pi = RTC_2 = \left(24.4 \frac{\text{liter-atm.}}{\text{mole}}\right)\left(\frac{0.00186 \text{ deg}}{1.86 \text{ deg/mole/liter}}\right)$$
$$= 0.0244 \text{ atm.} = 18.5 \text{ mm Hg} = 252 \text{ mm } H_2O.$$

12. (a) temperature
(b) uniform
(c) chemical's molar free energy
(d) uniform

13. $$-\log N_2 = \frac{4{,}440 \text{ cal/mole}}{2.303(1.987 \text{ cal/deg-mole})}\left(\frac{1}{298.15} - \frac{1}{353.20}\right) \text{deg}^{-1}$$
$$= 0.507$$
$$N_2 = 10^{-0.507} = 0.311$$

14. $$N_2 = e^{-\Delta H/RT} \times e^{\Delta H/RT_{mp}}$$

For given ΔH, the lower the melting point, the larger the solubility. From this one might expect that phenanthrene is more soluble than anthracene. At 25°C their solubilities are, respectively, 18.6 and 0.63 mole per cent.

15. (a) straight line
(b) $\Delta H/R$
(c) $-\Delta H/RT_{mp}$
(d) the melting temperature and the heat of fusion
(e) the heat of fusion and the solubility at one temperature.

16. $$\frac{n}{N+n} = e^{-N_0\varepsilon/RT} = 10^{-N_0\varepsilon/2.303RT}$$
$$= 10^{\overline{2.3(1.987)(1000)}}{}^{23{,}000} = 10^{-5}.$$

A concentration of vacancies of this magnitude may have a profound effect upon the chemical and physical properties of solids.

Chapter 35

1. $\Psi = -G/T.$ $\Delta\Psi = \Delta S_{\text{total}}$.

Chemical stability is associated with large values of Ψ. Ψ tends to be large when S is large and H is small, i.e. when the energy levels are closely spaced and the ground level is low-lying. The location of the ground level is particularly important at low temperatures.

Additional Problems

1. From the standpoint of the First Law of Thermodynamics, which states that energy is always conserved,

   Thermal energy could flow spontaneously from an icicle to your hand. But it doesn't.

   A brick could spring into the air spontaneously—if the brick, or the object upon which it rested, cooled slightly, to conserve the energy of the universe. But it doesn't.

   Water could dissociate spontaneously at room temperature into hydrogen and oxygen—at the expense of the thermal energy of its surroundings. But it doesn't.

   The air about you could spontaneously liquify, liberating thermal energy to its surroundings; or it could spontaneously separate into pure oxygen and nitrogen. But it doesn't.

   Why?

   Each hypothesized event, note, is the reverse of an actual event. An icicle held in your hand does get warmer (it may melt) and your hand gets colder; falling bricks do warmly come to rest—unless, perhaps, they are very very tiny; hydrogen and oxygen do form water (sometimes explosively), liberating thermal energy to the surroundings—unless the surroundings are very hot, or the gases are very dilute; and liquid air (an excellent coolant) does absorb thermal energy from its surroundings (and evaporates)—unless the surroundings are very cold, or the external pressure is very great.
2. Benjamin Thompson—later Count Rumford—noticed that cannons fired without shot got hotter than when shot was actually fired from them. How do you account for this behavior?
3. In "Sources of Heat Which is Excited by Friction," Count Rumford wrote, "I was struck with the very considerable heat which a brass

cannon acquires in a short time in being bored. . . ." Water around such cannons could be heated to and held at its boiling point for long periods of time. The boring equipment in these experiments was driven by horses. List the primary source(s) of "heat" in horse-driven boring operations that last for several minutes. Several months. Several centuries.

4. Discuss the relative merits of the following methods of using fodder to warm water: (1) Feed the fodder to a horse harnessed to a piece of friction machinery; (2) Feed the fodder to a horse immersed in a kettle of water; (3) Burn the fodder beneath a kettle of water.
5. Check Joule's estimate that water at the base of Niagara Falls should be about one-fifth of a degree warmer than water at the top of the Falls.
6. In one of his papers Joule makes an estimate of the mechanical equivalent of heat, using data reported earlier by Count Rumford, who had found that cannon-boring machinery that easily could be driven by one horse heated 26.58 lb. of ice-water to 180°F in $2\frac{1}{2}$ hours. Joule used also Watt's estimate that in one minute a horse can raise 330 lb. 100 ft. How well does the value of the mechanical equivalent of heat calculated from these data compare with Joule's determination?
7. A "dyne" is defined as the force that imparts to a 1 gram mass an acceleration of 1 cm/sec/sec. By comparison, the force of gravity imparts to objects at sea level an acceleration of 980 cm/sec/sec. The work involved when a force of one dyne operates through a distance of 1 cm is called an "erg." Estimate in ergs the work done by a flea during a push-up. Compare this figure with the energy of dissociation of a hydrogen molecule.
8. 10^7 ergs is called a "joule." Using Joule's experimental result that the heat produced when 772 lb. fall one foot is sufficient to raise the temperature of one pound of water 1° Fahrenheit, calculate the number of joules of energy required to raise the temperature of one gram of water 1° centigrade.
9. Determine by calculation or by experiment the approximate rate at which thermal energy is liberated by a burning candle. Using this figure, together with Rumford's statement that nine wax candles generate heat as fast as a single hard-working horse, and Watt's estimate of the rate horses do work, estimate the mechanical equivalent of heat.
10. Tait, a compatriot of Lord Kelvin, defines energy "as the power of doing work, or if we like to put it so, of doing mischief." Discuss this definition from the viewpoint of the First and Second Laws of Thermodynamics.
11. The sun, directly or indirectly, is the source of water power from waterfalls and dammed rivers, of wind power from convection currents in

the earth's atmosphere, and of steam power from combustion of wood and thc fossil fuels gas, oil, and coal. List several sources of terrestrial energy that are less directly dependent upon the sun.

12. Suppose the radiators in a cold room are turned up and the room temperature raised from 40° to 70°F. What effect has this on the room's content of thermal energy? (Disregard the increase in radiant energy density in the room.)
13. Why is the heat capacity of an ideal gas at constant pressure greater than its heat capacity at constant volume? Is C_P always greater than C_V?
14. Why is it that a quantity of water at 100°C is cooled to a lower temperature by a piece of iron at 0°C than by an equal weight of lead at 0°C? By approximately what factor do the degrees of cooling in the two cases differ?
15. The Law of Dulong and Petit as originally stated applies to low-melting metals. How might the law be stated for monatomic gases? For diatomic, linear triatomic, and non-linear triatomic gases at low temperatures? At high temperatures? For metallic chlorides, nitrates, and sulfates at room temperature?
16. Estimate the specific heat at 25°C of helium, hydrogen, methane, lithium, lead, diamond, rock salt, and calcium chloride.
17. Estimate the mechanical work obtained in the adiabatic expansion from 400° to 300°C of a mole of helium, hydrogen, and water vapor. What assumptions have you made? Which law(s) of thermodynamics have you used?
18. Thermodynamics has been described as a collection of statements of impotence. List several such statements. Can their physical content be formulated in a more positive manner?
19. Describe the random element introduced into the universe when Humpty Dumpty had his great fall.
20. While Dr. Black's work (Chapter 7) established the necessary condition for thermal equilibrium, it left unexplained the physical reasons for this condition. Why indeed do objects in thermal contact approach a common temperature?
21. Why does a marble rolling in a bowl eventually come to rest at the bottom of the bowl?
22. Why does a vibrating diving board eventually stop vibrating?
23. A struck tuning fork generates audible sound waves. If placed on a sounding board and struck in the same manner as before, it generates sound energy at a much greater rate, yet its vibrations do not die away very much faster than before. How do you account for this behavior?

24. Describe the energy transformations that occur in the universe during the operation of a pendulum clock.
25. P. G. Tait, in his 1876 Lectures on Some Recent Advances in Physical Science, asked, "Why is it that if I have a quantity of work or potential energy I can convert the whole of it, if I please, into heat; but when I have got it converted into heat, I cannot convert the heat back again, except in part, into work or potential energy?"
26. The important and very practical question, how much work can be obtained from heat? can be separated into two questions: what fraction of the heat can be converted to work? and to how much work does this fraction of the heat correspond? What men generally are credited with establishing the answers to these questions?
27. An analogy is sometimes drawn between the operations of heat engines and paddle wheels. In the latter case useful work is obtained from the descent of water from high to lower altitudes; in the former case—as Carnot noted—work is obtained from the descent of "caloric" from high to lower temperatures. In this analogy how should the term "caloric" be translated? Is it conserved, sometimes conserved, or never conserved?
28. Is it true that no heat engine can be more efficient that a reversible heat engine?
29. Why are solids and liquids generally not used as working substances in heat engines?
30. What is the mechanical equivalent of cold—for example, of a pound of ice on a hot summer day?
31. In a world threatened by over-population, which is the better use of cereal crops: bread production or beef production? Upon what physical law(s) is your argument based? (One alternative produces ten times more edible calories than the other.)
32. Is a man who transforms into work through muscular effort the energy of the food he eats and the air he breathes a more or less efficient machine than a reversible steam engine whose boiler temperature is normal body temperature and whose condenser temperature is the temperature of the surrounding air?
33. List two characteristics common to all natural processes.
34. What proceeds at a generally unpredictable rate, produces energywise precisely what it consumes, and never goes backward?
35. When are the initial and final states of a universe equivalent from the standpoint of the First Law? From the standpoint of the Second Law?
36. List three processes or events which if they were to occur would violate the Second Law of Thermodynamics but not the First.

37. Why aren't explosions reversible? Why, for example, couldn't the effects produced when the nitrate-laden freighter *Grandcamp* vanished in a thunderclap at 9:13 a.m. April 16, 1947, obliterating its crew and two hundred bystanders, flinging a steel barge fifty yards inland, knocking down two light planes circling overhead, breaking windows in every house of the neighboring town of Texas City, and destroying in an inferno a multimillion dollar chemical plant, be reversed in every respect? Why couldn't the atoms and molecules in the bits and pieces of this explosion, in the twisted wreckage, in the dismembered bodies, in the billowing smoke, and in the raging fires, spontaneously reassemble themselves into the original objects?
38. "What we experience in one moment," a philosopher has noted, "glides, in the next moment, into the past. There it remains forever, irretrievable, exempt from further change, inaccessible to further control by anything that the future will bring us—and yet enshrined in our memory as something that once filled our experience as an immediate present. Will it never come back? Why can it not be with us a second time?"*
39. Can the unidirectional flow of time be accounted for in terms of physical principles?
40. Is rotation of the moon about the earth an isentropic process?
41. Is it possible to decrease the entropy of a body without leaving behind changes in other bodies?
42. Is it possible to increase the entropy of a body without leaving behind changes in other bodies?
43. Is the following a valid statement? "When any actual process occurs it is impossible to invent a means of restoring every system concerned to its original condition."
44. Are all natural processes spontaneous?
45. Are all natural processes irreversible?
46. Josiah Willard Gibbs (1839–1903), founder of modern chemical thermodynamics and statistical mechanics and by many considered America's greatest theoretical physicist, described as follows in his famous paper of 1875, "On the Equilibrium of Heterogeneous Substances," the criterion of equilibrium for a material system that is isolated from all external influences.

For the equilibrium of any isolated system it is necessary and sufficient that in all possible variations of the state of the system which do not alter its energy, the variation of its entropy shall either vanish or be negative. If ε denotes the energy, and η the

* Hans Reichenbach, *The Direction of Time*, Berkeley, Calif.: University of California Press, 1956.

entropy of the system, and we use a subscript letter after a variation to indicate a quantity of which the value is not to be varied, the condition for equilibrium may be written

$$(\delta\eta)_\varepsilon \leqq 0.$$

Is this statement equivalent to the statement that the entropy of an isolated system tends to increase?

47. Important quantities encountered in classical mechanics are mass, length, time, area, volume, velocity, frequency, linear momentum, angular momentum, acceleration, force, pressure, torque, kinetic energy, potential energy, work, and the gravitational constant. What quantities not listed above are of central importance in applications of the First Law? The Second Law? Statistical mechanics? Quantum mechanics?
48. Which is the less likely event, the spontaneous flow of one calorie of thermal energy from a 300°K-reservoir to a 400°K-reservoir or the spontaneous conversion of one calorie of thermal energy at 25°C entirely to work?
49. Show with specific examples that for an irreversible event

$$\Delta S_\sigma \neq \frac{\Delta E_\sigma + P\Delta V_\sigma}{T_\sigma}. \qquad (W = 0)$$

Which side of the inequality is larger? Is this generally true?

50. List several quantities that might be called "additive invariants." Are all additive quantities invariant? Are all invariant quantities additive?
51. After many efforts to synthesize diamonds by compressing graphite, Bridgman wryly observed that "graphite is nature's best spring." What thermodynamic significance has this remark?
52. Why does a substance's entropy generally decrease as the substance is compressed isothermally? Might it be possible by a very great compression to reduce the entropy of a substance to zero?
53. A black body does not reflect light; however, it may emit light. In fact, of all bodies at a given temperature the perfectly black body must be the best emitter of light. Why? (Consider, at equilibrium, the emission and absorption of radiation within a hollow enclosure containing a perfectly black body and a body that is not perfectly black.)
54. P. G. Tait describes the following experiment. Polished platinum foil with ink letters on it is heated to deposit on the surface traces of iron oxide which tarnish the surface and make it absorb more light than the polished parts of the surface. When this foil is heated in a very hot flame there is seen on the original inked side faint traces of bright letters on a darker background; on the reverse side faint traces of dark letters

appear on a brighter background. How do you account for these observations?

55. Why does a red piece of glass look green when hot?
56. Show that the image of the sun formed by mirrors or lenses cannot be hotter than the sun.
57. The First Law may be considered a method for bookkeeping energy changes. Similarly, the Second Law may be considered a method for bookkeeping probability changes. Using Boltzmann's relation, the thermodynamic definition of temperature, the First Law, and the multiplicative property of Ω, show that for a universe composed of a chemical system σ and its thermal surroundings θ the statement

$$\Delta H_\sigma - T\Delta S_\sigma \leq 0$$

is equivalent to the statement

$$(\Omega_{\text{total}})_{\substack{\text{final}\\ \text{state}}} \geq (\Omega_{\text{total}})_{\substack{\text{initial}\\ \text{state}}} .$$

58. Show that the equation $G_{\text{reactants}} = G_{\text{products}}$ is equivalent to the equation $\Delta H_\sigma - T\Delta S_\sigma = 0$.
59. Because at equilibrium $\Delta H_\sigma - T\Delta S_\sigma = 0$, the statement often is made that equilibrium states are states in which there is a balance between energy and entropy effects. Energy and entropy have different physical units, however, and therefore the terms ΔH and ΔS cannot be compared with each other. Amend the original statement to make it dimensionally and physically correct.
60. What is the approximate value of ΔV_σ for the reversible evaporation of 1 mole of water at 0°C? At 100°C? At 374°C? At −272°C?
61. The enthalpies and entropies of melting at atmospheric pressure of the members of a homologous series often increase nearly linearly with increasing chain length and, insofar as this is so, can be represented by equations of the form

$$\Delta H = an + b$$

$$\Delta S = a'n + b' ,$$

where n stands for the number of carbon atoms in the molecule and a, b, a', b' are constants for a particular homologous series. Derive from these relations an expression for the normal melting points of the compounds of a homologous series. What form does this expression approach as n becomes very large?

For the normal paraffins it has been suggested that the parameters a, b, a', b' have, respectively, the values 0.6085 and −1.75 (in energy units) and 0.001491 and 0.00404 (in energy units per degree Kelvin).

Calculate from these numbers the normal melting point in degrees centigrade of linear polyethylene. (The actual melting point of a sample of polyethylene may differ from this calculated value owing to the presence in the sample of impurities, chain-branching, cross-linking, incomplete crystallization, strain at grain boundaries and, also, owing to deviations of ΔH and ΔS from strict linearity in n.)

62. Estimate from their boiling points the heats of vaporization of helium, hydrogen, neon, nitrogen, chlorine, benzene, n-octane, mercury, and zinc.
63. Why does cold water freeze and warm ice melt?
64. Describe the energy and entropy changes that occur in the universe in the combustion of coal.
65. Verify that the calculated value of the equilibrium partial pressures of hydrogen and iodine produced by the thermal dissociation of hydrogen iodide is the same whether one writes the dissociation reaction as $HI = (1/2)H_2 + (1/2)I_2$ or as $2HI = H_2 + I_2$.
66. Assuming that N_2, H_2, and NH_3 behave as ideal gases, show that the difference between ΔG and ΔG^0 for the reaction $N_2 + 3H_2 = 2NH_3$ is equal to

$$RT \ln (P^2_{NH_3}/P_{N_2}P^3_{H_2}).$$

Under what conditions does ΔG vanish?

What happens to the value of the equilibrium constant for this reaction as the total pressure on the system increases? As the temperature of the system increases?

67. For the reaction $aA + bB = cC + dD$

$$\Delta S_{\text{total}} = \Delta S^0 - R \ln \Pi - \frac{\Delta H^0}{T},$$

where
$$\Pi = \frac{a_C{}^c \cdot a_D{}^d}{a_A{}^a \cdot a_B{}^b}.$$

Interpret physically the three terms in the expression for ΔS_{total} and show that if the reaction is to occur spontaneously the activities a_A, a_B, a_C, a_D must be such that

$$\Pi < K_{eq},$$

where
$$K_{eq} = e^{-\Delta G^0/RT}.$$

Illustrate this condition with several specific examples.

The numerical value of the quotient Π can be diminished in two ways: (1) by decreasing the values of the activities that occur in the numerator, or (2) by increasing the values of the activities that occur in the denominator. What effects have (1) and (2) on the numerical values of the entropies of the reactants and products of the reaction? On ΔS_σ? On ΔS_θ? On ΔS_{total}?

68. The vapor pressure of many substances can be represented over modest temperature intervals by an equation of the form $P = A \exp(-B/T)$. What thermodynamic significance, if any, have the constants A and B?

69. List several compounds that at unit activity are thermodynamically unstable with respect to their constituent elements at unit activity.

70. Using the tabulated thermodynamic data in Appendix I, calculate (or estimate) for a temperature of 25°C,
    (a) The vapor pressure of water, methanol, bromine, iodine, sodium, and iron.
    (b) The solubility in water of lead sulfate, lead chloride, and lead sulfide (note hydrolysis of S^{--} and HS^-, and the self-ionization of water).
    (c) The equilibrium partial pressure of carbon dioxide over a mixture of the oxide and carbonate of magnesium, calcium, strontium, and barium.
    (d) The equilibrium partial pressure of water over the mixtures $BaCl_2$ - $BaCl_2 \cdot 2H_2O$, $MgCl_2$ - $MgCl_2 \cdot 6H_2O$, and CaO - $Ca(OH)_2$.
    (e) The equilibrium partial pressure of oxygen over the mixtures Ag - Ag_2O, Pb - PbO, and Al - Al_2O_3.

71. What is the equilibrium constant expression and the value of the equilibrium constant for the following reaction?

$$AgCl(c) + 2NH_3(aq) = Ag(NH_3)_2^+(aq) + Cl^-(aq)$$

How concentrated must the solution be in ammonia to dissolve one-tenth of a mole of silver chloride per liter?

72. For the following processes write down the equilibrium constant expression, calculate for a temperature of 25°C the value of the equilibrium constant, comment where possible on the physical significance of ΔS^0, and draw one concrete inference about each process.

The Haber process.
The thermite reaction.
The autoprotolysis of water.
The ionization of acetic acid.
The hydrolysis of sodium acetate.
The air oxidation of nitric oxide.
The dimerization of nitrogen dioxide.
The reaction of sodium with water.
The reaction of ammonia with hydrochloric acid.
The trimerization of acetylene to benzene.

The preparation of chlorine from hydrochloric acid and permanganate.
The reaction of copper with dilute and concentrated nitric acid.
The reaction of calcium carbide with water.
The oxidation of iodide ion to triiodide ion with bromine water.
The air oxidation of ferrous ion to ferric ion in acid solution.
The disproportionation in water of mercurous ion.
The disproportionation in water of cuprous ion.
The oxidation of lead sulfide with nitric acid.

73. For the reaction $H_2 = 2H$, it was found (Chapter 13) that

$$K(T = 10{,}000^\circ K) \approx 10^{2.4}$$

and that $T(K = 1) \approx 4{,}800^\circ K$. Can you improve upon these estimates?

74. List from memory the approximate values for as many molar entropies and enthalpies as you can.*

75. When can the spontaneity of a chemical reaction be judged solely by the sign of the heat of reaction?

76. List several exothermic reactions that are not spontaneous.

77. List several endothermic reactions that are spontaneous.

78. Discuss from several points of view the increase that occurs in the entropy of an ideal gas during an isothermal expansion.

79. What qualitative changes occur in the energy and entropy of an ideal gas and its thermal surroundings during a reversible isothermal expansion? During a reversible adiabatic expansion? During a free expansion?

80. What happens to the translational entropy of an ideal gas during a reversible adiabatic expansion if the gas is monatomic? If it is diatomic?

81. Show that in the reversible adiabatic expansion of an ideal monatomic gas $TV^{2/3}$ = constant.

82. What is physically unrealistic about the following graph?

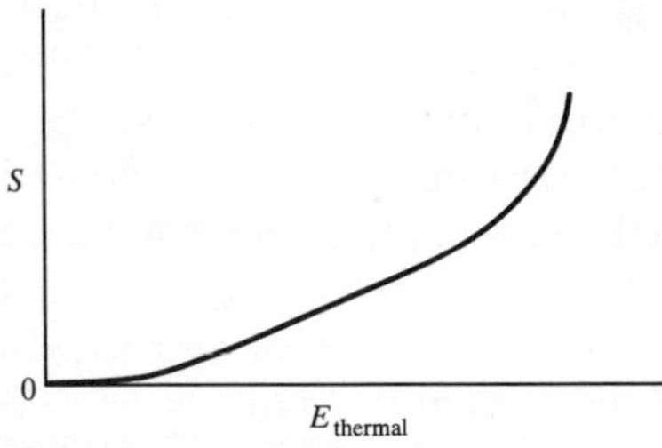

* A person who has frequently used the Second Law may remember the values of several typical entropies and enthalpies, just as a city dweller may remember the names of his city's important buildings and avenues, though he has not consciously tried to memorize them.

83. Plot against the thermal energy of a two-energy-level system the system's entropy, its absolute temperature, and the reciprocal of its absolute temperature—for which the symbol $\mathcal{L}$ has been suggested.
84. Which is hotter—that is, a better donor of thermal energy—a system with a small negative temperature or one with a large negative temperature?
85. Can a temperature of absolute zero be obtained by mixing a system whose absolute temperature is positive with a system whose absolute temperature is negative?
86. Show that if thermal energy can flow spontaneously from an object A to an object B, the fractional decrease produced thereby in Ω_A must be less than the fractional increase produced in Ω_B.
87. Using the thermodynamic definition of temperature and Boltzmann's relation, show that

$$T = \frac{1}{k} \cdot \frac{\Delta E}{\Delta \Omega} \cdot \Omega.$$

This equation shows why T tends to be small for objects that have little thermal energy, and large for objects that have a great deal of thermal energy. For when E is small, Ω is small, and when E is large, Ω is large.
88. The Third Law and spectroscopic entropies of O_3ClF differ by approximately 2.76 calories per deg per mole. Which value do you suppose is the larger? Does the magnitude of the discrepancy seem reasonable for a tetrahedral molecule isoelectronic with ClO_4^-?
89. Show that addition of a mole of solute to an ocean of solution in which the solute's mole fraction is N_2 increases the configurational entropy of the solution by the amount $-R \ln N_2$.
90. Are the phrases "pressure of the vapor" and "vapor pressure" synonymous?
91. List three ways the vapor pressure of a substance can be altered.
92. What are the activities of the solvent and the solute in an aqueous solution that produces in air a relative humidity of 98.2 per cent?
93. Using Raoult's Law and tables of vapor pressure data, estimate the molal boiling point constants of several common solvents.
94. Using Raoult's Law and tables of vapor pressure data, calculate the vapor pressure of a 1 molal aqueous sucrose solution at 0, -1, and -2°C. How do these numbers compare with the vapor pressure of pure ice at the same temperature? From a plot against temperature of the vapor pressure of each phase estimate the freezing point of a 1 molal sucrose solution.

95. Why do you suppose the first really good metal castings—made in Egypt around 2000 B.C.—were made of bronze?
96. At any point on a vapor pressure curve the slope of the tangent line is a function of the heat of vaporization, the absolute temperature, and the volume change during vaporization. Derive by dimensional arguments the form of the functional relation between the tangent line's slope and these three quantities.
97. Show that the quantity PV/T has the same dimensions as entropy.
98. Cite a substance whose entropy increases as its volume decreases. Is this an exception to the rule that entropies decrease as the volumes involved in the thermal motions of the constituent particles decrease?
99. List several phase transitions that have been recommended as fixed temperature points. What precautions must be observed to insure that the temperatures so determined are truly fixed?
100. In several of his papers Percy Bridgman, a Nobel prize winner in physics, refers to "fixed pressure points." What do you suppose he means by this phrase? What precautions must be observed to insure that the pressures at such points are truly fixed?
101. Why is thermal excitation of the vibrational degrees of freedom of water vapor almost negligible at room temperature? Note that thermal excitation of these degrees of freedom need not violate the First Law.
102. Using the information available in tables of thermodynamic data at 298°K, list the following molecules in order of increasing vibrational frequencies: hydrogen, oxygen, nitrogen, fluorine, chlorine, bromine, iodine.
103. Estimate from their heat capacities the vibrational frequencies of F_2 and Cl_2.
104. Is the energy of a photon of visible light greater or less than the average kinetic energy of a gas molecule at room temperature? Cite several familiar facts in support of your conclusion.
105. What modifications (if any) would occur in Boltzmann's factor and in the formula for the kinetic energy of an ideal gas if S were defined as simply $\ln \Omega$ and if T were defined (as now) as $(\partial E/\partial S)_V$?
106. How can one tell whether or not a reaction will go?
107. It has been said that the establishment of chemical equilibrium is a matter of manipulating entropies. Discuss this remark with the aid of several illustrative examples.
108. Entropy has been described as a measure of the disorder in a system. More precisely, Brillouin has suggested, *entropy measures the lack of information* about the actual structure of a system. This lack of information introduces the possibility of a great variety of microscopically distinct structures that in practice cannot be distinguished from each

other. Since any one of these different microstructures might be realized at a given time, the lack of information about a system corresponds to what we call the "disorder" in the system. This line of reasoning raises an interesting question.

Is there a quantitative connection between entropy and information?

More precisely, is it possible to set up between entropy and information a quantitative relation that has interesting, perhaps even useful, properties?

To answer this question we need a quantitative measure of information.

Consider a situation about which it is known before information on the situation is received that P_0 things might happen, and suppose that after a quantity of information I has been received it is known that only P_1 things might happen, P_1 being equal to or less than P_0 and equal to or greater than 1. Suggest for I a function of P_0 and P_1 that increases as P_0 increases and P_1 decreases, that is a maximum for given P_0 when $P_1 = 1$, that vanishes when $P_1 = P_0$, and that makes information received on independent situations additive.

Many of the texts listed in the bibliography contain problems. Denbigh's book has a very helpful set of answers and comments to approximately one hundred selected problems. Pitzer and Brewer's revision of Lewis and Randall lists answers to a like number of similarly chosen problems. Paul's text and the problem book of Sillén, Lange, and Gabrielson are especially noteworthy for problems based on the chemical literature.

Bibliography

Elementary treatments of chemical thermodynamics in paperback editions have been given by Mahan in *Elementary Chemical Thermodynamics*, (New York: Benjamin, 1963), and by Nash in *Elements of Chemical Thermodynamics* (Reading, Mass.: Addison-Wesley, 1962). Both books use some calculus.

Also available in paperback editions are Dover reissues of Planck's classic *Treatise on Thermodynamics* and Fermi's *Thermodynamics*.

A deservedly popular book that covers both classical and statistical thermodynamics is Wall's *Chemical Thermodynamics* (San Francisco: Freeman, 1958).

Two excellent books for physical insights into the meaning of entropy and free energy are Gurney's *Introduction to Statistical Mechanics* (New York: McGraw-Hill, 1949) and Fast's *Entropy* (New York: McGraw-Hill, 1962).

Another readable account of entropy and the Second Law considered from the micromolecular point of view is Rushbrooke's *Introduction to Statistical Mechanics* (Oxford: the Clarendon Press, 1949). More technical treatments have been given by Mayer and Mayer in *Statistical Mechanics* (New York: Wiley, 1940), a book with very useful appendices, and by Fowler and Guggenheim in *Statistical Thermodynamics* (Cambridge: Cambridge University Press, 1952).

Concise and mathematically elegant treatment of fundamentals have been given by Schrödinger in *Statistical Thermodynamics* (Cambridge: Cambridge University Press, 1948) and, at a more elementary level, by Guggenheim in *Boltzmann's Distribution Law* (New York: Interscience, 1955).

An interesting mathematical discussion of solutions and alloys is given by Guggenheim in *Mixtures* (Oxford: the Clarendon Press, 1952).

Classical thermodynamics has been the subject of many books. Notable for its readability and discussion of activity coefficients and standard states is Klotz's *Chemical Thermodynamics* (Englewood Cliffs, N.J.: Prentice-Hall, 1950). Pitzer and Brewer's outstanding revision of Lewis and Randall's great classic *Thermodynamics* (New York: McGraw-Hill, 1923, revised 1961) constitutes one of the most complete coverages of chemical thermodynamics available. In a similar category is Paul's *Principles of Chemical Thermodynamics* (New York: McGraw-Hill, 1951). Less comprehensive but highly recommended is Denbigh's *The Principles of Chemical Equilibrium* (Cambridge: Cambridge University Press, 1955). Well known for its formal treatment of classical thermodynamics is Guggenheim's *Thermodynamics* (New York: Interscience, 1949). Similar in spirit is A. H. Wilson's scholarly *Thermodynamics and Statistical Mechanics* (Cambridge: Cambridge University Press, 1957).

Excellent problems on classical thermodynamics are contained in Sillén, Lange, and Gabrielson's *Problems in Physical Chemistry* (Englewood Cliffs, N.J.: Prentice-Hall, 1952).

Extensive tabulations of thermodynamic data are given by Latimer in *Oxidation Potentials* (Englewood Cliffs, N.J.: Prentice-Hall, 1952) and by Kubaschewski and Evans in *Metallurgical Thermochemistry* (London: Butterworth, 1951). Interesting applications of thermodynamics to metallurgical problems are given by Wagner in *Thermodynamics of Alloys* (Reading, Mass.: Addison-Wesley, 1952).

Among many paperbacks recently issued on thermodynamics and related topics are the following.

Mendelssohn, *Cryophysics*, (New York: Interscience, 1960).

Bridgman, *The Thermodynamics of Electrical Phenomena in Metals* and *A Condensed Collection of Thermodynamic Formulas* (New York: Dover, 1961).

Carnot, *Reflections On the Motive Power of Fire* and other papers by Clausius and Clapeyron (New York: Dover, 1960).

Brown, *Count Rumford* (Garden City, N.Y.: Doubleday, Anchor Books, 1962).

Sandfort, *Heat Engines* (Garden City, N.Y.: Doubleday, Anchor Books, 1962).

Thirring, *Energy For Man* (New York: Harper and Row, 1962).

Cottrell, *Energy and Society* (New York: McGraw-Hill, 1955).

Schrödinger, *What is Life?* (Garden City, N.Y.: Doubleday, Anchor Books, 1956).

Numerous articles on thermodynamics have appeared in the *Journal of Chemical Education*.* The philosophy and pedagogical methods adopted in the present book have been described in that journal (Vol. 39, p. 491, 1962) under the title "The Second Law of Thermodynamics: Introduction for Beginners at Any Level."

* See particularly the bibliography on "Elementary Chemical Thermodynamics," Leonard K. Nash, Vol. 42, pp. 64–75, 1965.

Preface (and Postscript) to Appendix I

The tabulation of thermodynamic data in Appendix I represents a great heritage from the past—the distilled essence of the creative efforts of many minds and of millions of man-hours of careful and ingenious experimental work. This heritage cannot, however, be passively conveyed from one generation to the next. Each of us must learn over again what we have inherited from the past, Goethe has said, or it will not be ours. One purpose of the preceding pages has been to aid the reader in this historic process, to help him reach that point where with ease and confidence he may use the data of Appendix I to make useful predictions about the behavior of chemical systems.

Extensive as the data tabulation in Appendix I may appear, it represents hardly one page—one solvent at one temperature and one pressure—in that large book—the physical universe—which waits to be read by mankind. What are a substance's thermodynamic properties in liquid metals and in molten salts, for example? What are they at very low temperatures? And at very high pressures? Why is the enthalpy of formation of calcium carbonate -288.45 kcal/mole? And its entropy 22.2 cal/deg-mole? No one knows. These numbers cannot today be predicted from first principles. We know little, and understand less.

True, it has been said that with the discovery of quantum mechanics the possibility exists of making chemistry entirely theoretical. In fact, this possibility has not been realized. Precise calculations have been made for one-electron systems. Very accurate calculations have been made for two-electron systems. Beyond that the results are disappointing. Merely to print the solution to the Schrödinger equation for a single electronic state of a single atom as complex as iron would require for ink and paper more matter than is contained in the entire universe! Does this mean chemistry will forever remain essentially an experimental science? Or will some profoundly better way of making these calculations be found? That, perhaps, is for the younger generation to determine.

The data in Appendix I have been taken mainly from the definitive and much more comprehensive National Bureau of Standards Circular 500, "Selected Values of Chemical Thermodynamic Properties," compiled by Frederick D. Rossini, Donald D. Wagman, William H. Evans, Samuel Levine, and Irving Jaffe. The Circular can be obtained from the United States Government Printing Office, Washington, D.C.

Appendix I

Selected Values of Thermodynamic Properties at 298.15°K and 1 Atm.

H^0 and G^0 are expressed in kcal/mole

S^0 and $C_P{}^0$ are expressed in cal/deg-mole

c = crystalline, gas = gaseous, liq = pure liquid, aq = aqueous solution

For pure substances H^0 and G^0 are the molar enthalpies and free energies at 298.15°K and 1 atm.—with, for gases, small corrections for departures from ideal gas behavior—relative to the convention which assigns the value zero to these quantities for the elements in their standard states. The values tabulated for S^0 are the substances' molar entropies relative to the Third Law convention (p. 40).

For a solute in a solution the tabulated values are the solute's contribution to the solution's enthalpy, free energy, entropy, and heat capacity when the solute is present at unit concentration (more precisely, unit activity) relative to the convention which assigns the value zero to these quantities for the hydrogen ion, written H^+.

Elements are taken up generally by groups in the periodic table, VII through I.

Illustrative examples of some of the uses to which the data of Appendix I may be put are given in Chapter 28.

Formula	H^0	G^0	S^0	$C_P{}^0$
Hydrogen				
H(gas)	52.089	48.575	27.393	4.968
H^+(aq)	0.0	0.0	0.0	0.0
H_3O^+(aq)	−68.317	−56.690	16.716	17.996
H_2(gas)	0.0	0.0	31.211	6.892
Oxygen				
O(gas)	59.159	54.994	38.469	5.236
O_2(gas)	0.0	0.0	49.003	7.017
O_3(gas)	34.0	39.06	56.8	9.12
OH(gas)	10.06	8.93	43.888	7.141
OH^-(aq)	−54.957	−37.595	−2.52	−32.0
H_2O(gas)	−57.798	−54.635	45.106	8.025
H_2O(liq)	−68.317	−56.690	16.716	17.996
H_2O_2(liq)	−44.84	−27.240	(22)	
H_2O_2(aq)	−45.68	−31.470		
Fluorine				
F(gas)	18.3	14.2	37.917	5.436
F^-(aq)	−78.66	−66.08	−2.3	−29.5
F_2(gas)	0.0	0.0	48.6	7.52
HF(gas)	−64.2	−64.7	41.47	6.95
Chlorine				
Cl(gas)	29.012	25.192	39.457	5.2203
Cl^-(aq)	−40.023	−31.350	13.2	−30.0
Cl_2(gas)	0.0	0.0	53.286	8.11
ClO^-(aq)		−8.9	10.3	
ClO_2(gas)	24.7	29.5	59.6	
ClO_2^-(aq)	−16.5	−2.56	24.1	
ClO_3^-(aq)	−23.50	−0.62	39.	−18.0
ClO_4^-(aq)	−31.41	−2.	43.5	
Cl_2O(gas)	18.20	22.40	63.70	
HCl(gas)	−22.063	−22.769	44.617	6.96
HCl(aq)	−40.023	−31.350	13.2	−30.0
HClO(aq)	−27.83	−19.110	31.0	
ClF_3(gas)	−37.0	−27.2	66.61	15.33
Bromine				
Br(gas)	26.71	19.69	41.8052	4.9680
Br^-(aq)	−28.90	−24.574	19.29	−30.7
Br_2(gas)	7.34	0.751	58.639	8.60
Br_2(liq)	0.0	0.0	36.4	
Br_3^-(aq)	−32.0	−25.3	(40)	
BrO^-(aq)		−8.		
HBr(gas)	−8.66	−12.72	47.437	6.96
BrO_3^-(aq)	−9.6	10.9	38.9	−19.

Formula	H^0	G^0	S^0	$C_P{}^0$
Iodine				
I(gas)	25.482	16.766	43.184	4.9680
I^-(aq)	−13.37	−12.35	26.14	−31.0
I_2(gas)	14.876	4.63	62.280	8.81
I_2(c)	0.0	0.0	27.9	13.14
I_2(aq)	5.0	3.926		
I_3^-(aq)	−12.4	−12.31	41.5	
HI(gas)	6.2	0.31	49.314	6.97
IO^-(aq)	(−34)	−8.5		
HIO(aq)	(−38)	−23.5		
IO_3^-(aq)		−32.250	28.0	
ICl(gas)	4.2	−1.32	59.12	8.46
ICl_2^-(aq)		−38.35		
ICl_3(c)	−21.1	−5.40	41.1	
IBr(gas)	9.75	0.91	61.8	
Sulfur				
S(gas)	53.25	43.57	40.085	5.66
S(c, rhombic)	0.0	0.0	7.62	5.40
S(c, monoclinic)	0.071	0.023	7.78	5.65
S^{--}(aq)	10.0	20.0		
SO(gas)	19.02	12.78	53.04	
SO_2(gas)	−70.76	−71.79	59.40	9.51
SO_3(gas)	−94.45	−88.52	61.24	12.10
SO_3^{--}(aq)	−149.2	−118.8	10.4	
SO_4^{--}(aq)	−216.90	−177.34	4.1	4.0
$S_2O_3^{--}$(aq)	−154.0	−127.2	29.	
HS^-(aq)	−4.22	3.01	14.6	
H_2S(gas)	−4.815	−7.892	49.15	8.12
H_2S(aq)	−9.4	−6.54	29.2	
HSO_3^-(aq)	−150.09	−126.03	31.64	
HSO_4^-(aq)	−211.70	−179.94	30.32	
H_2SO_3(aq)	−145.5	−128.59	56.	
H_2SO_4(aq)	−216.90	−177.34	4.1	4.0
SF_6(gas)	−262.	−237.	69.5	
SO_2Cl_2(gas)		−73.6		
$(CH_3)_2SO$(liq)	−30.3	−18.9		
$(CH_3)_2SO_2$(liq)	−81.4	−64.7		
Selenium				
Se(gas)	48.37	38.77	42.21	4.968
Se(c, hex.)	0.0	0.0	10.0	5.95
Se^{--}(aq)	31.6	37.2	20.0	
SeO_3^{--}(aq)	−122.39	−89.33	3.9	
SeO_4^{--}(aq)	−145.3	−105.42	5.7	
HSe^-(aq)	24.6	23.57	42.3	
H_2Se(gas)	20.5	17.0	52.9	

Formula	H^0	G^0	S^0	$C_P{}^0$
Selenium (Cont.)				
H_2Se(aq)	18.1	18.4	39.9	
$HSeO_3^-$(aq)	−123.5	−98.3	(30.4)	
$HSeO_4^-$(aq)	−143.1	−108.2	22.0	
H_2SeO_3(aq)	−122.39	−101.8	45.7	
H_2SeO_4(aq)	−145.3	−105.42	5.7	
SeF_6(gas)	−246.0	−222.	75.10	
Tellurium				
Te(gas)	47.6	38.1	43.64	4.968
Te(c, II)	0.0	0.0	11.88	6.15
Te^{--}(aq)		52.7		
HTe^-(aq)		37.7		
H_2Te(gas)	36.9	33.1	56.0	
H_2Te(aq)		34.1		
TeO_2(c)	−77.69	−64.60	16.99	15.89
H_2TeO_3(c)	−144.7	−115.7	47.7	
TeF_6(gas)	−315.	−292.	80.67	
Nitrogen				
N(gas)	112.965	108.870	36.6145	4.968
N_2(gas)	0.0	0.0	45.767	6.960
N_3^-(aq)	60.3	77.7	(32)	
NO(gas)	21.600	20.719	50.339	7.137
NO_2(gas)	8.091	12.390	57.47	9.06
NO_2^-(aq)	−25.4	−8.25	29.9	
NO_3^-(aq)	−49.372	−26.43	35.0	
N_2O(gas)	19.49	24.76	52.58	9.251
$N_2O_2^{--}$(aq)	−2.59	33.0	6.6	
N_2O_4(gas)	2.309	23.491	72.73	18.90
N_2O_5(c)	−10.0	32.	27.1	
NH_3(gas)	−11.04	−3.976	46.01	8.523
NH_3(aq)	−19.32	−6.36	26.3	
NH_4^+(aq)	−31.74	−19.00	26.97	
N_2H_4(aq)	8.16	30.56	(33)	
HN_3(gas)	70.3	78.5	56.74	10.02
HN_3(aq)	60.50	71.30	(48)	
HNO_2(aq)	−28.4	−12.82		
HNO_3(liq)	−41.404	−19.100	37.19	26.26
HNO_3(aq)	−49.372	−26.43	35.0	
NH_2OH(aq)	−21.7	−5.60	(40)	
$H_2N_2O_2$(aq)	−13.7	8.6	52	
NOCl(gas)	12.57	15.86	63.0	
NOBr(gas)	19.56	19.70	65.16	
NH_4Cl(c, II)	−75.38	−48.73	22.6	20.1
$(NH_4)_2SO_4$(c)	−281.86	−215.19	52.65	44.81

Formula	H^0	G^0	S^0	$C_P{}^0$
Phosphorus				
P(gas)	75.18	66.71	38.98	4.9680
P(c, white)	0.0	0.0	10.6	5.55
P(c, red)	−4.4	−3.3	(7.0)	
P_4(gas)	13.12	5.82	66.90	16.0
PO_4^{---}(aq)	−306.9	−245.1	−52.	
P_4O_{10}(c)	−720.0			
PH_3(gas)	2.21	4.36	50.2	
HPO_3^{--}(aq)	−233.8	−194.0		
HPO_4^{--}(aq)	−310.4	−261.5	−8.6	
$H_2PO_3^-$(aq)		−202.35	(19.0)	
$H_2PO_4^-$(aq)	−311.3	−271.3	21.3	
H_3PO_3(aq)	−232.2	−204.8	(40.0)	
H_3PO_4(aq)	−308.2	274.2	42.1	
PCl_3(gas)	−73.22	−68.42	74.49	
PCl_5(gas)	−95.35	−77.57	84.3	
$POCl_3$(gas)	−141.5	−130.3	77.59	
PBr_3(gas)	−35.9	−41.2	83.11	
Arsenic				
As(gas)	60.64	50.74	41.62	
As(c, grey metal)	0.0	0.0	8.4	5.97
As_4(gas)	35.7	25.2	69.	
AsO^+(aq)		−39.1		
AsO_4^{---}(aq)	−208.0	−152.	−34.6	
$HAsO_4^{--}$(aq)	−214.8	−169.	0.9	
$H_2AsO_3^-$(aq)	−170.3	−140.4		
$H_2AsO_4^-$(aq)	−216.2	−178.9	28.	
H_3AsO_3(aq)	−177.3	−152.94	47.0	
H_3AsO_4(aq)	−214.8	−183.8	49.3	
AsF_3(gas)	−218.3	−214.7	69.08	30.3
AsF_3(liq)	−226.8	−215.5	43.31	
$AsCl_3$(gas)	−71.5	−68.5	78.2	
$AsCl_3$(liq)	−80.2	−70.5	55.8	
As_2S_3(c)	−35.0	−32.46	(26.8)	
Antimony				
Sb(gas)	60.8	51.1	43.06	4.968
Sb(c, metal)	0.0	0.0	10.5	6.08
SbO^+(aq)		−42.0	30.79	
Sb_4O_6(c)	−336.8	−298.0	58.8	48.46
Sb_2O_5(c)	−234.4	−200.5	29.9	28.1
SbH_3(gas)	34.	35.3	(53.0)	
$SbCl_3$(gas)	−75.2	−72.3	80.8	18.5
$SbCl_3$(c)	−91.34	−77.62	(44.5)	
$SbCl_5$(gas)	−93.9			
Sb_2S_3(c)	−36.0	−32.0	(30.3)	

Formula	H^0	G^0	S^0	$C_P{}^0$
Bismuth				
Bi(gas)	49.7	40.4	44.67	4.968
Bi(c)	0.0	0.0	13.6	6.1
BiO^+(aq)		−34.54	25.317	
Bi_2O_3(c)	−137.9	−118.7	36.2	27.2
$BiCl_3$(gas)	−64.7	−62.2	85.3	19.0
$BiCl_3$(c)	−90.61	−76.23	45.3	
BiOCl(c)	−87.3	−77.0	20.6	
Bi_2S_3(c)	−43.8	−39.4	35.3	
Carbon				
C(gas)	171.698	160.845	37.761	4.9803
C(c, diamond)	0.4532	0.6850	0.5829	1.449
C(c, graphite)	0.0	0.0	1.3609	2.066
CO(gas)	−26.4157	−32.8079	47.301	6.965
CO_2(gas)	−94.0518	−94.2598	51.061	8.874
CO_2(aq)	−98.69	−92.31	29.0	
CH_4(gas)	−17.889	−12.140	44.50	8.536
C_2H_2(gas)	54.194	50.0	47.997	10.499
C_2H_4(gas)	12.496	16.282	52.45	10.41
C_2H_6(gas)	−20.236	−7.860	54.85	12.585
C_6H_6(gas)	19.820	30.989	64.34	19.52
C_6H_6(liq)	11.718	41.30	29.756	
$(HCOOH)_2$(gas)	−187.7	−163.8	83.1	
HCOOH(gas)	−86.67	−80.24	60.0	
HCOOH(liq)	−97.8	−82.7	30.82	23.67
HCOOH(aq)	−98.0	−85.1	39.1	
$HCOO^-$(aq)	−98.0	−80.0	21.9	
H_2CO_3(aq)	−167.0	−149.00	45.7	
$HCO_3{}^-$(aq)	−165.18	−140.31	22.7	
$CO_3{}^{--}$(aq)	−161.63	−126.22	−12.7	
CH_3COOH(liq)	−116.4	−93.8	38.2	29.5
CH_3COOH(aq)	−116.743	−95.51		
CH_3COO^-(aq)	−116.843	−89.02		
$(COOH)_2$(c)	−197.6	−166.8	28.7	26.
$(COOH)_2$(aq)	−195.57	−166.8		
$HC_2O_4{}^-$(aq)	−195.7	−167.1	36.7	
$C_2O_4{}^{--}$(aq)	−197.0	−161.3	12.2	
HCHO(gas)	−27.7	−26.2	52.26	8.45
HCHO(aq)		−31.0		
CH_3OH(gas)	−48.10	−38.70	56.8	
CH_3OH(liq)	−57.036	−39.75	30.3	19.5
CH_3OH(aq)	−58.77	−41.88	31.63	
C_2H_5OH(gas)	−56.24	−40.30	67.4	
C_2H_5OH(liq)	−66.356	−41.77	38.4	26.64
CH_3CHO(gas)	−39.76	−31.96	63.5	15.0
CH_3CHO(aq)	−49.88			

Formula	H^0	G^0	S^0	$C_P{}^0$
Carbon (Cont.)				
$(CH_3)_2O$(gas)	−44.3	−27.3	63.72	15.76
C_2N_2(gas)	73.60	70.81	57.86	13.60
HCN(liq)	25.2	29.0	26.97	16.88
HCN(gas)	31.2	28.7	48.23	8.58
HCN(aq)	25.2	26.8	30.8	
CN^-(aq)	36.1	39.6	28.2	
HCNO(aq)	−35.1	−28.9	43.6	
CNO^-(aq)	−33.5	−23.6	31.1	
CH_3NH_2(gas)	−6.7	6.6	57.73	12.9
$CO(NH_2)_2$(c)	−79.634	−47.120	25.0	22.26
$CO(NH_2)_2$(aq)	−76.30	−48.72	41.55	
CH_3SH(gas)	−2.97	0.21	60.90	12.1
CS_2(gas)	27.55	15.55	56.84	10.91
CS_2(liq)	21.0	15.2	36.10	18.1
CNS^-(aq)	17.2	21.2	(36.0)	
CNCl(gas)	34.5	32.9	56.31	10.70
CF_4(gas)	−162.5	−151.8	62.7	
CCl_4(gas)	−25.50	−15.35	73.95	19.96
CCl_4(liq)	−33.34	−16.43	51.25	31.49
$COCl_2$(gas)	−53.30	−50.31	69.13	14.51
CH_3Cl(gas)	−19.58	−13.96	55.97	9.75
CH_3Br(gas)	−8.2	−5.9	58.74	10.18
$CHCl_3$(gas)	−24.	−16.	70.86	15.73
$CHCl_3$(liq)	−31.5	−17.1	48.5	27.8
Silicon				
Si(gas)	88.04	77.41	40.120	5.318
Si(c)	0.0	0.0	4.47	4.75
SiO(gas)	−26.72	−32.77	49.26	7.14
SiO_2(c, quartz)	−205.4	−192.4	10.00	10.62
SiH_4(gas)	−14.8	−9.4	48.7	10.24
SiF_4(gas)	−370.	−360.	68.0	18.2
SiF_6^{--}(aq)	−558.5	−511.	(−12.0)	
$SiCl_4$(gas)	−145.7	−136.2	79.2	21.7
$SiCl_4$(liq)	−153.0	−136.9	57.2	34.7
Germanium				
Ge(gas)	78.44	69.50	40.106	7.346
Ge(c)	0.0	0.0	10.14	6.24
GeO(gas)	−22.8	−28.2	52.56	7.39
Tin				
Sn(gas)	72.	64.	40.245	5.081
Sn(c, grey)	0.6	1.1	10.7	6.16
Sn(c, white)	0.0	0.0	12.3	6.30
SnO(c)	−68.4	−61.5	13.5	10.6

Formula	H^0	G^0	S^0	$C_P{}^0$
Tin (Cont.)				
SnO_2(c)	−138.8	−124.2	12.5	12.57
$HSnO_2^-$(aq)		−98.0		
$Sn(OH)_2$(c)	−138.3	−117.6	23.1	
$Sn(OH)_6^{--}$(aq)		−310.5		
$SnCl_2$(c)	−83.6	−72.2	(29.3)	
$SnCl_4$(liq)	−130.3	−113.3	61.8	39.5
SnS(c)	−18.6	−19.7	23.6	
Lead				
Pb(gas)	46.34	38.47	41.890	4.968
Pb(c)	0.0	0.0	15.51	6.41
Pb^{++}(aq)	0.39	−5.81	5.1	
PbO(c, red)	−52.40	−45.25	16.2	
PbO(c, yellow)	−52.07	−45.05	16.6	11.60
$HPbO_2^-$(aq)		−81.0?		
$Pb(OH)_2$(c)	−123.0	−100.6	21.	
PbO_2(c)	−66.12	−52.34	18.3	15.4
Pb_3O_4(c)	−175.6	−147.6	50.5	35.14
PbF_2(c)	−158.5	−148.1	29.0	
$PbCl_2$(c)	−85.85	−75.04	32.6	18.4
$PbBr_2$(c)	−66.21	−62.24	38.6	19.15
PbI_2(c)	−41.85	−41.53	42.3	
PbS(c)	−22.54	−22.15	21.8	11.83
$PbSO_4$(c)	−219.50	−193.89	35.2	24.9
$Pb(NO_3)_2$(c)	−107.35	−60.3	(50.9)	
$PbCO_3$(c)	−167.3	−149.7	31.3	20.9
PbC_2O_4(c)	−205.1	−180.3	(33.2)	
$PbCrO_4$(c)	−225.2	−203.6	(36.5)	
$PbMoO_4$(c)	−265.8	−231.7	(38.5)	
Thallium				
Tl(gas)	43.34	35.05	43.23	4.968
Tl(c)	0.0	0.0	15.4	6.35
Tl^+(aq)	1.38	−7.755	30.4	
Tl^{+++}(aq)	27.7	50.0	−106.	
Tl(OH)(c)	−56.9	−45.5	17.3	
$Tl(OH)_3$(c)	−156.	−123.0	(24.4)	
TlCl(gas)	−15.	−21.	61.1	8.66
TlCl(c)	−48.99	−44.19	25.9	
TlBr(gas)	−4.	−13.	63.8	8.81
TlBr(c)	−41.2	−39.7	28.6	
TlI(gas)	8.	−3.	65.6	8.86
TlI(c)	−29.7	−29.7	29.4	
Zinc				
Zn(gas)	31.19	22.69	38.45	4.968

Formula	H^0	G^0	S^0	$C_P{}^0$
Zinc (Cont.)				
Zn(c)	0.0	0.0	9.95	5.99
Zn^{++}(aq)	−36.43	−35.184	−25.45	
ZnO(c)	−83.17	−76.05	10.5	9.62
ZnO_2^{--}(aq)		−93.03		
$Zn(NH_3)_4^{++}$(aq)		−73.5		
$Zn(OH)_2$(c)	−153.5	−132.6	(19.9)	
ZnS(c, sphalerite)	−48.5	−47.4	13.8	10.8
ZnS(c, wurtzite)	−45.3	−44.2	(13.8)	
ZnS(c, precipt.)	(−44.3)	(−43.2)		
$ZnCl_2$(c)	−99.4	−88.255	25.9	18.3
$ZnSO_4$(c)	−233.88	−208.31	29.8	28.0
$ZnSO_4 \cdot H_2O$(c)	−310.6	−269.9	34.9	34.7
$ZnSO_4 \cdot 6H_2O$(c)	−663.3	−555.0	86.8	80.8
$ZnSO_4 \cdot 7H_2O$(c)	−735.1	−611.9	92.4	93.7
$Zn(CN)_2$(c)	18.4	29.	(22.9)	
$Zn(CN)_4^{--}$(aq)	82.0	100.4		
$ZnCO_3$(c)	−194.2	−174.8	19.7	19.16
Cadmium				
Cd(gas)	26.97	18.69	40.067	4.968
Cd(c)	0.0	0.0	12.3	6.19
Cd^{++}(aq)	−17.30	−18.58	−14.6	
CdO(c)	−60.86	−53.79	13.1	10.38
$Cd(OH)_2$(c)	−133.26	−112.46	22.8	
$CdCl_2$(c)	−93.00	−81.88	28.3	
$CdCl^+$(aq)		−51.8	5.6	
$CdCl_2$(aq, unionized)		−84.3	17.	
$CdCl_3^-$(aq)		−115.9	50.7	
CdS(c)	−34.5	−33.60	17.	
$CdSO_4$(c)	−221.36	−195.99	32.8	
$CdSO_4 \cdot H_2O$(c)	−294.37	−254.84	41.1	
$CdSO_4 \cdot \frac{8}{3} H_2O$(c)	−411.82	−349.63	57.9	
$Cd(CN)_2$(c)	39.0	49.7	(24.9)	
$Cd(CN)_4^{--}$(aq)		111.		
Mercury				
Hg(gas)	14.54	7.59	41.80	4.968
Hg(liq)	0.0	0.0	18.5	6.65
Hg^{++}(aq)	41.59	39.38	−5.4	
Hg_2^{++}(aq)		36.35		
HgO(c, red)	−21.68	−13.990	17.2	10.93
HgO(c, yellow)	−21.56	−13.959	17.5	
$HgCl_2$(c)	−55.0	−44.4	(34.5)	
Hg_2Cl_2(c)	−63.32	−50.35	46.8	24.3
$HgCl_4^{--}$(aq)		−107.7		
Hg_2Br_2(c)	−49.42	−42.714	50.9	

Formula	H^0	G^0	S^0	$C_P{}^0$
Mercury (Cont.)				
HgI_2(c, red)	−25.2	−24.07	(42.6)	
HgI_2(c, yellow)	−24.55	−23.1	(42.6)	
Hg_2I_2(c)	−28.91	−26.60	57.2	25.3
HgS(c, red)	−13.90	−11.67	18.6	
HgS(c, black)	−12.90	−11.05	19.9	
Hg_2SO_4(c)	−177.34	−149.12	47.98	
$Hg(CN)_2$(c)	62.5	74.3	(27.4)	
$Hg(CN)_4{}^{--}$(aq)	126.0	141.3		
Hg_2CrO_4(c)		−155.75		
Copper				
Cu(gas)	81.52	72.04	39.744	4.968
Cu(c)	0.0	0.0	7.96	5.848
Cu^+(aq)	(12.4)	12.0	(−6.3)	
Cu^{++}(aq)	15.39	15.53	−23.6	
CuO(c)	−37.1	−30.4	10.4	10.6
Cu_2O(c)	−39.84	−34.98	24.1	16.7
$Cu(OH)_2$(c)	−107.2	−85.3	(19.0)	
$CuF_2 \cdot 2H_2O$(c)	−274.5	−235.2	36.2	
CuCl(c)	−32.5	−28.2	20.2	
$CuCl_2$(c)	−52.3	−42.	(26.8)	
$CuCl_2{}^-$(aq)	−66.1	−57.9	49.4	
CuS(c)	−11.6	−11.7	15.9	11.43
Cu_2S(c)	−19.0	−20.6	28.9	18.24
$CuSO_4$(c)	−184.00	−158.2	27.1	24.1
$CuSO_4 \cdot H_2O$(c)	−259.00	−219.2	35.8	31.3
$CuSO_4 \cdot 3H_2O$(c)	−402.25	−334.6	53.8	49.0
$CuSO_4 \cdot 5H_2O$(c)	−544.45	−449.3	73.0	67.2
$Cu(NH_3)_4{}^{++}$(aq)	(−79.9)	−61.2	192.8	
$CuCO_3$(c)	−142.2	−123.8	21.	
Silver				
Ag(gas)	69.12	59.84	41.3221	4.9680
Ag(c)	0.0	0.0	10.206	6.092
Ag^+(aq)	25.31	18.430	17.67	9.
Ag^{++}(aq)		64.1		
Ag_2O(c)	−7.306	−2.586	29.09	15.67
AgCl(c)	−30.362	−26.224	22.97	12.14
$AgClO_3$(c)	−5.73	16.	(37.7)	
$AgClO_4$(c)	−7.75	21.	(38.8)	
AgBr(c)	−23.78	−22.930	25.60	12.52
$AgBrO_3$(c)		17.6		
AgI(c)	−14.91	−15.85	27.3	13.01
$AgIO_3$(c)	−41.7	−24.080	35.7	
Ag_2S(c, rhombic)	−7.60	−9.62	34.8	
Ag_2SO_4(c)	−170.50	−147.17	47.8	31.4

Formula	H^0	G^0	S^0	$C_P{}^0$
Silver (Cont.)				
$Ag(S_2O_3)_2{}^{---}(aq)$	(−285.5)	−247.6		
$Ag(SO_3)_2{}^{---}(aq)$		−225.4		
$AgNO_2(c)$	−10.605	4.744	30.62	18.8
$AgNO_3(c)$	−29.43	−7.69	33.68	22.24
$Ag(NH_3)_2{}^+(aq)$	−26.724	−4.16	57.8	
$Ag_2CO_3(c)$	−120.97	−104.48	40.0	26.8
$Ag_2C_2O_4(c)$	−159.1	−137.2	(48.0)	
$Ag(CH_3CO_2)(c)$	−93.41	−74.2	(33.8)	
$Ag_2MoO_4(c)$		−196.4		
$Ag_2CrO_4(c)$	−176.2	−154.7	51.8	
$Ag_4Fe(CN)_6(c)$		188.4		
$AgCN(c)$	34.94	39.20	20.0	
$Ag(CN)_2{}^-(aq)$	64.5	72.05	49.0	
Gold				
Au(gas)	82.29	72.83	43.12	4.9680
Au(c)	0.0	0.0	11.4	6.03
$Au^+(aq)$		39.0		
$Au^{+++}(aq)$		103.6		
$H_2AuO_3{}^-(aq)$		−45.8		
$HAuO_3{}^{--}(aq)$		−27.6		
$AuO_3{}^{---}(aq)$		−5.8		
$Au(OH)_3(c)$	−100.0	−69.3	29.0	
AuCl(c)	−8.4	−4.2	(24.0)	
$AuCl_3(c)$	−28.3	−11.6	(35.0)	
$AuCl_4{}^-(aq)$	−77.8	−56.2	61.	
$AuBr_4{}^-(aq)$	−45.5	−38.1	75.	
$Au(CN)_2{}^-(aq)$	58.4	51.5	99.	
Nickel				
Ni(gas)	101.61	90.77	43.502	5.5986
Ni(c)	0.0	0.0	7.20	6.21
$Ni^{++}(aq)$	(−15.3)	−11.53		
$Ni(OH)_2(c)$	−128.6	−108.3	19.	
$Ni(OH)_3(c)$	−162.1	−129.5	(19.5)	
$NiCl_2(c)$	−75.5	−65.1	25.6	18.6
NiS(c, α)		−17.7		
NiS(c, γ)		−27.3		
NiO(c)	−58.4	−51.7	9.22	10.60
$NiSO_4(c)$	−213.0	−184.9	18.6	33.4
$NiSO_4 \cdot 6H_2O$(c, blue)	−642.5	−531.0	73.1	82.
$Ni(NH_3)_6{}^{++}(aq)$		−60.1		
$Ni(CN)_4{}^{--}(aq)$	86.9	117.1	(33.0)	
Cobalt				
Co(gas)	105.	94.	42.881	5.5043

Formula	H^0	G^0	S^0	$C_P{}^0$
Cobalt (Cont.)				
Co(c)	0.0	0.0	6.8	6.11
Co^{++}(aq)	(−16.1)	−12.3	(−37.1)	
Co^{+++}(aq)		29.6		
$Co(OH)_2$(c)	−129.3	−109.0	(19.6)	
$Co(OH)_3$(c)	−174.6	−142.6	(20.0)	
CoF_2(c)	−157.	−146.45	(20.0)	
CoF_3(c)	−185.	−168.	(22.6)	
$CoCl_2$(c)	−75.8	−65.5	25.4	18.8
CoS(c, precipt.)	−19.3	−19.8	(16.1)	
$CoSO_4$(c)	−205.5	−180.1	27.1	
$Co(NO_3)_2$(c)	−100.9	−55.1	(46.0)	
$Co(NH_3)_6{}^{++}$(aq)		−57.7		
$Co(NH_3)_6{}^{+++}$(aq)		−55.2		
$Co(NH_3)_5Cl^{++}$(aq)	−162.1	−86.2	96.1	
Iron				
Fe(gas)	96.68	85.76	43.11	6.13
Fe(c)	0.0	0.0	6.49	6.03
Fe^{++}(aq)	−21.0	−20.30	−27.1	
Fe^{+++}(aq)	−11.4	−2.53	−70.1	
Fe_2O_3(c, hematite)	−196.5	−177.1	21.5	25.0
Fe_3O_4(c, magnetite)	−267.9	−242.4	35.0	
$Fe(OH)^{++}$(aq)	−67.4	−55.91	−23.2	
$Fe(OH)_2$(c)	−135.8	−115.57	19.	
$Fe(OH)_2{}^+$(aq)		−106.2		
$Fe(OH)_3$(c)	−197.0	−166.0	(23.0)	
$FeCl^{++}$(aq)	−42.9	−35.9	−22.	
$FeCl_3$(c)	−96.8	−80.4	(31.1)	
$FeBr^{++}$(aq)	−34.2	−27.9		
FeS(c)	−22.72	−23.32	16.1	
FeS_2(c)	−42.52	−39.84	12.7	
$FeSO_4$(c)	−220.5	−198.3	(27.6)	
$FeSO_4 \cdot 7H_2O$(c)	−718.7	−597.	(93.4)	
$FeNO^{++}$(aq)	−9.7	1.5	−10.6	
Fe_3C(c, cementite)	5.0	3.5	25.7	25.3
$Fe(CN)_6{}^{-4}$(aq)	126.7	170.4		
Fe_2SiO_4(c)	−343.7	−319.8	35.4	31.75
Manganese				
Mn(gas)	68.34	58.23	41.493	4.9680
Mn(c)	0.0	0.0	7.59	6.29
Mn^{++}(aq)	−52.3	−53.4	−20.	
MnO(c)	−92.0	−86.8	14.4	10.27
MnO_2(c)	−124.2	−111.4	12.7	12.91
$MnO_4{}^-$(aq)	−123.9	−101.6	45.4	
$MnO_4{}^{--}$(aq)		−120.4		

Formula	H^0	G^0	S^0	$C_P{}^0$
Manganese (Cont.)				
Mn_2O_3(c)	−232.1	−212.3	(22.1)	25.8
Mn_3O_4(c, I)	−331.4	−306.0	35.5	33.29
$Mn(OH)_2$(c, precipt.)	−166.8	−146.9	21.1	
$Mn(OH)_3$(c)	−212.	−181.	(23.8)	
MnF_2(c)	−189.	−179.	22.2	16.24
$MnCl_2 \cdot H_2O$(c)	−188.5	−164.5	(35.9)	
MnS(c, green)	−48.8	−49.9	18.7	11.94
MnS(c, precipt.)		−53.3		
$MnSO_4$(c)	−254.24	−228.48	26.8	23.94
$MnCO_3$(c)	−213.9	−195.4	20.5	19.48
Chromium				
Cr(gas)	80.5	69.8	41.637	4.9680
Cr(c)	0.0	0.0	5.68	5.58
Cr^{++}(aq)		−42.1		
Cr^{+++}(aq)		−51.5	−73.5	
Cr_2O_3(c)	−269.7	−250.2	19.4	28.38
$Cr_2O_7{}^{--}$(aq)	−349.1	−300.5	51.1	
$Cr(OH)^{++}$(aq)	−113.5	−103.0	−16.4	
$HCrO_4{}^-$(aq)	−220.2	−184.9	16.5	
$CrO_4{}^{--}$(aq)	−213.75	−176.1	9.2	
$Cr(OH)_2$(c)		−140.5		
$Cr(OH)_3$(c)	−247.1	−215.3	(19.2)	
CrF_3(c)	−265.2	−248.3	(22.2)	
CrF_2(c)	−181.0	−170.7	(19.6)	
$CrCl_2{}^+$(aq)	−130.0	−115.2		
$CrCl_2$(c)	−94.56	−85.15	27.4	16.87
$CrCl_3$(c)	−134.6	−118.0	30.0	21.53
Molybdenum				
Mo(gas)	155.5	144.2	43.462	4.9680
Mo(c)	0.0	0.0	6.83	5.61
$MoO_4{}^{--}$(aq)	−254.3	−218.8		
Tungsten (Wolfram)				
W(gas)	201.6	191.6	41.552	5.0903
W(c)	0.0	0.0	8.0	5.97
WO_3(c, yellow)	−200.84	−182.47	19.90	19.48
$WO_4{}^{--}$(aq)	−266.6	−220.		
Vanadium				
V(gas)	120.	109.	43.546	6.2166
V(c)	0.0	0.0	7.05	5.85
V^{++}(aq)		−54.7		
V^{+++}(aq)		−60.6		
$VO_4{}^-$(aq)	−210.9	−203.9	(48.0)	
V_2O_3(c)	−290.	−271.	23.58	24.83

Formula	H^0	G^0	S^0	$C_P{}^0$
Vanadium (Cont.)				
V_2O_4(c)	−344.	−318.	24.67	28.30
V_2O_5(c)	−373.	−344.	31.3	31.00
Titanium				
Ti(gas)	112.	101.	43.069	5.8385
Ti(c)	0.0	0.0	7.24	6.010
TiO_2(c, rutile III)	−218.0	−203.8	12.01	13.16
TiO^{++}(aq)		−138.		
Ti_2O_3(c)	−367.	−346.	18.83	23.27
Ti_3O_5(c)	−584.	−550.	30.92	37.00
TiF_2(c)	−198.	−187.1		
TiF_3(c)	−315.	−290.9		
TiF_4(c)	−370.	−346.3		
$TiCl_2$(c)	−114.	−96.		
$TiCl_3$(c)	−165.	−148.		
$TiCl_4$(liq)	−179.3	−161.2	60.4	37.5
Boron				
B(gas)	97.2	86.7	36.649	4.971
B(c)	0.0	0.0	1.56	2.86
B_2O_3(c)	−302.0	−283.0	12.91	14.88
B_2H_6(gas)	7.5	19.8	55.66	13.48
B_5H_9(gas)	15.0	39.6	65.88	19.
HBO_2(c)	−186.9	−170.5	11.	
$H_2BO_3^-$(aq)	−251.8	−217.6	7.3	
H_3BO_3(aq)	−255.2	−230.24	38.2	
H_3BO_3(c)	−260.2	−230.2	21.41	19.61
BF_3(gas)	−265.4	−261.3	60.70	12.06
BF_4^-(aq)	−365.	−343.	40.	
BCl_3(gas)	−94.5	−90.9	69.29	14.97
BCl_3(liq)	−100.0	−90.6	50.0	
BBr_3(gas)	−44.6	−51.0	77.49	16.25
BBr_3(liq)	−52.8	−52.4	54.7	
Aluminum				
Al(gas)	75.0	65.3	39.303	5.112
Al(c)	0.0	0.0	6.769	5.817
Al^{+++}(aq)	−125.4	−115.0	−74.9	
AlO_2^-(aq)		−204.7		
$H_2AlO_3^-$(aq)		−255.2		
Al_2O_3(c)	−399.09	−376.77	12.186	18.88
$Al(OH)_3$(amorphous)	−304.9	−271.9	(17.0)	
AlF_3(c)	−311.	−294.	23.	
AlF_6^{---}(aq)		−539.6		
$AlBr_3$(c)	−125.8	−120.7	44.	24.5
$AlCl_3$(c)	−166.2	−152.2	40.	21.3

Formula	H^0	G^0	S^0	$C_P{}^0$
Aluminum (Cont.)				
$(NH_4)Al(SO_4)_2 \cdot 12H_2O$(c)	−1419.40	−1179.02	166.6	163.3
$Al_2(SO_4)_3$(c)	−820.98	−738.99	57.2	62.00
$Al_2(SO_4)_3 \cdot 6H_2O$(c)	−1268.14	−1105.14	112.1	117.8
Beryllium				
Be(gas)	76.63	67.60	32.545	4.9680
Be(c)	0.0	0.0	2.28	4.26
Be^{++}(aq)	−93.	−85.2		
BeO(c)	−146.0	−139.0	3.37	6.07
Magnesium				
Mg(gas)	35.9	27.6	35.504	4.9680
Mg(c)	0.0	0.0	7.77	5.71
Mg^{++}(aq)	−110.41	−108.99	−28.2	
MgO(c)	−143.84	−136.13	6.4	8.94
$Mg(OH)_2$(c)	−221.00	−199.27	15.09	18.41
MgF_2(c)	−263.5	−250.8	13.68	14.72
$MgCl_2$(c)	−153.40	−141.57	21.4	17.04
$MgCl_2 \cdot 6H_2O$(c)	−597.42	−505.65	87.5	75.46
$Mg(ClO_4)_2$(c)	−140.6	−79.4		
$MgSO_4$(c)	−305.5	−280.5	21.9	23.01
$Mg(NO_3)_2$(c)	−188.72	−140.63	39.2	33.94
$MgCO_3$(c)	−266.	−246.	15.7	18.05
$MgSiO_3$(c)	−357.9	−337.2	16.2	19.56
$MgNH_4PO_4$(c)		−390.		
Calcium				
Ca(gas)	46.04	37.98	36.99	4.968
Ca(c)	0.0	0.0	9.95	6.28
Ca^{++}(aq)	−129.77	−132.18	−13.2	
CaO(c)	−151.9	−144.4	9.5	10.23
CaH_2(c)	−45.1	−35.8	10.	
$Ca(OH)_2$(c)	−235.8	−214.33	18.2	20.2
CaF_2(c)	−290.3	−277.7	16.46	16.02
$CaBr_2$(c)	−161.3	−156.8	31.	
CaI_2(c)	−127.8	−126.6	34.	
CaC_2(c)	−15.0	−16.2	16.8	14.90
$CaCO_3$(c, calcite)	−288.45	−269.78	22.2	19.57
$CaCO_3$(c, aragonite)	−288.49	−269.53	21.2	19.42
$CaC_2O_4 \cdot H_2O$(c, precipt.)	−400.4	−361.9	37.28	36.40
$Ca(HCO_2)_2$(c)	−323.5	−300.8		
$Ca(CN)_2$(c)	−44.2	−33.1		
$CaSiO_3$(c)	−378.6	−358.2	19.6	20.38
$CaSO_4$(c, anhydrite)	−342.42	−315.56	25.5	23.8
$CaSO_4 \cdot \frac{1}{2}H_2O$(c, α)	−376.47	−343.02	31.2	28.6
$CaSO_4 \cdot 2H_2O$(c)	−483.06	−429.19	46.36	44.5

Formula	H^0	G^0	S^0	$C_P{}^0$
Calcium (Cont.)				
Ca_3N_2(c)	−103.2	−88.1	25.	22.5
$Ca(NO_3)_2$(c)	−224.00	−177.34	46.2	35.69
$Ca_3(PO_4)_2$(c, β)	−988.9	−932.0	56.4	54.45
$CaHPO_4$(c)	−435.2	−401.5	21.	
$CaHPO_4 \cdot 2H_2O$(c)	−576.0	−514.6	40.	
Strontium				
Sr(gas)	39.2	26.3	39.325	4.9680
Sr(c)	0.0	0.0	13.0	6.0
Sr^{++}(aq)	−130.38	−133.2	−9.4	
$Sr(OH)_2$(c)	−229.3	−207.8	(21.0)	
SrO(c)	−141.1	−133.8	13.0	10.67
SrF_2(c)	−290.3	−277.8		
$SrCl_2$(c)	−198.0	−186.7	28.	18.9
$SrBr_2$(c)	−171.1	−166.3		
SrI_2(c)	−135.5	−135.		
$SrSO_4$(c)	−345.3	−318.9	29.1	
$Sr(NO_3)_2$(c)	−233.25	−186.		
$SrCO_3$(c)	−291.9	−271.9	23.2	19.46
Barium				
Ba(gas)	41.96	34.60	40.699	4.9680
Ba(c)	0.0	0.0	16.	6.30
Ba^{++}(aq)	−128.67	−134.0	3.	
BaO(c)	−133.4	−126.3	16.8	11.34
BaO_2(c)	−150.5	−135.8		
BaF_2(c)	−286.9	−274.5	23.0	17.02
$BaCl_2$(c)	−205.56	−193.8	30.	18.0
$BaCl_2 \cdot H_2O$(c)	−278.4	−253.1	40.	28.2
$BaCl_2 \cdot 2H_2O$(c)	−349.35	−309.8	48.5	37.10
Lithium				
Li(gas)	37.07	29.19	33.143	4.9680
Li(c)	0.0	0.0	6.70	5.65
Li^+(aq)	−66.554	−70.22	3.4	
Li_2(gas)	47.6	37.6	47.06	8.52
Li_2O(c)	−142.4	−133.9	9.06	
LiH(gas)	30.7	25.2	40.77	7.06
LiH(c)	−21.61	−16.72	5.9	8.3
LiOH(c)	−116.45	−105.9	12.	
$LiOH \cdot H_2O$(c)	−188.77	−163.3	17.07	
LiF(c)	−146.3	−139.6	8.57	10.04
LiCl(c)	−97.70	−91.7	(13.2)	
LiBr(c)	−83.72	−81.2	(16.5)	
LiI(c)	−64.79	−64.		
Li_2SO_4(c)	−342.83	−316.6	(27.0)	

Formula	H^0	G^0	S^0	$C_P{}^0$
Lithium (Cont.)				
Li_2CO_3(c)	−290.54	−270.66	21.60	23.28
$LiBH_4$(c)	−44.6			
$LiAlH_4$(c)	−24.2			
Sodium				
Na(gas)	25.98	18.67	36.715	4.9680
Na(c)	0.0	0.0	12.2	6.79
Na^+(aq)	−57.279	−62.589	14.4	
Na_2(gas)	33.97	24.85	55.02	
NaO_2(c)	−61.9	−46.5		
Na_2O(c)	−99.4	−90.0	17.4	16.3
Na_2O_2(c)	−120.6	−102.8	(16.0)	
NaH(gas)	29.88	24.78	44.93	7.002
NaH(c)	−13.7			
NaOH(c, II)	−101.99	−90.1	(12.5)	19.2
$NaOH \cdot H_2O$(c)	−175.17	−149.00	20.2	
NaF(c)	−136.0	−129.3	14.0	11.0
NaCl(c)	−98.232	−91.785	17.3	11.88
$NaClO_4$(c)	−92.18	−61.4		
NaBr(c)	−86.030	−83.1		
NaI(c)	−68.84	−56.7		
Na_2S(c)	−89.2	−86.6		
Na_2SO_3(c)	−260.6	−239.5	34.9	28.7
Na_2SO_4(c)	−330.90	−302.78	35.73	30.50
$Na_2SO_4 \cdot 10H_2O$(c)	−1033.48	−870.93	141.7	140.4
$NaNO_3$(c, II)	−111.54	−87.45	27.8	22.24
Na_2CO_3(c)	−270.3	−250.4	32.5	26.41
$Na_2C_2O_4$(c)	−314.3	−308.		
$NaHCO_3$(c)	−226.5	−203.6	24.4	20.94
NaCN(c)	−21.46	−14.7		
Na_2SiO_3(c)	−363.	−341.	27.2	26.72
$NaBH_4$(c)	−43.82	−28.57	25.02	20.7
Na_2SiF_6(c)	−677.	−610.4		
Potassium				
K(gas)	21.51	14.62	38.296	4.968
K(c)	0.0	0.0	15.2	6.97
K^+(aq)	−60.04	−67.46	24.5	
K_2(gas)	30.8	22.1	59.69	
K_2O(c)	−86.4	−76.2		
K_2O_2(c)	−118.	−100.1		
KH(gas)	30.0	25.1	47.3	
KOH(c)	−101.78	−89.5		
KF(c)	−134.46	−127.42	15.91	11.73
$KF \cdot 2H_2O$(c)	−277.00	−242.7	36.	
KHF_2(c)	−219.98	−203.73	24.92	18.37

Formula	H^0	G^0	S^0	$C_P{}^0$
Potassium (Cont.)				
KCl(c)	−104.175	−97.592	19.76	12.31
KCl(gas)	−51.6	−56.2	57.24	8.66
$KClO_3$(c)	−93.50	−69.29	34.17	23.96
$KClO_4$(c)	−103.6	−72.7	36.1	26.33
KBr(c)	−93.73	−90.63	23.05	12.82
$KBrO_3$(c)	−79.4	−58.2	35.65	25.07
KI(c)	−78.31	−77.03	24.94	13.16
KI_3(c)	−76.6	−73.5		
KIO_3(c)	−121.5	−101.7	36.20	25.42
K_2SO_4(c, II)	−342.66	−314.62	42.0	31.1
KNO_3(c)	−117.76	−93.96	31.77	23.01
KH_2PO_4(c)	−374.9	−339.2		
K_2CO_3(c)	−273.93	−255.5		
$K_2C_2O_4$(c)	−320.8	−296.7		
$KHCO_3$(c)	−229.3	−205.7		
KCN(c)	−26.90	−20.		
$K_3Fe(CN)_6$(c)	−41.4	−3.3		
$K_4Fe(CN)_6$(c)	−125.1	−84.0		
$KMnO_4$(c)	−194.4	−170.6	41.04	28.5
$KCr(SO_4)_2$(c)	−562.	−510.		
$KCr(SO_4)_2 \cdot 12H_2O$(c)	−1383.1	−1164.		
$KAl(SO_4)_2$(c)	−589.24	−534.29	48.9	46.12
$KAl(SO_4)_2 \cdot 12H_2O$(c)	−1447.74	−1227.8	164.3	155.6
Rubidium				
Rb(gas)	20.51	13.35	40.628	4.9680
Rb(c)	0.0	0.0	16.6	7.27
Rb^+(aq)	−58.9	−67.45	29.7	
Rb_2O(c)	−78.9	−69.5		
RbOH(c, II)	−98.9	−87.1		
RbF(c)	−131.9	−124.3	27.2	
RbCl(c)	−102.91	−96.8		12.3
RbBr(c)	−93.03	−90.38	25.88	12.68
RbI(c)	−78.5	−77.8	28.21	12.50
Rb_2SO_4(c)	−340.50	−312.8		
$RbNO_3$(c)	−117.04	−93.3		
Cesium				
Cs(gas)	18.83	12.24	41.944	4.9680
Cs(c)	0.0	0.0	19.8	7.42
Cs^+(aq)	−59.2	−67.41	31.8	
Cs_2O(c)	−75.9	−65.6		
CsOH(c)	−97.2	−84.9		
CsF(c)	−126.9	−119.5		
CsCl(c, II)	−103.5	−96.6		
CsBr(c)	−94.3	−91.6	29.	12.4

Formula	H^0	G^0	S^0	$C_P{}^0$
Cesium (Cont.)				
$CsI(c)$	−80.5	−79.7	31.	12.4
$Cs_2SO_4(c)$	−339.38	−310.7		
$CsNO_3(c)$	−118.11	−94.0		
$CsAl(SO_4)_2 \cdot 12H_2O(c)$	−1449.5	−1281.5	164.	148.1

Appendix II

Physical Constants, Numerical Factors, and Energy Conversion Factors

Absolute temperature of ice point (0°C) .... 273.150°K
Absolute temperature of triple point of H_2O, by definition .... 273.160°K
Thermochemical calorie, by definition .... 4.1840 joules
Standard gravity, by definition .... 980.665 cm/sec^2
Atmosphere (by definition 760 mm Hg at s.c.) .... 1,013,250 dynes/cm^2

Avogadro's number: $N_0 = 6.0232 \times 10^{23}$/mole

Boltzmann's gas constant: $k = 1.38045 \times 10^{-16}$ erg/deg

Planck's constant: $h = 6.6252 \times 10^{-27}$ erg sec

Electronic charge: $e = 4.8029 \times 10^{-10}$ esu
$= 1.60206 \times 10^{-19}$ coulomb

Velocity of light: $c = 2.99793 \times 10^{10}$ cm/sec

Faraday's constant: $\mathscr{F} = 96{,}493.5$ coul/*g-eq* or joules/volt *g-eq*
$= 23{,}062.3$ cal/volt *g-eq*

Molar gas constant: $R = 1.98726$ cal/deg mole
$= 8.31470$ joules/deg mole
$= 82.0597$ cc atm./deg mole

$$\ln(\ ) = 2.302585 \log(\ )$$

$$R \ln(\ ) = 4.5758 \log(\ ) \text{ cal/deg mole}$$

$$298.15R \ln(\ ) = 1364.3 \log(\ ) \text{ cal/mole}$$

$$\frac{298.15R}{\mathscr{F}} \ln(\ ) = 0.05916 \log(\ ) \text{ volt } g\text{-}eq\text{/mole}$$

$$hc/k = 1.4388 \text{ cm deg}$$

ENERGY CONVERSION FACTORS

	cal/mole	cc atm./mole	cm^{-1}	electron volt	erg/molecule
1 cal/mole	1	41.292	0.34974	4.3361×10^{-5}	6.9465×10^{-17}
1 cc atm./mole	0.024218	1	8.470×10^{-3}	1.0501×10^{-6}	1.6823×10^{-18}
1 cm^{-1}	2.8593	118.07	1	1.2398×10^{-4}	1.9862×10^{-16}
1 electron volt	23,062	9.523×10^{5}	8065.7	1	1.6020×10^{-12}
1 erg/molecule	1.4396×10^{16}	5.944×10^{17}	5.0348×10^{15}	6.2422×10^{11}	1

Index